Microorganisms for Sustainability

Volume 2

Series editor

Naveen Kumar Arora, Environmental Microbiology, School for Environmental Science, Babasaheb Bhimrao Ambedkar University, Lucknow, Uttar Pradesh, India

More information about this series at http://www.springer.com/series/14379

Samina Mehnaz
Editor

Rhizotrophs: Plant Growth Promotion to Bioremediation

Springer

Editor
Samina Mehnaz
Department of Biological Sciences
Forman Christian College
Lahore, Pakistan

Microorganisms for Sustainability
ISBN 978-981-10-4861-6 ISBN 978-981-10-4862-3 (eBook)
DOI 10.1007/978-981-10-4862-3

Library of Congress Control Number: 2017944613

Printed on acid-free paper

This Springer imprint is published by Springer Nature
The registered company is Springer Nature Singapore Pte Ltd.
The registered company address is: 152 Beach Road, #21-01/04 Gateway East, Singapore 189721, Singapore

Contents

About the Series Editor

Dr. Naveen Kumar Arora, Ph.D. in microbiology, associate professor in the Department of Environmental Microbiology, Babasaheb Bhimrao Ambedkar University (a central university), Lucknow, Uttar Pradesh, India, is a renowned researcher in the field of environmental microbiology and biotechnology. His specific area of research is rhizosphere biology and PGPRs. He has more than 50 research papers published in premium international journals and several articles published in magazines and dailies. He is editor of three books, published by Springer. He is member of several national and international societies and reviewer of several international journals. He has delivered lectures in conferences and seminars around the globe. He has a long-standing interest in teaching at the PG level and is involved in taking courses in bacteriology, microbial physiology, environmental microbiology, agriculture microbiology, and industrial microbiology. He has been advisor to 57 postgraduate and 8 doctoral students. Recently, he was awarded for excellence in research by the honorable governor of Uttar Pradesh. Although an academician and researcher by profession, he has a huge obsession for the wildlife and its conservation and has authored a book, *Splendid Wilds*. He has a dedicated website www.naveenarora.co.in for the cause of wildlife and environment conservation.

About the Editor

Dr. Samina Mehnaz, Ph.D. in microbiology, is a professor and chairperson of the Department of Biological Sciences, Forman Christian College (a chartered university), Lahore, Punjab, Pakistan. She is a researcher in the field of microbiology since 1991. Her specific area of interest is rhizosphere microbiology, more specifically PGPRs, bioformulations, and microbial metabolites. She has been an Alexander von Humboldt (AvH) fellow (Germany), National Science and Engineering Research Council (NSERC) fellow (Canada), International Atomic Energy Agency (IAEA) fellow (Belgium), and a postdoctorate fellow (USA). She has 39 research papers published in international journals and 4 book chapters (as a single author) published by Springer. She is editor of a book, *Bioformulations: For Sustainable Agriculture*, recently published by Springer, India. She is editor of an international journal, *Archives of Phytopathology and Plant Protection*. She has reviewed more than 65 research articles for more than 30 international journals and is a regular reviewer for *Microbial Ecology* and *International Journal of Systematic and Evolutionary Microbiology*. She has been a winner of Research Productivity Award (RPA) from 2010 to 2014, awarded by the Pakistan Council of Science and Technology (PCST), and a Productive Scientist of Pakistan since 2006, as listed by PCST. In January 2016, she was appointed as "Humboldt Ambassador Scientist" of Pakistan (a representative of Alexander von Humboldt (AvH) Foundation, Germany) by the president of AvH Foundation and is currently serving at this position.

Gluconacetobacter azotocaptans: A Plant Growth-Promoting Bacteria

1

Samina Mehnaz and George Lazarovits

Abstract

Gluconacetobacter diazotrophicus is a very well-known and well-studied member of family *Acetobacteraceae*. Strains of this species have been isolated from all over the world. On the other side, only two strains of *Gluconacetobacter azotocaptans* have been isolated, i.e., one from coffee plant in Mexico (type strain) and another from corn plant in Canada; therefore, not much information is available about it. This chapter is an effort to bring this species in the limelight and show its significance as plant growth promoter. Authors have not only isolated the strain DS1, they have also evaluated its plant growth-promoting potential in four varieties of corn, radish, cucumber, tomato, pepper, and potato, under greenhouse, field, and tissue culture conditions, and data is presented in this chapter. Inoculation effect of DS1 was compared with *G. diazotrophicus* PAL5, its *nif* mutant, and strains of other genera including *Pseudomonas*, *Azospirillum*, *Enterobacter*, *Burkholderia*, and *Sphingobacterium*. Work on wheat was done by another group in Canada, and their data has also been added. It is a nitrogen-fixing, indole acetic acid-producing strain with phosphate solubilization ability. It has promoted the plant growth in most cases. Its growth-promoting effect varied in tissue culture medium depending on sucrose concentration as it is linked with IAA and ethylene production. Based on data presented here, authors are recommending its use as a biofertilizer.

S. Mehnaz (✉)
Department of Biological Sciences, Forman Christian College (A Chartered University), Ferozepur Road, Lahore 54600, Pakistan

A & L Research Laboratories, London, ON, Canada
e-mail: saminamehnaz@fccollege.edu.pk

G. Lazarovits
A & L Research Laboratories, London, ON, Canada

S. Mehnaz (ed.), *Rhizotrophs: Plant Growth Promotion to Bioremediation*, Microorganisms for Sustainability 2, DOI 10.1007/978-981-10-4862-3_1

1.1 Introduction

Gluconacetobacter is a gram-negative, acetic acid-producing bacterium. It was initially categorized as sub-genus of *Acetobacter*; however, Yamada et al. (1997) suggested to raise its status to genus level; it was accepted and validated in 1998. This genus has 24 species, and among these *G. diazotrophicus* is the one that is widely isolated from all over the world and extensively studied due to the same reason. *G. diazotrophicus* is an endophytic bacterial species that occurs predominantly in vegetatively propagated plants. It has been isolated from numerous types of plant tissues including the internal tissues of sucrose-accumulating plants such as sugarcane, washed roots and aerial parts of *Pennisetum purpureum*, sweet potato stems and roots, rhizosphere soil of coffee plants, as well as the surface-sterilized stems and roots and inner tissues of *Eleusine coracana*, pineapple plants, and wetland rice varieties (Munoz-Rojaz and Caballero-Mellado 2003; Muthukumarasammy et al. 2005). Recently, a review published by Eskin et al. (2014) listed all reported hosts of *G. diazotrophicus*.

G. diazotrophicus was the only known nitrogen-fixing species of this genus until Jimenez-Salgado et al. (1997) isolated two other acetic acid-producing, diazotrophic bacteria from rhizosphere of coffee plants. These diazotrophs shared features with the genus *Gluconacetobacter* but differ from *G. diazotrophicus* with respect to morphological and biochemical traits as well as genetic and molecular features. Results of intensive taxonomic analysis by Fuentes-Ramirez et al. (2001) led to the recommendation that these new isolates be assigned to novel species within the family *Acetobacteraceae*. These isolates were named as *G. azotocaptans* and *G. johannae*. Like other species of *Gluconacetobacter* (except *G. diazotrophicus*), reports of these two species are also rare.

Mehnaz et al. (2006) reported the isolation of *G. azotocaptans* strain DS1 from corn rhizosphere. Till now there is no other report about the isolation of this species from any other host or other parts of the world. Therefore, work done on this species is also very limited. Focus of this chapter is to throw light on the significance of this species as a plant growth-promoting bacteria.

1.2 *Gluconacetobacter azotocaptans*

It is a gram-negative, nitrogen-fixing bacterium isolated from coffee and corn rhizosphere (Fuentes-Ramirez et al. 2001; Mehnaz et al. 2006). Authors have extensively worked on this bacterium; part of this work has been published (Mehnaz and Lazarovits 2006; Mehnaz et al. 2006). This chapter is presenting the published and unpublished work done on this strain by authors and the University of Saskatchewan. This strain was explored for its plant growth-promoting potential in lab studies, field experiments, and tissue culture conditions. Corn, wheat, potato, and some vegetables were used as host to observe the effect of this bacterium. Detailed information about this work is provided below.

Mehnaz and Lazarovits (2006) screened this strain for plant growth-promoting traits and its potential to be used as biofertilizer. Authors performed qualitative and quantitative assays and reported the nitrogenase activity (40 nmol ethylene/hr/mg protein), indole acetic acid production (106 μg/l), phosphate solubilization, and antifungal activity of this strain, against *Fusarium solani*, *F. solani phaseoli*, *F. moniliforme*, and *F. sambucinum*.

1.3 Growth Promotion of Cereals Inoculated with *G. azotocaptans* DS1

1.3.1 Corn

Authors had worked on the project of bioformulation for corn at Agriculture and Agri-Food Canada, London, Ontario, Canada. Data presented here is part of that project. Corn plants were inoculated with DS1 and other bacterial strains, grown in sterilized sand and unsterilized soil, in pot experiment. DS1 was also used as inoculum for field trials.

For pot experiment, four Pioneer varieties of corn, i.e., 39D82, 39H84, 39M27, and 39T68, were inoculated with *G. azotocaptans* DS1 (corn isolate) and other strains of *Azospirillum* (N7 and N8; corn isolates), *Pseudomonas*, and *G. diazotrophicus* PAL5 and its *nifD* mutant; 10^8 cells per plant at the time of transplantation. Three-days-old seedlings were grown for 4 weeks, in 250 g sterilized sand and NPK fertilizers (20:10:50 kg/ha, respectively). DS1 inoculated plants showed significant and maximum increase (27%) in root weight of 39T68 as compared to all other treatments and other corn varieties used in this study (Table 1.1). For 39T68, *A. brasilense* N8 and *P. putida* CQ179 also increased the root weight, but it was slightly less than DS1. With two varieties, 39H84 and 39M27, inoculation effect for all

Table 1.1 Effect of bacterial isolates on root weight of four corn varieties after 30 days growth, in sterilized sand

	Corn varieties			
Treatments	39D82 (mg/plant)	39H84 (mg/plant)	39M27 (mg/plant)	39T68 (mg/plant)
Control	260 ± 40 bc	210 ± 56 a	270 ± 50 a	220 ± 60 c
A. zeae N7	300 ± 42 a	220 ± 36 a	270 ± 47 a	255 ± 52 abc
A. brasilense N8	290 ± 41 ab	200 ± 59 a	265 ± 63 a	270 ± 64 ab
P. putida CQ179	250 ± 34 c	215 ± 39 a	250 ± 46 a	270 ± 64 ab
***G. azotocaptans* DS1**	**250 ± 32 c**	**220 ± 65 a**	**260 ± 41 a**	**280 ± 90 a**
G. diazotrophicus PAL5 Wt	260 ± 33 bc	210 ± 57 a	290 ± 42 a	230 ± 49 bc
*G. diazotrophicus PAL5 nif*D	250 ± 40 c	200 ± 42 a	250 ± 42 a	240 ± 50 abc

Values are average of 12 replicates. Letters indicate a statistically significant difference between treatments according to Duncan's multiple range test (DMRT) at $P \leq 0.05$. Mean separation within a column followed by the same letters does not differ significantly

Data is taken from Mehnaz and Lazarovits (2006)

bacterial strains on root weight was non-significant as compared to uninoculated plants. For 39D82, *Azospirillum* strains, N7 and N8, showed maximum increase in root weight, 15% and 11%, respectively, as compared to uninoculated and all other inoculated plants. DS1 and all other inoculated strains did not show any difference with uninoculated plants. Interestingly, *G. diazotrophicus* strains did not show significant difference in root weight of any variety.

Shoot weight of all varieties increased when inoculated with DS1 (Table 1.2); however, 39M27 and 39T68 showed significantly high shoot weight, ranging from 23% to 29% (Mehnaz and Lazarovits 2006). *Azospirillum brasilense* N8 and *A. zeae* N7 (Mehnaz et al. 2007a) also significantly increased shoot weight of these two varieties; however, it was less than *G. azotocaptans* DS1, i.e., 15.7%. *G. diazotrophicus* PAL5 strains, i.e., Wt and *nif*D, could not contribute significantly in shoot weight of these varieties. For variety 39H84, increase in shoot weight was non-significant by all bacterial strains. For variety 39D82, shoot weight was significantly increased by both strains of *G. diazotrophicus*, i.e., 12.5–13.7%. *G. azotocaptans* DS1 showed 11.3% increase in shoot weight that was slightly less than *G. diazotrophicus*. *Azospirillum* strains N7 and N8 showed 7.5% and 12.5% non-significant increase, respectively. However, *P. putida* CQ179 showed lowest shoot weight as compared to all other inoculated and control plants.

Same varieties were used for plant experiment in unsterilized soil collected from corn field and inoculated with DS1 and strains of *Azospirillum*, *Pseudomonas*, and *G. diazotrophicus*. DS1 significantly increased the root weight of 39H84, i.e., 19% (Table 1.3), as compared to other varieties and other bacterial strains. None of the bacterial strains could increase the root weight of variety 39T68 as compared to control plants. *P. putida* CQ179 was the only strain that significantly increased the root weight, i.e., 20.8%, of variety 39M27 plants. Performance of DS1 was very poor with this variety as lowest root weight was recorded for plants inoculated with

Table 1.2 Effect of bacterial isolates on shoot weight of four corn varieties, after 30 days growth in sterilized sand

	Corn varieties			
Treatments	39D82 (mg/plant)	39H84 (mg/plant)	39M27 (mg/plant)	39T68 (mg/plant)
Control	400 ± 62 bcd	525 ± 108 ab	310 ± 60 b	510 ± 98 d
A. zeae N7	430 ± 41 abc	510 ± 124 ab	315 ± 54 b	590 ± 66 abc
A. brasilense N8	450 ± 58 ab	490 ± 124 b	330 ± 66 b	590 ± 130 ab
P. putida CQ179	360 ± 54 d	560 ± 100 a	350 ± 60 b	570 ± 74 abcd
***G. azotocaptans* DS1**	**445 ± 71 ab**	**540 ± 86 ab**	**400 ± 71 a**	**630 ± 122 a**
G. diazotrophicus PAL5 Wt	455 ± 65 a	510 ± 98 ab	330 ± 57 b	530 ± 88 cd
*G. diazotrophicus PAL5 nif*D	450 ± 55 a	535 ± 105 ab	320 ± 51 b	560 ± 85 bcd

Values are average of 12 replicates. Letters indicate a statistically significant difference between treatments according to Duncan's multiple range test (DMRT) at $P \leq 0.05$. Mean separation within a column followed by the same letters does not differ significantly
Data is taken from Mehnaz and Lazarovits (2006)

Table 1.3 Effect of bacterial isolates on root weight of four corn varieties, after 30 days growth, in non-sterilized corn field soil

Treatments	Corn varieties			
	39D82 (mg/plant)	39H84 (mg/plant)	39 M27 (mg/plant)	39 T68 (mg/plant)
Control	240 ± 30 c	290 ± 45 bc	240 ± 45 bc	320 ± 61 a
A. zeae N7	290 ± 48 ab	310 ± 53 abc	260 ± 47 abc	340 ± 74 a
A. brasilense N8	250 ± 34 bc	310 ± 70 abc	255 ± 33 abc	330 ± 79 a
P. putida CQ179	300 ± 39 a	330 ± 50 ab	290 ± 52 a	335 ± 63 a
***G. azotocaptans* DS1**	**250 ± 57 bc**	**345 ± 53 a**	**220 ± 69 c**	**320 ± 65 a**
G. diazotrophicus PAL5 Wt	240 ± 37 c	280 ± 66 c	260 ± 64 abc	340 ± 75 a
*G. diazotrophicus PAL5 nif*D	240 ± 25 c	290 ± 75 bc	280 ± 53 ab	340 ± 63 a

Values are average of 12 replicates. Letters indicate a statistically significant difference between treatments according to Duncan's multiple range test (DMRT) at $P \leq 0.05$. Mean separation within a column followed by the same letters does not differ significantly
Data is taken from Mehnaz and Lazarovits (2006)

Table 1.4 Effect of bacterial isolates on shoot weight of four corn varieties, after 30 days growth, in non-sterilized corn field soil

Treatments	Corn varieties			
	39D82 (mg/plant)	39H84 (mg/plant)	39M27 (mg/plant)	39T68 (mg/plant)
Control	500 ± 67 b	570 ± 87 b	450 ± 111 b	690 ± 99 a
A. zeae N7	570 ± 64 a	610 ± 92 b	560 ± 66 a	710 ± 127 a
A. brasilense N8	520 ± 46 ab	590 ± 120 b	450 ± 109 b	670 ± 97 a
P. putida CQ179	560 ± 59 a	675 ± 88 a	560 ± 63 a	720 ± 131 a
***G. azotocaptans* DS1**	**480 ± 42 b**	**570 ± 149 b**	**470 ± 111 b**	**700 ± 144 a**
G. diazotrophicus PAL5 Wt	500 ± 72 b	595 ± 130 b	505 ± 81 ab	710 ± 118 a
*G. diazotrophicus PAL5 nif*D	520 ± 49 ab	580 ± 117 b	590 ± 99 a	710 ± 120 a

Values are average of 12 replicates. Letters indicate a statistically significant difference between treatments according to Duncan's multiple range test (DMRT) at $P \leq 0.05$. Mean separation within a column followed by the same letters does not differ significantly
Data is taken from Mehnaz and Lazarovits (2006)

this strain; however, the difference was non-significant. The rest of the inoculated strains showed non-significant increase in root weight.

G. azotocaptans DS1 could not significantly increase the shoot weight of plants of any variety as compared to uninoculated plants (Table 1.4). *A. zeae* N7 and *P. putida* CQ179 significantly increased the shoot weight of variety 39D82, i.e., 14% and 12%, respectively. The rest of the strains contributed non-significantly in shoot weight. Lowest but non-significant root weight was recorded for DS1 inoculated plants. *P. putida* CQ179 also significantly increased the shoot weight, i.e., 18.4%, of variety 39H84 plants. Shoot weight of the rest of the inoculated plants was far below as compared to CQ179 inoculated plants. For variety 39M27, *G. diazotrophicus nif*D, *P. putida* CQ179, and *A. zeae* N7 significantly increased shoot weight of the plants as compared to all other treatments. CQ179 and N7 both increased 24.4% shoot weight, and *G. diazotrophicus nif*D showed 31% increase in this parameter.

For field experiment, corn variety 39D82 was used. Bacterial cultures, i.e., *G. azotocaptans* DS1 and *Azospirillum canadense* DS2 (another isolate from corn rhizosphere; Mehnaz et al. 2007b), were individually used as inoculum. Bacterial cell pellets were suspended in 0.85% saline with 1% polyvinyl pyrrolidone K30 (PVP; sticker). Approximately 10^6 cells per seed were applied by soaking them in inoculum. For control, seeds were coated only with sticker. Replicated field plots of corn variety Pioneer 39D82 were established in Southwestern Ontario at the Delhi research farm. All plots received a broadcast application of granular fertilizer containing 55 kg N + 20 kg P_2O_5 + 110 K_2O/ha incorporated to a depth of about 10 cm prior to seeding. In the fall, grains were harvested, the moisture content was determined, and the 15% moisture content (MC) yield was calculated. Data was analyzed by using SAS statistical software (ver.9.1). ANOVA was carried out in SAS, and comparison among treatments was done by using Duncan's multiple range test (DMRT).

Grain yield for *G. azotocaptans* DS1 inoculated plants was 9 t/ha as compared to 8.7 and 8.6 t/ha for *A. canadense* DS2 and control plants, respectively. Another field experiment was performed with corn variety, NK N-35 B8, and inoculated with the same organisms (DS1 and DS2) and the same conditions. Grain yield for DS1 was 11.1 t/ha as compared to10.7 and 10.2 t/ha for DS2 and control plants, respectively. For both experiments, DS1 inoculated plants showed highest grain yield; however, the difference was non-significant.

1.3.2 Wheat

The detailed studies with strain DS1 of *G. azotocaptans* on wheat were done at the University of Saskatchewan, Canada, by Morley in 2013. He used DS1 and a strain of *Azospirillum zeae* N7 (Mehnaz et al. 2007a) to inoculate wheat plants cv. Lillian under lab and field conditions and observed their effect on dry matter, %Ndfa, and survival of these strains. Survival of strains was observed on sterilized, non-sterilized, and fungicide-coated seeds. It was reported that DS1 and N7 survived well in the presence of fungicide, Dividend® XLRTA®, on the seed coat, showing their resistance to this product. Survival was better on non-sterilized and fungicide-coated seeds as compared to sterilized seeds as it declined at fast rate on sterilized ones.

DS1 and N7 are nitrogen fixers, and this ability was observed in these experiments. It is known that presence of nitrogen fertilizer reduces the biological nitrogen fixation (BNF). Interestingly, in this study both strains contributed more through BNF, in the presence of nitrogen fertilizer. Nitrogen uptake and %Ndfa increased with the increase of nitrogen fertilizer. Inoculated plants grown in growth chamber with 12.2–24.5 μg N/g had highest %Ndfa, i.e., 25.5%. Inoculated plants fertilized with higher amount showed highest nitrogen uptake, 1.3 g/pot, at maturity, as well. Similar results were obtained in field study as inoculated plants provided with 80 kg N/ha showed highest nitrogen uptake of 47 kg N/ha and significantly higher ($P < 0.05$) %Ndfa, 10.5%, as compared to other dosage of chemical fertilizer. Accumulation of nitrogen varied in different parts of the plants with bacterial strains.

For pot experiments, *Azospirillum zeae* N7 inoculated plants accumulated significant amount of nitrogen in spikes, and *G. azotocaptans* DS1 inoculated plants had highest amount in stem. This trend was not observed in field experiment. In pot experiments, plants were harvested at 40, 65, and 102 days after sowing. Significant increase in dry matter of *G. azotocaptans* DS1 inoculated plants provided with fertilizer, was observed after 40 days of sowing. At other stages and inoculation with *Azospirillum* N7, it was not as effective as DS1. Among dry matter, the weight of stem for DS1 inoculated plants with different fertilizer doses was higher as compared to uninoculated plants and plants without any fertilizer. However, dry matter for spikes was highest for *Azospirillum* inoculated plants as compared to other treatment.

Field experiments with these two strains were conducted at three sites of Saskatchewan. Field soil was provided with different doses of nitrogen fertilizer. It was observed that presence of nitrogen fertilizers did not inhibit the nitrogen fixation by DS1 and N7, as determined by %Ndfa. Increase in dry matter or grain yield was observed with increase in fertilizer dose. However, increase in yield or dry matter was non-significant in inoculated plants.

1.4 Growth Promotion of Vegetables Inoculated with *G. azotocaptans* DS1

Authors have worked on the *G. azotocaptans* DS1 to observe its growth-promoting effects on different vegetables. For cereals, work was done in greenhouse and field as well. However, for vegetables, experiments were done only in greenhouse under controlled temperature and light conditions. In addition to *G. azotocaptans* DS1, four other bacterial strains, *Azospirillum zeae* N7, *A. brasilense* N8, *A. canadense* DS2, and *Pseudomonas putida* CQ179, isolated from corn rhizosphere were used to inoculate the vegetable crops. Inoculum was applied to cucumber, pepper, radish, and tomato seedlings, at the time of transplantation in promix (mixture of peat and vermiculite) and grown under greenhouse conditions. Plants were harvested after 4 weeks, and root, shoot, and whole plant weights were recorded.

For each crop, 12 replicates were used for each treatment. All experiments were repeated three times. The repeated experiments showed similar trends, and there were non-significant differences between the same treatments in each experiment. The data was analyzed by using SAS statistical software. One-way analysis of variance (ANOVA) was done with the ANOVA procedure in SAS, and comparison among treatments was done by using Duncan's multiple range test (DMRT). All analyses were performed at the $P = 0.05$ level.

Table 1.5 Effect of nitrogen fixers on sweet pepper cultivar "California wonder" and cucumber cultivar "Marketmore 76" plants after 30 days growth in promix, under greenhouse conditions

	Sweet pepper			Cucumber		
Treatments	Total weight (mg/plant)	Shoot weight (mg/plant)	Root weight (mg/plant)	Total weight (mg/plant)	Shoot weight (mg/plant)	Root weight (mg/plant)
Control	113 bc	58 b	53 b	773 bc	603 bc	175 ab
A. zeae N7	103 c	54 b	49 b	756 bc	583 bc	168 b
A. brasilense N8	142 ab	82 a	71 a	829 ab	659 ab	160 b
***G. azotocaptans* DS1**	**161 a**	**89 a**	**72 a**	**931 ab**	**730 a**	**201 a**
A. canadense DS2	177 a	97 a	78 a	635 c	504 c	123 c
P. putida CQ179	163 a	86 a	74 a	844 ab	666 ab	167 b

Values are average of 12 replicates. Letters indicate a statistically significant difference between treatments according to Duncan's multiple range test (DMRT) at $P \leq 0.05$. Mean separation within a column followed by the same letters does not differ significantly

Fig. 1.1 Growth-promoting effect of *G. azotocaptans* DS1 on cucumber plant, under greenhouse conditions

1.4.1 Sweet Pepper

It was observed that *G. azotocaptans* DS1 significantly promoted root, shoot, and total plant weight. Significantly high total weight was recorded for *G. azotocaptans* DS1, *A. canadense* DS2, and *P. putida* CQ179 inoculated plants as compared to uninoculated and two other strains. Although *A. canadense* DS2 inoculated plants showed highest weight as compared to *G. azotocaptans* DS1 and *P. putida* CQ179, the difference was non-significant (Table 1.5; Fig. 1.1). Root and shoot weight for all inoculated plants except *A. zeae* N7 were significantly higher than control plants. N7 inoculated plants had non-significant difference with control. Among other

Fig. 1.2 Growth-promoting effect of *G. azotocaptans* DS1 on pepper plant, under greenhouse conditions

strains, highest root and shoot weights were also observed with *A. canadense* DS2 inoculated plants; difference with DS1, N8, and CQ179 was non-significant.

1.4.2 Cucumber

G. azotocaptans DS1 significantly promoted the total plant weight and shoot weight of cucumber plants as compared to uninoculated and plants inoculated with other strains (Table 1.5). Increase in shoot weight was quite visible in DS1 inoculated plants (Fig. 1.2). *A. brasilense* N8 and *P. putida* CQ179 also showed total weight and shoot weight higher than control; however, the difference was non-significant with *G. azotocaptans* DS1 and uninoculated plants. Root weight of DS1 inoculated plants was highest among all inoculated and uninoculated plants, but difference was non-significant with uninoculated plants.

1.4.3 Tomato

G. azotocaptans DS1, *A. canadense* DS2, and *P. putida* CQ179 significantly enhanced the root, shoot, and total weight of tomato plants as compared to uninoculated and rest of the inoculated plants (Table 1.6). Highest shoot weight and total plant weight were recorded for DS2, and root weight was highest for DS1 inoculated plants, however, difference among these three strains was non-significant. Significantly lowest weights among all inoculated and uninoculated plants were recorded for *A. zeae* N7 inoculated plants. *A. brasilense* N8 inoculated plants showed total plant and root weight higher than control plants and N7 but lower than other inoculated plants.

Table 1.6 Effect of nitrogen fixers on radish cultivar "Cherry belle" and tomato cultivar "Bellstar 409" plants after 30 days growth in promix, under greenhouse conditions

	Radish			Tomato		
Treatments	Total weight (mg/plant)	Shoot weight (mg/plant)	Root weight (mg/plant)	Total weight (mg/plant)	Shoot weight (mg/plant)	Root weight (mg/plant)
Control	965 b	308 c	619 bc	320 c	257 b	64 c
A. zeae N7	1,124 a	408 a	711 abc	240 d	190 c	59 c
A. brasilense N8	1,161 a	340 abc	823 a	364 b	282 b	84 b
***G. azotocaptans* DS1**	**1,114 a**	**383 ab**	**680 bc**	**460 a**	**361 a**	**103 a**
A. canadense DS2	925 b	361 abc	585 c	511 a	418 a	102 a
P. putida CQ179	1,041 ab	326 bc	732 ab	472 a	383 a	93 ab

Values are average of 12 replicates. Letters indicate a statistically significant difference between treatments according to Duncan's multiple range test (DMRT) at $P \leq 0.05$. Mean separation within a column followed by the same letters does not differ significantly

1.4.4 Radish

A. brasilense N8, *A. zeae* N7, and *G. azotocaptans* DS1 significantly enhanced total weight of radish plants, as compared to rest of the treatments including control plants (Table 1.6). Among these three, highest weight was recorded for *A. brasilense* N8 inoculated plants; however, difference with *A. zeae* N7 and *G. azotocaptans* DS1 was non-significant. Highest shoot weight was recorded with N7 inoculated plants. DS1 inoculated plants were second highest and had non-significant difference with N7. N8 showed highest significant increase in root weight. The rest of the strains, except *A. canadense* DS2, showed higher root weight as compared to control, but difference was non-significant.

1.5 Use of *G. azotocaptans* DS1 as Inoculum for Potato Plants in Tissue Culture

Effect of *G. azotocaptans* DS1, on potato cultivar "Kennebec," was observed in tissue culture conditions. Plantlets were grown in MS medium after individual inoculation of *G. azotocaptans* DS1, *E. cloacae* CR1, *S. maltophilia* CR3, *P. putida* CR7, and *S. canadense* CR11. After 8 weeks, it was observed that DS1 significantly reduced the shoot height, shoot weight, and total biomass; root weight was better than control, but difference was not significant (Table 1.7). CR3, CR7, and CR11 significantly enhanced shoot weight and total biomass; however, root weight and shoot height were non-significantly higher than control. CR1 significantly reduced all parameters.

Response of DS1 was quite discouraging in tissue culture. To investigate this effect, three different concentrations of sucrose were used in MS medium, i.e., 7, 15, and 30 g/l. Regular amount of sucrose in MS medium is 30 g/l. Plantlets of same cultivar, "Kennebec," were inoculated with DS1, CR1, and *B. phytofirmans* E24. E24 is

Table 1.7 Effect of bacterial isolates on the "in vitro" growth of potato cultivar "Kennebec" after 8 weeks growth in MS medium

Treatments	Shoot height (cm/plant)	Total biomass (mg/plant)	Root weight (mg/plant)	Shoot weight (mg/plant)
Uninoculated	12.8 ± 1.52 a	57.2 ± 10.3 c	9.05 ± 4.7 b	45.1 ± 12.4 b
E. cloacae CR1	3.0 ± 1.02 c	15.2 ± 6.6 d	0 ± 0 c	15.2 ± 6.6 c
***G. azotocaptans* DS1**	**3.2 ± 0.3 c**	**22.6 ± 5.0 d**	**13.8 ± 2.7 b**	**13.3 ± 3.8 c**
S. maltophilia CR3	13.7 ± 1.99 a	79.5 ± 16.2 b	12.9 ± 5.2 b	66.7 ± 12.0 a
P. putida CR7	14.0 ± 1.32 a	72.1 ± 9.7 b	12.9 ± 4.3 b	59.2 ± 7.3 a
S. canadense CR11	13.5 ± 1.73 a	70.2 ± 15.4 b	9.9 ± 5.5 b	59.7 ± 11.1 a
B. phytofirmans E24	10.9 ± 1.24 b	110 ± 14.5 a	58.4 ± 8.1 a	46.9 ± 12.5 b

Values are average of 10 replicates. Letters indicate a statistically significant difference between treatments according to Duncan's multiple range test (DMRT) at $P \leq 0.05$. Mean separation within a column followed by the same letters does not differ significantly

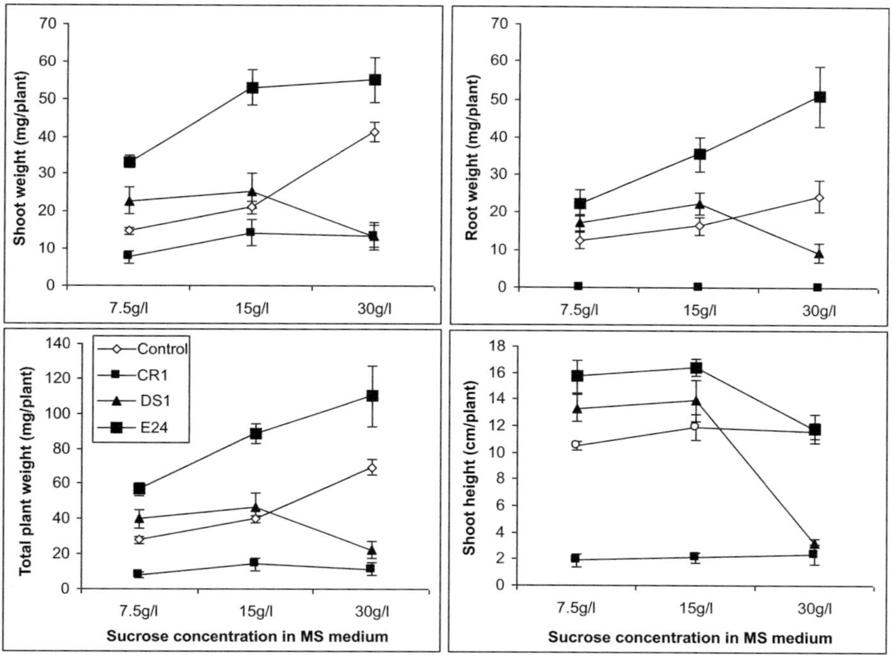

Fig. 1.3 Effect of sucrose concentration in MS medium on different growth parameters of potato cultivar "Kennebec"

known to be a plant growth promoter for corn, potato, tomato, and pepper (Lazarovits and Novak 1997; Mehnaz et al. 2010); it was used as positive control. After 8 weeks, it was observed that at 7 and 15 g/l, DS1 promoted all the parameters, i.e., root weight, shoot weight, total biomass, and shoot height, but it was drastically reduced at 30 g/l (Fig. 1.3). E24 increased all parameters at all concentrations, with maximum positive effect at 30 g/l. CR1 decreased all parameters at all concentrations.

1.6 *Gluconacetobacter azotocaptans* DS1 as Plant Growth Promoter

Role of *G. azotocaptans* DS1 as plant growth promoter is exclusively evaluated by authors. The strain improved the growth of corn, wheat, radish, pepper, tomato, and cucumber, under greenhouse and/or field conditions. Data mentioned above, support that DS1 has potential to be used as biofertilizer. As strain has ability of nitrogen fixation and IAA production, it seems that these two mechanisms are playing major role in plant growth promotion. Ability of this strain was evaluated under greenhouse and field conditions for cereals and vegetables, and in most of the cases results were positive even if non-significant. In soil, sand, and promix, DS1 strain has performed well.

Under tissue culture conditions, results were positive for lower concentration of sucrose. At higher concentration, it almost killed the plantlets. Effect of sucrose on plant growth is reported in literature. Studies have suggested extensive connections between sugar signaling and phytohormone pathways, in which abscisic acid acts positively and ethylene acts negatively (Kozuka et al. 2005). It is known that sucrose increases ethylene production in plant tissues (Meir et al. 1985) and also enhances the sensitivity to auxin (de Klerk et al. 1999). Calamar and Klerk (2002) studied the effect of sucrose concentration on adventitious root regeneration in apple and noticed strong reduction of rooting at higher sucrose concentration. Kozuka et al. (2005) examined the role of photoreceptors and sucrose on differential growth of leaf blade and petiole. They observed the inhibition in leaf blade expansion with increasing sucrose concentration in white light.

Researchers who isolated and named the species, i.e., *G. azotocaptans*, did not analyze the potential of this strain as PGPR; at least authors could not find any report. Therefore, all the information is based on one strain. Work done by Morley (2013) on this strain also validated its plant growth potential.

Unfortunately, only two strains of this species have been reported by now. It seems that more strains might exist but not discovered yet. One strong reason can be that 16S rRNA of *G. azotocaptans* has 98.5% similarity with *G. diazotrophicus* (Fuentes-Ramirez et al. 2001; Mehnaz et al. 2006) that is strong enough to declare it as *G. diazotrophicus*. To confirm it as *G. azotocaptans,* one has to get complete sequence of 1.5 kb, and that is not a regular practice; mostly researchers used a smaller fragment. Another way is to use specific primers designed for this species (Fuentes-Ramirez et al. 2001).

1.7 Conclusions

In this chapter, authors have discussed the role of *G. azotocaptans* in plant growth promotion. Most of the work was carried out by authors on isolate DS1, and data provided here strongly suggests that it should be used as biofertilizer. Now, DS1 strain is with an international company known for bioformulation production, and authors expect that may be soon we will see the commercially available bioformulation based on this strain alone or as a part of consortium.

In addition, authors recommend that researchers finding new strains of *G. diazotrophicus* should also make sure by doing the sequencing of longer fragment of 16S rRNA that they have the right species.

Acknowledgments Authors are grateful to Brian Weselowski, Tom Kowalik, and Linda Li for lab work. This work was supported by grant from Agriculture and Agri-Food Canada.

References

Calamar A, de Klerk GJ (2002) Effect of sucrose on adventitious root regeneration in apple. Plant Cell Tissue Organ Cult 70:207–212

de Klerk GJ, Paffen A, Jasik J, Haralampieva V (1999) A dual effect of ethylene during rooting of apple microcuttings. In: Altman A, Ziv M, Izhar S (eds) Plant biotechnology and in vitro biology in 21st century. Kluwer Publishers, Dordrecht, pp 41–44

Eskin N, Vessey K, Tian L (2014) Research progress and perspectives of nitrogen fixing bacterium, *Gluconacetobacter diazotrophicus*, in monocot plants.Int J Agron, *Vol 2014, Article ID 208383*

Fuentes-Ramirez LE, Bustillos-Cristales R, Tapia-Hernandez A et al (2001) Novel nitrogen fixing acetic acid bacteria, *Gluconacetobacter johannae* sp. nov. and *Gluconacetobacter azotocaptans* sp. nov., associated with coffee plants. Int J Syst Evol Microbiol 51:1305–1314

Jimenez-Salgado T, Fuentes-Ramirez LE, Tapia-Hernandez A et al (1997) *Coffea arabica* L., a new host plant for *Acetobacter diazotrophicus*, and isolation of other nitrogen fixing acetobacteria. Appl Environ Microbiol 63:3676–3683

Kozuka T, Horiguchi G, Kim GT et al (2005) The different growth responses of the *Arabidopsis thaliana* leaf blade and the petiole during shade avoidance are regulated by photoreceptors and sugar. Plant Cell Physiol 46(1):213–223

Lazarovits G, Nowak J (1997) Rhizobacteria for improvement of plant growth and establishment. Hortic Sci 32:188–192

Mehnaz S, Lazarovits G (2006) Inoculation effects of *Pseudomonas putida, Gluconacetobacter azotocaptans* and *Azospirillum lipoferum* on corn plant growth under green house conditions. Microb Ecol 51:326–335

Mehnaz S, Weselowski B, Lazarovits G (2006) Isolation of *Gluconacetobacter azotocaptans* from corn rhizosphere. Syst Appl Microbiol 29(6):496–501

Mehnaz S, Weselowski B, Lazarovits G (2007a) *Azospirillum zeae* sp. nov., diazotrophic bacteria isolated from rhizosphere soil of *Zea mays*. Int J Syst Evol Microbiol 57(12):2805–2809

Mehnaz S, Weselowski B, Lazarovits G (2007b) *Azospirillum canadense* sp. nov., a nitrogen fixing bacterium isolated from corn rhizosphere. Int J Syst Evol Microbiol 57(3):620–624

Mehnaz S, Kowalik T, Reynold B, Lazarovits G (2010) Growth promoting effects of corn (*Zea mays*) bacterial isolates under greenhouse and field conditions. Soil Biol Biochem 2(10):1848–1856

Meir S, Philosophos-Hidas S, Epstein E, Aharoni N (1985) Carbohydrates stimulate ethylene production in tobacco leaf discs: interaction with auxin and the relation to auxin metabolism. Plant Physiol 78:131–138

Morley RE (2013) Impact of free-living diazotrophs, *Azospirillum lipoferum* and *Gluconacetobacter azotocaptans*, on growth and nitrogen utilization by wheat (*Triticum aestivum* cv. *Lillian*). Dissertation, University of Saskatchewan

Munoz-Rojaz J, Caballero-Mellado J (2003) Population dynamics of *Gluconacetobacter diazotrophicus* in sugarcane cultivars and its effect on plant growth. Microbiol Ecol 46:454–464

Muthukumarasamy R, Cleenwerck I, Revathi G et al (2005) Natural association of *Gluconacetobacter diazotrophicus* and diazotrophic *Acetobacter peroxydans* with wetland rice. Syst Appl Microbiol 28:277–286

Yamada Y, Hoshino K, Ishikawa T (1997) The phylogeny of acetic acid bacteria based on the partial sequences of 16S ribosomal RNA: the elevation of the subgenus Gluconacetobacter to generic level. Biosci Biotechnol Biochem 61:1244–1251 [validation list no. 64. Int J SystBacteriol (1998) 48: 327–328]

Modifying the Rhizosphere of Agricultural Crops to Improve Yield and Sustainability: *Azospirillum* as a Model Rhizotroph

2

María Alejandra Pereyra and Cecilia M. Creus

Abstract

The chemicals used in agriculture to increase yields, and to kill pathogens, pests, and weeds, may have a harmful impact on the ecosystem. Current public concerns about the side effects of agrochemicals imposed scientists a new challenge in improving the understanding of cooperative activities among plants and rhizosphere microbial populations. The future goal is the gradual reduction in the use of chemicals without affecting yield or quality of the crops. A new generation of technologies must be developed focusing on the favorably partitioning of the biomolecules produced during the interaction between plants and microbes. The objective of this chapter is to review the current knowledge about the effects of plant growth-promoting rhizobacteria and their potential use as innovative tools for the sustainability of agroecosystems, with emphasis on the *Azospirillum*, and their use in Argentina.

2.1 Introduction

Agricultural intensification has greatly increased the productive capacity of agroecosystems, though it also has unintended environmental consequences including degradation of soil and water resources and alteration of biogeochemical cycles (Drinkwater and Snapp 2007; Lehman et al. 2015). In modern cultivation processes, indiscriminate use of fertilizers, particularly the nitrogenous and phosphorus ones, has led to substantial pollution of soil, air, and water (Gupta et al. 2015). The application of fertilizers on a long-term basis often leads to reduction in soil pH and in

M.A. Pereyra • C.M. Creus (✉)
Facultad de Ciencias Agrarias, Universidad Nacional de Mar del Plata,
Ruta 226 Km 73.5, 7620 Balcarce, Buenos Aires, Argentina
e-mail: creus.cecilia@inta.gob.ar

S. Mehnaz (ed.), *Rhizotrophs: Plant Growth Promotion to Bioremediation*,
Microorganisms for Sustainability 2, DOI 10.1007/978-981-10-4862-3_2

exchangeable bases, making nutrients unavailable to crops and declining their productivity (Gudzic et al. 2015). To obviate this problem and obtain higher plant yields, farmers have become increasingly dependent on chemical sources of nitrogen (N) and phosphorus (P). Besides being costly, the production of chemical fertilizers depletes nonrenewable resources and poses human and environmental hazards (Joshi et al. 2006). It has also been reported that the excessive use of chemicals not only affects the fertility status of soil and pollutes the environment but also might exert deleterious effects on soil microorganisms (Youssef and Eissa 2014).

By the other side, pathogenic microorganisms that affect plant health are also a major and chronic threat to sustainable agriculture and ecosystem stability, worldwide. Pesticides still represent one of the main pressures from agriculture on human health and ecosystems, though chemical substances placed on the market tend to be less harmful in response to the requirements of the directives and regulations in force, as they undergo more risk assessment (Eurostat. Statistics Explained 2012).

All these setbacks were the consequence of centering agricultural production and soil conservation to cover human needs, without considering either soil stability or soil health (Welbaum et al. 2004). The importance of soil health and quality in relation to sustainable land management is an actual concern (Doran 2002; Karlen 2012).

2.2 The Concept of Soil Health

Soil is a dynamic living matrix, and it is a critical resource in agricultural and food security. Soil health is defined as the sustained capacity of soil to function as a vital living system. This concept is based on that it contains biological elements that are key to ecosystem function within land-use boundaries (Doran and Zeiss 2000; Karlen et al. 2001). These functions are able to sustain biological productivity of soil, maintain the quality of surrounding air and water environments, as well as promote plant, animal, and human health (Doran et al. 1996). Keeping this definition in mind, the quality of a specific kind of soil has been defined as "the capacity to function, within natural or managed ecosystem boundaries, to sustain biological productivity, promote environmental quality, and maintain plant and animal health" (Blanco and Lal 2012). Soil erosion, atmospheric pollution, extensive soil cultivation and grazing, high irrigation, salinization, and desertification not only decrease the productivity of an agricultural land but also perturb or degrade its health. A balance of chemical, physical, and biological components contributes toward maintaining soil quality (Das and Varma 2010). Agroecosystem functioning is governed largely by soil microbial dynamics (Kennedy and Smith 1995). Sustainable and productive agriculture depends on a healthy community of soil microbes. These decompose organic matter and contribute to the biological recycling of chemical nutrients that affect soil fertility. The functioning of agricultural ecosystems, including the health of soil, mostly depends on the interaction between the diversity of primary producers (plants) and decomposers (microbes), which are the two key functional groups that form the basis of all soil ecosystems (Elmqvist et al. 2010).

Microbes balance soil ecology being an integral part of every soil ecosystem. Only 1 g of soil may contain billions of microbes with thousands of different species. Metabolic activities of the microbes, such as plant growth-promoting rhizobacteria (PGPR), cyanobacteria, and fungal organisms (mycorrhiza), and of soil fauna (nematodes, worms, protozoans, etc.) promote soil health and crop productivity. Several biotic or abiotic factors lead to the alteration of the microbial community structure and composition, which may influence directly or indirectly the soil ecosystem, nutrient cycling activity, and crop production (Chaudhry et al. 2012).

In addition, anthropogenic intervention for the management and treatment of soil, involving fertilizers, pesticides, manure, or genetically modified microorganisms and plants, also influences microbial diversity (Nautiyal 2012). Different studies have shown that the application of chemical fertilizers (NPK) enhanced crop yield but affected the diversity in microbial population and their enzymatic activities (Zhang et al. 2015). By the other side, it has been shown that chemical fertilizers could increase the soil microbial biomass, exerting no significant changes in the microbial characteristics of the soil (Nautiyal 2012). Evidences linking direct impact of chemical fertilizers on microbial diversity, function, and phylogeny are still not well documented (Nautiyal 2012).

Changes in microbial parameters are correlated with the soil organic carbon content, as all the soil organisms essentially need a carbon source for their survival, and not to the application of P and N. This is the reason why soils poor in organic matter are usually poor in microbial activities. Crop productivity greatly depends upon the amount of available nutrients in the soil, which is governed by transformations of soil microbial biomass. Thus, the growth and activity of microorganisms are functions of soil properties, such as nutrition, texture, pH, temperature, and moisture content, and they are sensitive indicators of changes in these soil properties. The optimal functioning of each organism usually appears as a part of small well-structured communities carrying species which are interdependent on each other. In this context, augmentative approaches study the possibility of isolating organisms from the vast pool of biodiversity, with any special enhanced activity, and introducing them in the ecosystem. These activities can be also enhanced by manipulation such as drainage (aeration) or crop rotation. This approach can involve the use of selected wild-type organisms or genetically modified organisms which have their function introduced or enhanced by the use of recombinant DNA. Nevertheless, such constructed organisms may not fit ecologically as the comparable wild types (Prakash et al. 2011). Future marketing of transgenic bioinoculant products and their release into the environment as eco-friendly alternations to agrochemicals will depend on the generation of the biosafety data required for the registration of PGPR agents.

2.3 The Rhizosphere

The rhizosphere, the interface between growing roots and the mineral world in the soil, provides a particular ecosystem where ecological feedbacks, chemical interactions, and inter-organism communication take place. According to root nearness, soil can be divided in three main zones: (1) the rhizoplane or root surface, (2) the ectorhizosphere that portion of the soil under root influence, and (3) soil that is devoid of plants (Manthey et al. 1994). The ectorhizosphere, the rhizoplane, and the root cortex are together called the rhizosphere.

Rhizosphere affects and even transforms a large soil environment, including all of the so-called bulk soil (Richter et al. 2007). Plants alter the rhizobacterial community by releasing different substrates, which can vary from single sugar components to complex aromatic structures, and therefore selecting for increased numbers of certain taxa and/or functional groups of bacteria (Kravchenko et al. 2003). Microorganisms can also influence the plant by promoting or inhibiting growth (Glick et al. 1998). Many of the interactions between microbes and plants are still unknown.

During the past years, there has been an increased recognition of the role that biological processes play in soil function and in sustainable crop production (Nautiyal 2012). A recent major strategy to counteract the rapid decline in environmental quality is to promote sustainable agriculture. The objective is to sustain high production with the gradual reduction in the use of fertilizers and pesticides and the greater use of the biological and genetic potential of plant and microbial species. In this sense, the strategies that relay on sustainable agricultural techniques do not harm the environment or health, not only for animals and human beings but also for soil.

2.4 Using Microbes to Attain Higher Benefits in Sustainable Agriculture

2.4.1 Beneficial Microbial Modes of Action on Plants

The application of PGPR as biofertilizers and biocontrol agents is being considered as an alternative or supplemental way of reducing the use of chemicals in crop production (Kloepper et al. 1989; Vessey 2003; Maheshwari 2011). In addition to the plants ability to modify their physiology and metabolism, certain rhizosphere microorganisms can help plants to either avoid or partially overcome environmental stresses (Govindasamy et al. 2008). PGPR that were isolated from heavy metal polluted soils are able to enhance plant growth and development under heavy metal stress conditions, such as in the presence of arsenic (Reichman 2014), cadmium (Guo and Chi 2014), or both zinc and cadmium (Pereira et al. 2015). This line of research is oriented to soil bioremediation attaining in addition a boost effect on crop growth.

Sustainable approaches are those that not only aim to improve short-term crop yields but also to assure the maximum long-term performance, protecting the ecology of agricultural systems and the interests of the farmers. The so-called microbial technologies deal with restructuring the crop rhizosphere by inoculating crops with beneficial microorganisms and using cultural practices that enrich indigenous beneficial ones. The long-term sustainability of agricultural systems is highly dependent on effective handling of the indigenous resources of agroecosystems. Actual methods, used to investigate microbial structure and composition, include culture-dependent and molecular methods. The first ones can detect less than 1 % of the microorganisms present in soil. The second ones include high-throughput DNA-sequencing techniques with the potential to detect, cost-effectively, low abundant uncultivable microbial species (Roesch et al. 2007). Another tool is the phospholipid fatty acids profiles determination that provides a wide-ranging measurement of microbial communities at the phenotypic level. Although these profiles do not give information on species composition, they reveal the fingerprint of community structure. Actually, they are considered as a robust tool that consistently discriminates between different communities (Kaur et al. 2005). During the last 20 years, the new biotechnologies have opened new scenes for the enhancement of sustainable agriculture production. These advances have made possible to take advantage of soil microorganisms for improving crop productivity. They also offer an economically attractive and ecologically viable choice to reduce external inputs.

When applied as inoculants, the so-called PGPR enhances plant growth by a wide variety of mechanisms that have been classified by their direct or indirect modes of action. The first comprises the production of bacterial metabolites, mainly phytohormones, that stimulate plants (Dobbelaere et al. 2003) but also include volatiles (Ping and Boland 2004; Santoro et al. 2015) and signal molecules like nitric oxide (Creus et al. 2005; Molina Favero et al. 2008). Lowering the ethylene level in plants (Glick et al. 2007), improving the plant nutrient status by mobilizing nutrients in soils or fixing atmospheric N (Hayat et al. 2010), and stimulating disease resistance in plants by triggering induced systemic resistance response (Van Wees et al. 1997) are also direct modes of action. Indirect effects are originated on the ability of some PGPR to constraint other soil microbes thus giving pathogenic ones less chance to develop (Lugtenberg and Kamilova 2009).

Both symbiotic and nonsymbiotic associations between organisms in the rhizosphere rely on interacting factors and chemical signals that operate on time and space scales. Among these, compounds of hormonal nature play major roles. To make the picture more complex, all these factors vary with water content, temperature, nutrients and soil structure, and others (Molina Favero et al. 2007).

2.4.2 Microbial Inoculants as Components in Current Agronomical Practices

The progress made in the last three decades in the understanding of the diversity of PGPR, along with their colonization abilities, and modes of action has facilitated their application as a new component in the management of sustainable agricultural systems. The practical application of living bacteria as inoculants was quite controversial from the beginning, because the response of crops is not completely predictable and depends on many ecological and agro-technological factors. Nevertheless, much progress has been made in this field, leading to an ever-growing and successful application of rhizobacteria in several regions of the world, especially in South and Central America. Diverse symbiotic (*Rhizobium*, *Bradyrhizobium*, *Mesorhizobium*) and nonsymbiotic (*Pseudomonas*, *Bacillus*, *Klebsiella*, *Azotobacter*, *Azospirillum*) rhizobacteria are now being used worldwide as inoculants to promote plant growth and/or to protect crops and attain higher yields.

There are a number of studies revealing the benefits of bacterial inoculation even though chemical application is accomplished as a current agronomical practice. Beneficial bacteria showed certain and variable grade of compatibility with herbicides (Ahemad and Khan 2010, 2011a), insecticides (Ahemad and Khan 2011b, c), and fungicides (Pereyra et al. 2009; Ahemad and Khan 2011d, 2012) even under stress conditions like soil heavy metal contamination (Wani and Khan 2010; Ma et al. 2011), salinity (Mayak et al. 2004), or water stress (Creus et al. 2004; Pereyra et al. 2009).

The leading countries in field applications of PGPR are Mexico with estimated 300,000 ha inoculated fields in 2007 and Argentina where over 220,000 ha of wheat and corn were commercially inoculated with *Azospirillum* in 2008 (Bashan and Hartmann 2009). Particularly, *Azospirillum* sp. has been commercially used on a relative large scale in Argentina, Mexico, Europe, South Africa, and India, mainly on cereals but also on other crops (Fuentes-Ramirez and Caballero-Mellado 2005; Bashan and Hartmann 2009; Díaz-Zorita and Fernández-Canigia 2009). Many of these studies showed promising results as "Microbial Technologies."

Regarding on the benefits they induce on crops, PGPR can be classified as phytostimulators, biofertilizers, and biocontrol agents, depending on the proposed use of the commercial product. Nevertheless, certain groups of bacteria show overlapping applications. Here, we describe proved cases of the application of PGPR on crops that can be ascribed to one or more of these groups.

2.4.2.1 Phytostimulators

A major goal to improve agricultural performance and increase food production is to attain high yields, even at low soil fertility or without intensive fertilization. To achieve this goal, root development and physiology appear to be central. Roots are dynamic anchorages of plants. They not only support the whole plant growth but also its physiological activity. Treatments to achieve greater adventitious rooting, increased number of lateral roots, higher length, and density of root hairs are targets of many research projects in plant biology (Molina Favero et al. 2007). It is generally accepted that root activity modifies the habitat of microorganisms and these, in turn, could trigger changes in the overall plant behavior. Some PGPR stimulate plant growth directly by the production of substances that mimic plant hormones. However, to produce these stimulatory effects on plants, the first step is the proper root colonization by bacteria that attach to root surface forming clumps or biofilms on it (Salcedo et al. 2015). As a primary target, root is the organ that shows the first stimulating bacterial effects. This is particularly remarkable in plants inoculated with *Azospirillum* spp. (Okon 1985).

The production of plant growth regulators, mainly auxins, cytokinins, and gibberellins, is the most commonly invoked mechanism for plant growth promotion exerted by PGPR (Bashan and de Bashan 2010). Auxin, cytokinin, gibberellin, abscisic acid, and jasmonate production has been reported in several associative and endophytic diazotrophic species of many genera such as *Azospirillum*, *Klebsiella*, *Gluconacetobacter*, *Azoarcus*, *Herbaspirillum*, *Enterobacter*, *Bacillus*, *Achromobacter*, *Acetobacter*, *Burkholderia*, *Pseudomonas*, *Serratia*, *Xanthomonas*, and *Azotobacter* (Ping and Boland 2004; Tsavkelova et al. 2006; Baca and Elmerich 2007). Auxins and cytokinins are important regulators of plant development, regulating processes involved in the determination of the root architecture (Overvoorde et al. 2010). Nevertheless, auxins are thought to play the major role in stimulating root growth by rhizobacteria (Dobbelaere et al. 2003). Cytokinins play important roles for plant developmental processes from seed germination to senescence, including maintenance of stem cell systems in shoots and roots, organogenesis, leaf senescence, and interacting with auxins both participate in root vascular development and the control of shoot branching (Castillo et al. 2015). Early work from Barea et al. (1976) found that at least 90 % of the bacteria isolated from the rhizosphere of important crops were able to produce cytokinin-type compounds in chemically defined medium.

Gibberellins play an important role in the early stages of plant development by enhancing shoot and root growth and increasing root hair density, though they also regulate many aspects of reproductive growth in plants. Bottini et al. (1989) were the first to confirm the ability of *Azospirillum* sp. to produce gibberellins in chemically defined culture medium. Gibberellic acid production and their conjugates metabolism by *Azospirillum* sp. were summarized by Bottini et al. (2004).

Experiments with IAA-attenuated mutant bacteria inoculated on wheat (Barbieri and Galli 1993; Dobbelaere et al. 1999) or those carried with dwarf rice or maize deficient in the production of physiologically active gibberellins (Castillo et al. 2015) are strong evidence that the production of phytohormones by associated bacteria accounts for the phytostimulatory effects. In addition, the study of the expression profiles of inoculated plants would help to understand the complex metabolic changes produced upon inoculation. The transcript profile of in vitro grown sugarcane inoculated with *G. diazotrophicus* and *H. rubrisubalbicans* revealed differentially expressed genes related to auxins, gibberellins, and ethylene (Nogueira et al. 2001). The transcriptional profile of rice plants inoculated with *H. seropedicae* identified expressed sequence tags (ESTs) involved in auxins and ethylene pathways that are regulated during the association (Brusamarello-Santos et al. 2012).

Apart from the production of plant growth regulators, the decrease in the levels of ethylene in inoculated plants is another proposed phytostimulatory effect (Glick 2004). Some rhizosphere and endophytic bacteria produce the enzyme 1-aminocyclopropane-1-carboxylate deaminase (ACC deaminase). The activity of ACC deaminase can reduce ACC content from the ethylene biosynthesis pathway in plants (Desbrosses et al. 2009). ACC deaminase-containing rhizobacteria bound to a plant act as a sink for ACC, thereby lowering ethylene levels in plant tissues. The result of the functioning of this enzyme is an increase in the growth of plant roots and shoots and a reduction of the inhibitory effects of ethylene synthesis especially during stressful conditions (Glick 2004).

The phytostimulatory effects of PGPR on plants were also studied in greenhouse and field conditions. Experiments involving *Azospirillum* inoculation during the 1990s were carried out in many countries including Israel, France, Belgium, Argentina, Uruguay, Mexico, and South Africa. These experiments' results were summarized in two interesting reviews by Okon and Labandera-González (1994) and Dobbelaere et al. (2001). They concluded that inoculation with *Azospirillum* resulted in significant yield increases in the magnitude of 5–30 % in about 60–70 % of the experiments. The beneficial effects were mainly observed in lighter soils under intermediate levels of fertilizer (N, P, and K) and water regimes. These pioneer reviews on the application of bacteria at the field lead to establish the first basis for designing larger experiments to assess the conditions and management practices for attaining positive and reproducible results.

Veresoglou and Menexes (2010) concluded in an excellent report on a meta-analysis conducted on 59 available articles to evaluate the extent, to which *Azospirillum* might contribute to wheat growth properties, that a mean increase of 8.9 % in seed yield and of 17.8 % in aboveground dry weight resulted from inoculation of wheat with *Azospirillum*. Other crops like corn, sorghum, rice, and legume showed yield increases in ranges from 5 % to 30 % over non-inoculated controls (Dobbelaere et al. 2001; Díaz-Zorita and Fernández-Canigia 2009; Hungria et al. 2010; Helman et al. 2011).

2.4.2.2 Biofertilizers

Apart from the described increasing effects on growth and yield, PGPR application might also enhance nutrient uptake from soils, thus reducing the need for fertilizers and preventing water contamination with nitrate and phosphate accumulation in agricultural soils (Bashan and de Bashan 2010). The expanded root system can improve the efficiency of the uptake of soil nutrients and fertilizers. This can be accomplished either by increasing the volume of explored soil or by enhancing nutrients uptake rates per root surface unit. The first possibility is the best established for PGPR as the better exploration of soil allows a major accessibility to micro-sites where low mobile nutrients could be enriched. A reduction in fertilizer application would lessen the effects of water contamination from fertilizers and lead to economical savings for farmers. These savings would increase the cost/benefit ratio, a crucial aspect for sustainable agriculture in many developing countries. Symbiotic processes leading to enhanced N fixed by *Rhizobium* and *Bradyrhizobium* species or P availability by mycorrhizal fungi have been the most studied up today. These two genera of microorganisms are well known, and many researchers have shown the contribution in N and P, respectively.

Though in less magnitude, associative nonsymbiotic bacteria can contribute for fixing N (Welbaum et al. 2004) or remobilizing of nonmobile P sources by acid production (Ahemad and Kibret 2014). When applied from outside as inoculants, PGPR facilitate resource acquisition (N, P, and essential minerals). In this sense, the theoretical needs of chemical inputs could be decreased. Nevertheless, the balance of P and other elements is a long-term variable that must be analyzed to determine if the system is sustainable. Although the cumuli of knowledge in PGPR effects and their modes of action are very large, actually the information on practical biofertilization techniques to lesser chemical inputs is still scarce.

A two-season field study was performed in the south of Vietnam to assign the effects of a product containing a pseudomonad, two bacilli and soil yeast, on rice. Results indicated that application improved significantly the N use efficiency by rice, saving 43 kg N ha^{-1} with an additional yield of 270 kg ha^{-1} in the two consecutive seasons (Cong et al. 2009). The extra efficiency was shown by the fact that both treatments, biofertilizer with the application of about 40 and 60 kg less N-fertilizer and urea alone full dose, reached the same maximum yields in two successive harvests on the same plots (Cong et al. 2009).

Results obtained from a 3-year field research conducted to test whether microbial inoculants could be used to increase maize yield and to enhance nutrient uptake were published by Adesemoye et al. (2008). They showed that inoculated plots removed higher amounts of N, P, or K from the soil, potentially reducing nutrient losses to the environment.

In a large study conducted during 2002–2006 growing seasons, the performance of a commercial inoculant based on INTA Az-39 strain of *A. brasilense* was evaluated in 297 experimental field trials in the Pampas region of Argentina (Díaz-Zorita and Fernández-Canigia 2009). At all sites, the sown wheat varieties were regionally adapted and recommended for high yielding environmental and crop management conditions. N and P fertilization were applied when necessary according to recommendations based on chemical soil analysis and suggested protocols for each local site. Wheat grain yield from those 297 experimental sites varied in a range from 850 to 8050 kg ha^{-1} according to the management. The yield average increase was 260 kg ha^{-1}, equivalent to 8.0 % of the mean wheat yield attained under the dry land farming conditions found in the region. Positive responses were determined in about 70 % of the sites, depending mostly on the attainable yield, and independently of fertilization and other crop, and soil management practices. This is in agreement with the reported efficiency estimated from green house and field studies conducted in different parts of the world (Okon and Labandera-González 1994; Dobbelaere et al. 2001). The interaction between inoculation and N and/or P fertilization was also analyzed. As it was expected, fertilized wheat yield was enhanced with respect to that of unfertilized crop. However, regardless of the fertilization practice, inoculation significantly and positively affected yield, with mean yield responses of 259 and 260 kg ha^{-1} for unfertilized and fertilized wheat, respectively.

Sugarcane inoculation with *G. diazotrophicus* also resulted in improved N uptake (Suman et al. 2005). Studies on rice inoculated with ten different associative and endophytic diazotrophs, including *Paenibacillus* sp., *Bacillus* sp., *Burkholderia* sp., *Herbaspirillum* sp., and *Azorhizobium* sp., indicated that bacterial inoculation had a significant positive impact on N uptake and on shoot and root growth (Islam et al. 2009).

Despite of the large reservoir of P on earth, the amount of soil available P-forms to plants is generally low because the majority of soil P is found in insoluble forms, while plants absorb it only as soluble ionic phosphates (Bhattacharyya and Jha 2012). Organisms coupled with phosphate-solubilizing activity often termed as phosphate-solubilizing microorganisms may provide the available forms of P to the plants and hence comprise a viable substitute to chemical P fertilizers (Khan et al. 2006). Bacteria of the genera *Azotobacter*, *Bacillus*, *Beijerinckia*, *Burkholderia*, *Enterobacter*, *Erwinia*, *Flavobacterium*, *Microbacterium*, *Pseudomonas*, *Rhizobium*, and *Serratia* are reported as the most significant phosphate-solubilizing bacteria (Bhattacharyya and Jha 2012).

The increment of plant P nutrition by inoculation with PGPR can be explained by different mechanisms. One is the capability these soil bacteria have to solubilize inorganic P as a consequence of the action of low molecular weight organic acids they are able to synthesize and excrete (Zaidi et al. 2009). Another different mechanism is the mineralization of organic P that occurs through the synthesis of a variety of different phosphatases, catalyzing the hydrolysis of phosphoric esters. Nevertheless, both mechanisms can coexist in the same bacterial strain (Ahemad and Kibret 2014).

Many strains of *Pseudomonas* are able to solubilize P in soil and increase its availability to plants (Sundara et al. 2002). The beneficial effects of the inoculation with phosphate-solubilizing bacteria as single or combined inoculants are well documented (Ahemad and Kibret 2014). Positive results of inoculation with phosphate-solubilizing bacteria of soybean (Fernández et al. 2007), sorghum (Vikram 2007), wheat (Afzal and Bano 2008), and many other crops were reported.

Seed inoculation with *H. seropedicae* enhanced the N content in leaf of maize (12 %) under soil acidity conditions without N supply (Inagaki et al. 2015). Inoculation of PGPR in acidic sandy soil (4.5–5.0 pH) resulted in higher P concentration in the leaf tissue of maize, indicating increase of P solubilization promoted by the diazotrophic bacteria (Inagaki et al. 2015).

Microorganisms such as those of the genera *Aspergillus*, *Bacillus*, and *Clostridium* were found to be efficient in potassium solubilizing by excreting organic acids that directly dissolve rock potassium or chelate silicon ions to solubilize and mobilize it in different crops (Mohammadi and Sohrabi 2012; Parmar and Sindhu 2013). Nevertheless, little research has been done on potassium solubilization, which is the third major essential macronutrient for plant growth.

By the other side, iron is an essential compound for most living organisms. However, despite its abundance on earth, and the micromolar concentrations required for cell growth, it is biologically unavailable in most environments. Its availability in nature is limited by the rapid oxidation of the ferrous form to the very insoluble ferric form, which aggregates into insoluble oxy-hydroxide polymers. Also reduced ferrous form might induce Fenton reaction producing free radicals which are deleterious to cellular macromolecules (Halliwell and Gutteridge 1984). To fulfill their iron needs, bacteria have multiple iron acquisition systems. One of relevant importance for rhizobacteria relies on molecules (siderophores and hemophores) synthesized and released by bacteria into the extracellular medium; these molecules scavenge iron or heme from various sources (Wandersman and Delepelaire 2004). Siderophores are low molecular weight compounds produced by microorganisms under limited availability of iron. These compounds are able to bind iron from the environment and to transfer it into the bacterial cell (Stintzi et al. 2000). Studies with PGPR showed that siderophore-mediated iron uptake systems present in these microorganisms exert a strong influence on the whole microbial community that can be quite beneficial to the plant (Kloepper et al. 1980). Yet in spite of many researches, it remains to be elucidated if the true effect relies on a better plant iron nutrition or if it is a biocontrol on the pathogenic bacteria for the quest of iron in the rhizosphere. The probable implication of siderophores produced by PGPR has been considered as a potential way to improve plant growth, nodulation, and N_2 fixation in iron-deficient conditions (Fernández-Scavino and Pedraza 2013). Plants assimilate iron from bacterial siderophores by means of different mechanisms, for instance, chelate and release of iron, the direct uptake of siderophore-Fe complexes, or a ligand exchange reaction. Pandey et al. (2005) observed that *Pseudomonas aeruginosa* GRC1, a prolific producer of hydroxamate-type siderophores in iron-deficient conditions enhanced the growth of *Brassica campestris* in field trials. More recently, Radzki et al. (2013) showed that siderophores produced by *Chryseobacterium* C138 provided iron to iron-starved tomato plants in hydroponics culture.

2.4.2.3 Biological Control Agents

Some root-colonizing bacteria are able to both suppress disease in host plants by the production of inhibitory compounds that restrain soil pathogen growth and, at the same time, stimulate growth and defense responses in host plants. Biological control is, thus, considered as an alternative or a supplemental way of reducing the use of chemicals in agriculture (Gerhardson 2002; Lugtenberg and Kamilova 2009).

The mechanism of pathogen growth inhibition is due to diverse metabolic abilities of biocontrol bacteria that produce inhibitory allelochemicals. These compounds include iron-chelating siderophores, antibiotics, and antifungal metabolites like HCN, phenazines, pyrrolnitrin, 2,4-diacetylphloroglucinol (DAPG), pyoluteorin, viscosinamide, and tensin (Compant et al. 2005; Bhattacharyya and Jha 2012). Biocidal volatiles, lytic enzymes, and detoxification enzymes produced by some bacteria are also other metabolic biocontrol compounds (Compant et al. 2005).

As it was stated in the preceding section, iron is an essential element with low bioavailability. Under iron-limited conditions, the strongly producers of siderophores are correlated with an enhanced capacity to niche occupancy required by many rhizobacteria. Although various bacterial siderophores differ in their abilities to bind iron, in general, they deprive pathogenic fungi of this essential element since the fungal siderophores have lower affinity (Compant et al. 2005).

Pseudomonas spp. are ubiquitous bacteria in agricultural soils and have many traits that make them well suited as biocontrol agents of soilborne pathogens (Weller 2007). Weller (2007) proposed that pseudomonads are able to improve plant growth by suppressing either "major" pathogens (produce well-known root or vascular diseases with obvious symptoms) or "minor" pathogens (parasites or saprophytes that damage mainly juvenile tissue such as root hairs and tips and cortical cells). In addition to pseudomonads, the genus *Bacillus* is widely recognized as a powerful biocontrol agent. *B. subtilis* and other bacilli are potentially useful as biocontrol agents due to their broad host range and their ability to form endospores and to produce different biologically active compounds with a broad spectrum of activity (Nagorska et al. 2007).

On the other hand, colonization of roots with PGPR can lead to systemic resistance in the plant providing protection against several types of pathogenic diseases.

The protection is typically manifested as both a reduction in disease symptoms and an inhibition of pathogen growth. This process appears to be phenotypically similar to pathogen-induced systemic acquired resistance. This effect of rhizobacteria is referred to as an induced systemic resistance (ISR) and has been demonstrated in different plant species, e.g., bean, carnation, cucumber, radish, tobacco, and tomato, and in the model plant *Arabidopsis thaliana* (Van Wees et al. 1997). Elicitation of ISR by plant-associated bacteria was initially demonstrated using *Pseudomonas* spp. and other Gram-negative bacteria (van Loon and Glick 2004). Various specific strains of *Bacillus* also elicit a significant reduction in the incidence or severity of various diseases on a diversity of hosts (Kloepper et al. 2004). Many individual bacterial components induce ISR, such as lipopolysaccharides, flagella, siderophores, cyclic lipopeptides, DAPG, homoserine lactones, and volatiles like acetoin and 2,3-butanediol (Lugtenberg and Kamilova 2009).

Root inoculation of *A. thaliana* ecotype Columbia with *Pseudomonas fluorescens* CHA0 partially protected leaves from the oomycete *Peronospora parasitica* (Iavicoli et al. 2003). Using mutants derived from strain CHA0 (pyoverdine deficient, exoprotease deficient, HCN deficient, pyoluteorin deficient, or DAPG deficient), it was demonstrated that DAPG production in *Pseudomonas fluorescens* is required for the induction of ISR in *Peronospora parasitica* (Iavicoli et al. 2003). Although DAPG is a known antifungal compound, it was also demonstrated to affect the physiology of plants and other eukaryotes (Keel et al. 1992). Although the mode of action of DAPG is not fully understood, it inhibited primary root growth and stimulated lateral root production in tomato seedlings (Brazelton et al. 2009). In this sense, there seems to be overlapping effects of some secondary metabolites. Apart from exerting direct biocontrol by competing with pathogens for the resources or the niche occupancy, they can also stimulate the host growth or induce ISR.

Whatever the mechanism involved in the control, the ability of biocontrol agents to compete in the rhizosphere is crucial to produce their benefits when commercial products are applied in the field. This competence comprises effective root colonization combined with the ability to survive and proliferate along growing plant roots over a considerable period, in the presence of the indigenous microflora (Parke 1991; Lugtenberg et al. 2001). Given the importance of rhizosphere competence as a prerequisite for an effective biological control, understanding root-microbe interaction, and the genetic and environmental factors that affect rhizosphere colonization, will significantly contribute to improve the efficacy of products based on biocontrol agents.

2.5 *Azospirillum* as a Model Rhizotroph

Among PGPR, the species of *Azospirillum* have gained the reputation of being the most studied plant-associative bacteria (Bashan et al. 2004). When present in plants in proper amounts, they stimulate the density and length of root hairs, the rate of appearance of lateral roots, and root surface area (Okon and Labandera-González 1994). These effects cause roots to take up more water and mineral nutrients resulting in faster plant growth. Under appropriate agronomic conditions, these processes would increase crop yield (Creus et al. 2004). In general, they contribute to reduce the burden of soil nutrient loss in arable lands, to counteract part of the negative effects of water and saline stresses on plant growth, and to help plants avoid or minimize contaminant uptake (Barassi et al. 2007). Moreover, in view of the high input of agrochemicals in contemporary crop production and the likelihood that *Azospirillum* inoculation could be regularly used in the near future in regular crop production, studies on the interaction of *Azospirillum* inoculants with common pesticides are essential (Pereyra et al. 2009).

Azospirillum spp. are included into the alpha subclass of *Proteobacteria* belonging to the IV rRNA superfamily (Xia et al. 1994). After the recent reclassification of *Azospirillum irakense* to *Niveispirillum irakense* and *Azospirillum amazonense* to *Nitrospirillum amazonense* based on their polyphasic taxonomic characteristics, at present this genus encompasses 15 valid species (Young et al. 2015). *A. brasilense* is the most used as biofertilizer. *Azospirillum* is not a plant-specific bacterium but a general root colonizer. Although it has first been isolated from and studied on cereal crops, at present there are more non-cereal species successfully inoculated with *Azospirillum* (Bashan and de Bashan 2010).

Azospirillum congregates several characteristics present in different microorganisms that make it a valuable PGPR. The very first studies on *Azospirillum*-inoculated subtropical grasses (*Z. mays*, *O. sativa*, and forages such as *Digitaria* spp.) attributed the growth promotion effects primarily to the biological N_2 fixation exerted by the bacteria (Döbereiner and Day 1976). Even though this characteristic could be extremely valuable in agriculture, later field studies failed to demonstrate a significant N_2 fixation in *Azospirillum*-inoculated crops (Vande Broek et al. 2000). Further studies ascribed the positive bacterial effects on plants to morphological and physiological changes in the inoculated roots, which would lead to an enhancement of water and mineral uptake (Okon and Kapulnik 1986). *Azospirillum brasilense* produces plant growth regulators mainly IAA, which is associated with the beneficial effects observed after inoculation (Baca and Elmerich 2007). Several mechanisms have been postulated to explain how *Azospirillum* enhances growth and development of plants (Bothe et al. 1992; Bashan and Holguin 1997; Steenhoudt and Vanderleyden 2000; Bashan and de Bashan 2010). Nevertheless, to date no unique mechanism had been established to explain the growth promotion capability of these bacteria. Instead, the most accepted hypothesis postulates that a sum of events accounts for the general plant growth promotion effect (Bashan and Holguin 1997).

It was previously reported that the cell wall is a target for *A. brasilense* growth promotion (Creus et al. 2004; Pereyra et al. 2010). Plant cell growth is constrained by the primary cell wall which consists of cellulosic microfibrils embedded in a matrix of interwoven noncellulosic polysaccharides and proteins. Cucumber seeds inoculated with *Azospirillum* resulted in seedlings presenting larger hypocotyls. Cell wall dynamics of these inoculated plants was affected including greater acid-induced cell wall extension and lower activity of two important enzymes from the cell wall metabolism, NADH oxidase and ferulic acid peroxidase (Pereyra et al. 2010). These lesser activities, coupled with a lesser content of ferulic acid, responsible of the stiffening of the cell wall, could be another mechanism accounting for the growth promotion induced by *Azospirillum* (Dal Lago et al. 2015).

These and other physiological changes observed in the inoculated plants subjected to abiotic stresses were reported. *Azospirillum*-inoculated wheat (*T. aestivum*) seedlings subjected to mild osmotic stress developed significant higher coleoptiles, with higher fresh weight and better water status than non-inoculated seedlings (Alvarez et al. 1996; Creus et al. 1998). A larger root system was evident in *Azospirillum*-inoculated wheat seedlings growing either under well-irrigated or water stress conditions (Pereyra et al. 2006). It was also proved that part of the negative effects, mild and severe salt stresses would cause on wheat seedlings were significantly reversed in *Azospirillum*-inoculated roots (Creus et al. 1997). Fresh weight, fresh weight/dry weight, water content, and relative water content were higher in shoots from inoculated plants than in stressed controls (Creus et al. 1997). Vessel size has an important role in the adaptation to water stress environmental conditions. *A. brasilense*-inoculated wheat seedlings showed wider xylem vessels and less negative water potential in their coleoptiles when grown exposed to osmotic stress (Pereyra et al. 2012). The induction of wider xylem vessels by inoculation might imply an enhanced coleoptile hydraulic conductance which in turn could explain the better water status observed in plants. Indeed, field experiments carried out with different *Azospirillum* strains in *S. bicolor*, *Z. mays*, and *T. aestivum* have shown significantly increased yields, enhanced mineral uptake, and less canopy temperature (Sarig et al. 1988; Okon and Labandera-González 1994; Casanovas et al. 2003; Creus et al. 2004). In this sense, inoculation technology with *Azospirillum* could be extended to arid soils to protect crops against drought. Under drought conditions, inoculated plants responded in a different way to water stress compared to non-inoculated ones. They showed significantly higher water content, relative water content, water potential, apoplastic water fraction, and lower cell wall modulus of elasticity values (Creus et al. 2004).

Although *Azospirillum* is not considered to be a classic biocontrol agent of soil-borne plant pathogens, there have been reports on moderate capabilities of *A. brasilense* in biocontroling crown gall-producing *Agrobacterium* (Bakanchikova et al. 1993), bacterial leaf blight of mulberry (Sudhakar et al. 2000), and bacterial leaf and/or vascular tomato diseases (Bashan and de-Bashan 2002a, b). In addition, *A. brasilense* can restrict the proliferation of other nonpathogenic rhizosphere bacteria (Holguin and Bashan 1996).

It is agreed that the benefits *Azospirillum* imposes on plants rely upon root colonization. In this sense, the formation of complex bacterial communities on the roots, known as biofilms, is crucial. Previous studies showed that nitric oxide production by *A. brasilense* Sp245 was responsible, at least in part, of the effects on root growth and proliferation (Creus et al. 2005). Nitric oxide is also a signaling molecule implicated in biofilm formation and was shown to regulate the formation of biofilm in *A. brasilense* Sp245 (Arruebarrena di Palma et al. 2013). Biofilm dynamics is of enormous importance for *Azospirillum* to exert beneficial effects on plants. So, the mechanisms operating in these phenomena are intensively and actively investigated.

Finally, in the last years, biofertilizers composed of mixed species are being used. They showed a better impact in crop yields than single species ones.

Ruiz-Sánchez et al. (2011) reported that rice inoculation with *Glomus intraradices* and *A. brasilense* increased growth under water stress. Combined formulations based on pseudomonads, *Azospirillum*, and many other PGPR microorganisms are also available for agronomic purposes. However, still little information is available about interspecies and multispecies interactions. There are several field experiments of single species seed inoculation with *Azospirillum* or *Pseudomonas*, but there is really very limited agronomical data regarding co-inoculation with both microorganisms (Valverde et al. 2015). A series of field trials with dual inoculation was conducted in Argentina during the seasons 2010–2013. The performance of *A. brasilense* single inoculation was compared to a combined formulation containing also *P. fluorescens* in wheat and corn fields. In all cases, *A. brasilense* alone or in combination with *P. fluorescens* had a positive effect on plant biomass at all three N fertilization levels essayed. The potential yield predicted was higher for dual inoculation, mainly when N-fertilizer was applied.

Thought, the number of spikes per m^2, a good predictor of the potential yield, was higher for dual inoculation, mainly when N-fertilizer was applied. This effect was translated into higher grain yield for dual inoculation than for single *A. brasilense* application (Valverde et al. 2015). The same effect on maize yield was observed in many locations, with or without N fertilization, though always major effect is observed under slight N fertilization. In a study conducted during 2009 growing season, the performance of a commercial combined inoculant based on INTA Az-39 strain of *A. brasilense* and *P. fluorescens* was evaluated in the presence of different levels of N, in an experimental field trial in Balcarce, Argentina. The sown maize variety was regionally adapted and recommended for high yielding environmental and crop management conditions. P fertilization was not necessary, and N (urea) was applied at sown, in three treatments: no N, half the dose (125 kg ha^{-1}), and complete dose (250 kg ha^{-1}), according to recommendation based on chemical soil analysis. Grain yield varied in a range from 9930 to 11,995 kg ha^{-1} according to the treatment (Fig. 2.1). The percentage of yield average increases due to inoculant application were 3.3 %, 16.6 %, and 2.5 % when no N, half dose, and complete dose were applied (Table 2.1). This result is a clear picture of the benefit inoculation exerts, in view of the achievement of greater yields reducing N applications. Co-inoculation is a promising field to be considered in the development of new biofertilizers.

2.6 Outlook and Conclusion

Organic farming differs from conventional agriculture in the production process, and it relies on techniques such as crop rotation, green manure, and biological pest control to maintain the soil productivity instead of chemical fertilizers and pesticides (Zhengfei et al. 2005). Several researchers have demonstrated that organic farming leads to improved soil quality with higher microbiological activity (Nautiyal 2012). Research must be focused in exploring bacterial structure, including PGPR consortium changes under different cropping practices and systems, and get a better

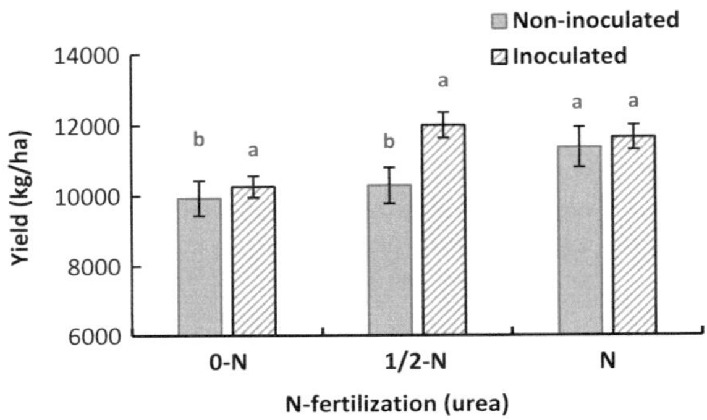

Fig. 2.1 Yield (kg ha^{-1}) of *Zea mays* inoculated with a formulation containing *A. brasilense* and *P. fluorescens* with three different doses of N-fertilizer applied as urea: 0-N, no urea applied; ½-N, half dose; and N, complete dose according to recommendation based on chemical soil analysis. Bars represent averages of four blocks as replicas. Small bars are standard deviation. Different letters within each level of N fertilization represent significant differences ($p < 0.05$) according to Duncan test

Table 2.1 Percentage of yield increase over non-inoculated control for each N-level

	N fertilization urea		
	0-N	½ -N	N
Increase over non-inoculated	3.3 %	16.6 %	2.5 %

understanding on how to build soil holistic ecology to maintain the health and productivity of plants. Long-term experiences have shown that neither the organic manure nor the chemical fertilizers alone can achieve sustained high yields. Integrated use of organic manures, biofertilizers, and chemical fertilizers, therefore, remains the only promising option in improving crop productivity.

Sustainable agriculture strategies should maintain the biodiversity of PGPR in the soil which might be affected by agricultural practices (Mäder et al. 2002; Esperschutz et al. 2007; Sugiyama et al. 2010). Studies aimed to understand and integrate plant responses during association, based on the profiling of plant gene expression, are a great help. New alternatives should be taken in mind for the use of bioinoculants. Some of them might include extending the technology to other valuable crops such as fruits, vegetables, and flowers, developing new formulations, and including multistrain bacterial consortia. Also the optimization of growth conditions, self-life of PGPR products, and application alternatives should be considered.

References

Adesemoye AO, Torbert HA III, Kloepper JW (2008) Enhanced plant nutrient use efficiency with PGPR and AMF in an integrated nutrient management system. Can J Microbiol 54:876–886

Afzal A, Bano A (2008) *Rhizobium* and phosphate solubilizing bacteria improve the yield and phosphorus uptake in wheat (*Triticum aestivum* L.) IJAB 10:85–88

Ahemad M, Khan MS (2010) Ameliorative effects of *Mesorhizobium* sp. MRC4 on chickpea yield and yield components under different doses of herbicide stress. Pestic Biochem Physiol 98:183–190

Ahemad M, Khan MS (2011a) Response of greengram [*Vigna radiata* (L.) Wilczek] grown in herbicide-amended soil to quizalafop-p-ethyl and clodinafop tolerant plant growth promoting *Bradyrhizobium* sp. (vigna) MRM6. J Agric Sci Technol 13:1209–1222

Ahemad M, Khan MS (2011b) Insecticide-tolerant and plant growth promoting *Bradyrhizobium* sp. (vigna) improves the growth and yield of greengram [*Vigna radiata* (L.) Wilczek] in insecticide-stressed soils. Symbiosis 54:17–27

Ahemad M, Khan MS (2011c) *Pseudomonas aeruginosa* strain PS1 enhances growth parameters of greengram [*Vigna radiata* (L.) Wilczek] in insecticide-stressed soils. J Pestic Sci 84:123–131

Ahemad M, Khan MS (2011d) Plant growth promoting fungicide-tolerant *Rhizobium* improves growth and symbiotic characteristics of lentil (*Lens esculentus*) in fungicide-applied soil. J Plant Growth Regul 30:334–342

Ahemad M, Khan MS (2012) Productivity of greengram in tebuconazole-stressed soil, by using a tolerant and plant growth-promoting *Bradyrhizobium* sp. MRM6 strain. Acta Physiol Plant 34:245–254

Ahemad M, Kibret M (2014) Mechanisms and applications of plant growth promoting rhizobacteria: current perspective. JKSUS 26:1–20

Alvarez MI, Sueldo RJ, Barassi CA (1996) Effect of *Azospirillum* on coleoptile growth in wheat seedlings under water stress. Cereal Res Commun 24:101–107

Arruebarrena Di Palma A, Pereyra CM, Moreno-Ramirez L et al (2013) Denitrification derived nitric oxide modulates biofilm formation in *Azospirillum brasilense*. FEMS Microbiol Lett 338:77–85

Baca BE, Elmerich C (2007) Microbial production of plant hormones. In: Elmerich C, Newton W (eds) Associative and endophytic nitrogen-fixing bacteria and cyanobacterial associations. Springer, Dordrecht, pp 113–143

Bakanchikova TI, Lobanok EV, Pavlova-Ivanova LK et al (1993) Inhibition of tumor formation process in dicotyledonous plants by *Azospirillum brasilense* strains. Mikrobologiya 62:515–523

Barassi CA, Sueldo RJ, Creus CM et al (2007) *Azospirillum* spp., a dynamic soil bacterium favoring vegetable crop production. DSDP 1(2):68–82

Barbieri R, Galli E (1993) Effect on wheat root development of inoculation with an *Azospirillum brasilense* mutant with altered indole-3-acetic acid production. Res Microbiol 144:69–75

Barea J, Navarro E, Montoya E (1976) Production of plant growth regulators by rhizosphere phosphate-solubilizing bacteria. J Appl Bacteriol 40:129–134

Bashan Y, de Bashan LE (2002a) Protection of tomato seedlings against infection by *Pseudomonas syringae* pv. Tomato by using the plant growth-promoting bacterium *Azospirillum brasilense*. Appl Environ Microbiol 68:2637–2643

Bashan Y, de Bashan LE (2002b) Reduction of bacterial speck (*Pseudomonas syringae* pv. Tomato) of tomato by combined treatments of plant growth-promoting bacterium, *Azospirillum brasilense*, streptomycin sulfate, and chemo-thermal seed treatment. Eur J Plant Pathol 108:821–829

Bashan Y, de Bashan LE (2010) How the plant growth-promoting bacterium *Azospirillum* promotes plant growth – a critical assessment. Adv Agron 108:77–136

Bashan Y, Hartmann A (2009) Ecology and application of *Azospirillum* and other plant growth-promoting bacteria (PGPB) – special issue. Eur J Soil Biol 45:1–2

Bashan Y, Holguin G (1997) *Azospirillum* -plant relationships: environmental and physiological advances. Can J Microbiol 43:103–112

Bashan Y, Holguin G, de Bashan LE (2004) *Azospirillum*–plant relationships: physiological, molecular, agricultural, and environmental advances (1997–2003). Can J Microbiol 50:521–577

Bhattacharyya PN, Jha DK (2012) Plant growth-promoting rhizobacteria (PGPR): emergence in agriculture. World J Microbiol Biotechnol 28:1327–1350

Blanco H, Lal L (2012) Erosion control and soil quality. In: Principles of soil conservation and management, Springer Science and Business Media, Dordrecht, pp 477–492

Bothe H, Körsgen H, Lehmacher T et al (1992) Differential effects of *Azospirillum*, auxin and combined nitrogen on the growth of the roots of wheat. Symbiosis 13:167–179

Bottini R, Fulchieri M, Pearce D et al (1989) Identification of gibberellins A1, A3 and iso-A3 in culture of *Azospirillum lipoferum*. Plant Physiol 89:1–3

Bottini R, Cassán F, Piccoli P (2004) Gibberellin production by bacteria and its involvement in plant growth promotion and yield increase. Appl Microbiol Biotechnol 65(5):497–503

Brazelton JN, Pfeufer EE, Sweat TA et al (2009) 2,4-Diacetyl phloroglucinol alters plant root development. MPMI 21(10):1349–1358

Brusamarello-Santos LCC, Pacheco F, Aljanabi SMM et al (2012) Differential gene expression of rice roots inoculated with the diazotroph *Herbaspirillum seropedicae*. Plant Soil 356:113–125

Casanovas EM, Barassi CA, Andrade FH et al (2003) *Azospirillum*-inoculated maize plant responses to irrigation restraints imposed during flowering. Cereal Res Commun 31:395–402

Castillo P, Molina R, Andrade A et al (2015) Phytohormones and other plant growth regulating substances. In: Cassán et al (eds) Phytohormones and other plant growth regulators produced by PGPR: the genus *Azospirillum*, handbook for *Azospirillum*: technical issues and protocols. Springer, International Publishing, Cham, pp 115–138

Chaudhry V, Rehman A, Mishra A et al (2012) Changes in bacterial community structure of agricultural land due to long term organic and chemical amendments. Microbiol Ecol 64(2):450–460

Compant S, Duffy B, Nowak J et al (2005) Use of plant growth-promoting bacteria for biocontrol of plant diseases: principles, mechanisms of action, and future prospects. Appl Environ Microbiol 71(9):4951–4959

Cong PT, Dung TD, Hien TM et al (2009) Inoculant plant growth-promoting microorganisms enhance utilisation of urea-N and grain yield of paddy rice in southern Vietnam. Eur J Soil Biol 45:52–61

Creus CM, Sueldo RJ, Barassi CA (1997) Shoot growth and water status in *Azospirillum*-inoculated wheat seedlings grown under osmotic and salt stresses. Plant Physiol Biochem 35:939–944

Creus CM, Sueldo RJ, Barassi CA (1998) Water relations in *Azospirillum* inoculated wheat seedlings under osmotic stress. Can J Bot 76:238–244

Creus CM, Sueldo RJ, Barassi CA (2004) Water relations and yield in *Azospirillum*-inoculated wheat exposed to drought in the field. Can J Bot 82:273–281

Creus CM, Graziano M, Casanovas E et al (2005) Nitric oxide is involved in the *Azospirillum brasilense*-induced lateral root formation in tomato. Planta 221:297–303

Dal Lago C, Creus CM, Pereyra MA (2015) *Azospirillum* growth promotion is related to changes in ferulate and dehydrodiferulate contents in cell wall of inoculated cucumber seedlings hypocotyls. Am J Plant Sci 6:1650–1661

Das SK, Varma A (2010) Role of enzymes in maintaining soil health. In: Shukla G, Varma A (eds) Soil enzymology, soil biology 22. Springer, Berlin/Heidelberg, pp 25–42

Desbrosses G, Contesto C, Varoquaux F et al (2009) PGPR–*Arabidopsis* interactions is a useful system to study signaling pathways involved in plant developmental control. PS&B 4:321–323

Díaz-Zorita M, Fernández-Canigia MV (2009) Field performance of a liquid formulation of *Azospirillum brasilense* on dryland wheat productivity. Eur J Soil Biol 45:3–11

Dobbelaere S, Croonenborghs A, Thys A et al (1999) Phytostimulatory effect of *Azospirillum brasilense* wild type and mutant strains altered in IAA production on wheat. Plant Soil 212:155–164

Dobbelaere S, Croonenborghs A, Thys A et al (2001) Responses of agronomically important crops to inoculation with *Azospirillum*. Aust J Plant Physiol 28:871–879

Dobbelaere S, Vanderleyden J, Okon Y (2003) Plant growth-promoting effects of diazotrophs in the rhizosphere. Crit Rev Plant Sci 22:107–149

Döbereiner J, Day L (1976) Associative symbiosis in tropical grasses: characterization of microorganisms and dinitrogen fixing sites. In: Newton WE, Nyman CJ (eds) Proceedings 1st Int symposium on N_2 fixation. Washington State University Press, Pullman, pp 518–538

Doran JW (2002) Soil health and global sustainability: translating science into practice. Agric Ecosyst Environ 88:119–127

Doran JW, Zeiss MR (2000) Soil health and sustainability: managing the biotic component of soil quality. Appl Soil Ecol 15:3–11
Doran JW, Sarrantonio M, Liebig M (1996) Soil health and sustainability. In: Sparks DL (ed) Advances in agronomy, vol 56. Academic Press, San Diego, pp 1–54
Drinkwater LE, Snapp SS (2007) Nutrients in agroecosystems: re-thinking the management paradigm. Adv Agron 92:163–186
Elmqvist T, Matby E, Barker T et al (2010) In: Kumar et al (eds) Biodiversity, ecosystems and ecosystem services, the economics of ecosystems and biodiversity: ecological and economic foundations. Earthscan, London, pp 1–96
Esperschutz J, Gattinger A, Mader P et al (2007) Response of soil microbial biomass and community structures to conventional and organic farming systems under identical crop rotations. FEMS Microbiol Ecol 61:26–37
Eurostat. Statistics Explained (2012) Agri-environmental indicator – pesticide risk. http://ec.europa.eu/eurostat/statistics-explained/index.php/Agri-environmental_indicator_-_pesticide_risk#Agri-environmental_context. ISSN 2443-8219. Accessed 8 Feb 2016
Fernández L, Zalba P, Gómez M et al (2007) Phosphate-solubilization activity of bacterial strains in soil and their effect on soybean growth under greenhouse conditions. Biol Fertil Soils 43:805–809
Fernández-Scavino A, Pedraza RO (2013) The role of siderophores in plant growth-promoting bacteria. In: Maheshwari DK, Saraf M, Aeron A (eds) Bacteria in agrobiology: crop productivity. Springer, Heidelberg, pp 265–285
Fuentes-Ramirez LE, Caballero-Mellado J (2005) Bacterial biofertilizers. In: Siddiqui ZA (ed) PGPR: biocontrol and Biofertilization. Springer, Dordrecht, pp 143–172
Gerhardson B (2002) Biological substitutes for pesticides. Trends Biotechnol 20:338–343
Glick BR (2004) Bacterial ACC deaminase and the alleviation of plant stress. Adv Appl Microbiol 56:291–312
Glick BR, Penrose DM, Li J (1998) A model for the lowering of plant ethylene concentrations by plant growth-promoting bacteria. J Theor Biol 190:63–68
Glick BR, Todorovic B, Czarny J et al (2007) Promotion of plant growth by bacterial ACC deaminase. Crit Rev Plant Sci 26:277–242
Govindasamy V, Senthilkumar M, Gaikwad K et al (2008) Isolation and characterization of ACC deaminase gene from two plant growth promoting rhizobacteria. Curr Microbiol 57:312–317
Gudzic N, Aksic M, Djikic A et al (2015) Soil fertility changes trends and available Fe, Mn and Zn in a long-term system user nitrogen fertilizers. IJAAR 7(2):14–20
Guo J, Chi J (2014) Effect of Cd-tolerant plant growth-promoting *Rhizobium* on plant growth and Cd uptake by *Lolium multiflorum* Lam and *Glycine max* (L.) Merr in Cd-contaminated soil. Plant Soil 375:205–214
Gupta G, Parihar SS, Ahirwar NK et al (2015) Plant growth promoting rhizobacteria (pgpr): current and future prospects for development of sustainable agriculture. J Microbiol Biotechnol 7:96–102
Halliwell B, Gutteridge MC (1984) Oxygen toxicity, oxygen radicals, transition metals and disease. Biochem J 219:1–14
Hayat R, Ali S, Amara U et al (2010) Soil beneficial bacteria and their role in plant growth promotion: a review. Annu Rev Microbiol 60:579–598
Helman Y, Burdman S, Okon Y (2011) Plant growth promotion by rhizosphere bacteria through direct effects. In: Rosenberg E, Gophna U (ed) Beneficial microorganisms in multicellular life form. Heidelberg: Springer, pp 89–103
Holguin G, Bashan Y (1996) Nitrogen-fixation by *Azospirillum brasilense* Cd is promoted when co-cultured with a mangrove rhizosphere bacterium (*Staphylococcus* sp.) Soil Biol Biochem 28:1651–1660
Hungria M, Campo RJ, Souza EM et al (2010) Inoculation with selected strains of *Azospirillum brasilense* and *A. lipoferum* improves yields of maize and wheat in Brazil. Plant Soil 331:413–425

Iavicoli A, Boutet E, Buchala A et al (2003) Induced systemic resistance in *Arabidopsis thaliana* in response to root inoculation with *Pseudomonas fluorescens* CHA0. MPMI 16(10):851–858
Inagaki AM, Guimarães VF, Lana MC et al (2015) Maize initial growth with the inoculation of plant growth-promoting bacteria (PGPB) under different soil acidity levels. AJCS 9(4):271–280
Islam M, Madhaiyan M, Boruah HPD et al (2009) Characterization of plant growth-promoting traits of free-living diazotrophic bacteria and their inoculation effects on growth and nitrogen uptake of crop plants. J Microbiol Biotechnol 19:1213–1222
Joshi KK, Kumar V, Dubey RC et al (2006) Effect of chemical fertilizer adaptive variants, *Pseudomonas aeruginosa* GRC2 and *Azotobacter chroococcum* AC1 on *Macrophomina phaseolina* causing charcoal rot of *Brassica juncea*. Korean J Environ Agric 25:228–235
Karlen DL (2012) Soil health: the concept, its role, and strategies for monitoring. Diana H Wall et al Soil ecology and ecosystem services, 1st Edition, Oxford University Press, Oxford 331–336
Karlen DL, Andrews SS, Doran JW (2001) Soil quality: current concepts and applications. Adv Agron 74:1–40
Kaur A, Chaudhary A, Kaur A et al (2005) Phospholipid fatty acid – a bioindicator of environment monitoring and assessment in soil ecosystem. Curr Sci India 89:1103–1112
Keel C, Schnider U, Maurhofer M et al (1992) Suppression of root disease by *Pseudomonas fluorescens* CHA0: importance of the bacterial secondary metabolite 2,4-diacetyl phloroglucinol. MPMI 5:4–13
Kennedy AC, Smith KL (1995) Soil microbial diversity and the sustainability of agricultural soils. Plant Soil 170:75–86
Khan MS, Zaidi A, Wani PA (2006) Role of phosphate solubilizing microorganisms in sustainable agriculture – a review. Agron Sustain Dev 27:29–43
Kloepper JW, Leong J, Teintze M et al (1980) Enhanced plant growth by siderophores produced by plant growth promoting rhizobacteria. Nature (London) 286:885–886
Kloepper JW, Lifshitz R, Zablotowicz RM (1989) Free-living bacterial inocula for enhancing crop productivity. Trends Biotechnol 7:39–43
Kloepper JW, Ryu CM, Zhang S (2004) Induced systemic resistance and promotion of plant growth by *Bacillus* spp. Phytopathology 94:1259–1266
Kravchenko AN, Thelen KD, Bullock DG et al (2003) Relationship among crop grain yield, topography, and soil electrical conductivity studied with cross-correlograms. Agron J 95:1132–1139
Lehman RM, Cambardella CA, Stott DE et al (2015) Understanding and enhancing soil biological health: the solution for reversing soil degradation. Sustainability 7:988–1027
Lugtenberg B, Kamilova F (2009) Plant-growth-promoting rhizobacteria. Annu Rev Microbiol 63:541–556
Lugtenberg BJJ, Dekkers L, Bloemberg GV (2001) Molecular determinants of rhizosphere colonization by *Pseudomonas*. Annu Rev Phytopathol 39:461–490
Ma Y, Prasad MNV, Rajkumar M et al (2011) Plant growth promoting rhizobacteria and endophytes accelerate phytoremediation of metalliferous soils. Biotechnol Adv 29:248–258
Mäder P, Fließbach A, Dubois D et al (2002) Soil fertility and biodiversity in organic farming. Science 296:1694–1697
Maheshwari DK (2011) Plant growth and health promoting bacteria. In: Maheshwari DK (ed) Microbiology monographs. Springer Series, Heidelberg, pp 99–116
Manthey JA, McCoy DL, Crowley DE (1994) Stimulation of rhizosphere iron reduction and uptake in response to iron deficiency in citrus rootstocks. Plant Physiol Biochem 32:211–215
Mayak S, Tirosh T, Glick BR (2004) Plant growth-promoting bacteria that confer resistance to water stress in tomatoes and peppers. Plant Sci 166:525–530
Mohammadi K, Sohrabi Y (2012) Bacterial biofertilizers for sustainable crop production: a review. J Agric Biol Sci 7:307–316
Molina Favero C, Creus CM, Lanteri ML et al (2007) Nitric oxide and plant growth promoting rhizobacteria: common features influencing root growth and development. Adv Bot Res 46:1–33
Molina Favero C, Creus CM, Simontachi M et al (2008) Aerobic nitric oxide production by *Azospirillum brasilense* Sp245 and its influence on root architecture in tomato. MPMI 21:1001–1009

Nagorska K, Bikowski M, Obuchowski M (2007) Multicellular behaviour and production of a wide variety of toxic substances support usage of *Bacillus subtilis* as a powerful biocontrol agent. Acta Biochim Pol 54(3):495–508

Nautiyal CS (2012) Microbes for soil sustainability and crop productivity. International society of environmental botanists 18 (3) This article has been reproduced from the archives of EnviroNews – Newsletter of ISEB India

Nogueira EDM, Vinagre F, Masuda HP et al (2001) Expression of sugarcane genes induced by inoculation with *Gluconacetobacter diazotrophicus* and *Herbaspirillum rubrisubalbicans*. Genet Mol Biol 24:199–206

Okon Y (1985) *Azospirillum* as a potential inoculant for agriculture. Trends Biotechnol 3:223–228

Okon Y, Kapulnik Y (1986) Development and function of *Azospirillum*-inoculated roots. Plant Soil 90:3–16

Okon Y, Labandera-González CA (1994) Agronomic applications of *Azospirillum*: an evaluation of 20 years worldwide field inoculation. Soil Biol Biochem 26:1591–1601

Overvoorde P, Fukaki H, Beeckman T (2010) Auxin control of root development. CSH. Perspectives 2(6.) 2):a001537

Pandey P, Kang SC, Gupta CP et al (2005) Rhizosphere competent *Pseudomonas aeruginosa* GRC1 produces characteristic siderophore and enhances growth of Indian mustard (*Brassica campestris*). Curr Microbiol 51:303–309

Parke JL (1991) Root colonization by indigenous and introduced microorganisms. In: Keister DL, Gregan PB (eds) The rhizosphere and plant growth. Kluwer Academic Publishers, Dordrecht, pp 33–42

Parmar P, Sindhu SS (2013) Potassium solubilization by rhizosphere bacteria: influence of nutritional and environmental conditions. J Microbiol Res 3:25–31

Pereira SIA, Barbosa L, Castro PML (2015) Rhizobacteria isolated from a metal-polluted area enhance plant growth in zinc and cadmium-contaminated soil. IJEST 12:2127–2142

Pereyra MA, Zalazar CA, Barassi CA (2006) Root phospholipids in *Azospirillum*-inoculated wheat seedlings exposed to water stress. Plant Physiol Biochem 44:873–879

Pereyra MA, Ballesteros FM, Creus CM et al (2009) Seedlings growth promotion by *Azospirillum brasilense* under normal and drought conditions remains unaltered in Tebuconazole-treated wheat seeds. Eur J Soil Biol 45:20–27

Pereyra CM, Ramella NA, Pereyra MA et al (2010) Changes in cucumber hypocotyl cell wall dynamics caused by *Azospirillum brasilense* inoculation. Plant Physiol Biochem 48:62–69

Pereyra MA, García P, Colabelli MN et al (2012) A better water status in wheat seedlings induced by *Azospirillum* under osmotic stress is related to morphological changes in xylem vessels of the coleoptile. Appl Soil Ecol 53:94–97

Ping L, Boland W (2004) Signals from the underground: bacterial volatiles promote growth in *Arabidopsis*. Trends Plant Sci 9(6):263–266

Prakash D, Verma S, Bhatia R et al (2011) Risks and precautions of genetically modified organisms. ISRN Ecology, Article ID 369573, 13 pages

Radzki W, Gutiérrez-Mañero FJ, Algar E et al (2013) Bacterial siderophores efficiently provide iron to iron-starved tomato plants in hydroponics culture. Ant Van Leeuw J Microb 104:321–330

Reichman SM (2014) Probing the plant growth-promoting and heavy metal tolerance characteristics of *Bradyrhizobium japonicum* CB1809. Eur J Soil Biol 63:7–13

Richter DD, OhN-H FR et al (2007) The rhizosphere and soil formation. In: Cardon ZG, Whitbeck JL (eds) The rhizosphere: an ecological perspective. Elsevier Academic Press, Burlington, pp 179–200

Roesch LF, Fulthorpe RR, Riva A et al (2007) Pyrosequencing enumerates and contrasts soil microbial diversity. ISME J 1(4):283–290

Ruiz-Sánchez M, Armada E, Muñoz Y et al (2011) *Azospirillum* and arbuscular mycorrhizal colonization enhance rice growth and physiological traits under well-watered and drought conditions. J Plant Physiol 168:1031–1037

Salcedo F, Pereyra C, Arruebarrena Di Palma A et al (2015) Methods for studying biofilms in Azospirillum and other PGPRs. In: Cassán et al (eds) Handbook for *Azospirillum*: technical issues and protocols. Springer, Cham, pp 199–229
Santoro M, Capellari L, Giordano W et al (2015) Production of volatile organic compounds in PGPR. In: Cassán et al (eds) Handbook for *Azospirillum*: technical issues and protocols. Springer International Publishing, Cham, pp 307–317
Sarig S, Blum A, Okon Y (1988) Improvement of the water status and yield of field-grown grain sorghum (*Sorghum bicolor*) by inoculation with *Azospirillum brasilense*. J Agric Sci 110:271–277
Steenhoudt O, Vanderleyden J (2000) *Azospirillum* a free-living nitrogen fixing bacterium closely associated with grasses: genetic, biochemical and eco-logical aspects. FEMS Microbiol Rev 24:487–506
Stintzi A, Barnes C, Xu J et al (2000) Microbial iron transport via a siderophore shuttle: a membrane ion transport paradigm. Proc Natl Acad Sci USA 97:10691–10696
Sudhakar P, Gangwar SK, Satpathy B et al (2000) Evaluation of some nitrogen fixing bacteria for control of foliar diseases of mulberry (*Morus alba*). Indian J Seric 39:9–11
Sugiyama A, Vivanco JM, Jayanty S et al (2010) Pyrosequencing assessment of soil microbial communities in organic and conventional potato farms. Plant Dis 94:1329–1335
Suman A, Gaur A, Shrivastava AK et al (2005) Improving sugarcane growth and nutrient uptake by inoculating *Gluconacetobacter diazotrophicus*. Plant Growth Regul 47:155–162
Sundara B, Natarajan V, Hari K (2002) Influence of phosphorous solubilizing bacteria on the changes in soil available phosphorous and sugarcane and sugar yields. Field Crop Res 77:43–49
Tsavkelova E, Klimova S, Cherdyntseva T et al (2006) Microbial producers of plant growth stimulators and their practical use: a review. Appl Biochem Microbiol 42:117–126
Valverde C, González Anta G, Ferraris G (2015) *Azospirillum* and *Pseudomonas*. In: Cassán et al (eds) Handbook for *Azospirillum*: technical issues and protocols. Springer, International Publishing, Cham, pp 389–409
van Loon LC, Glick GR (2004) Increased plant fitness by rhizobacteria. In: Sandermann H (ed) Molecular ecotoxicology of plants, 170. Springer, Berlin, pp 177–205
Van Wees SCM, Pieterse CMJ, Trijssenaar A et al (1997) Differential induction of systemic resistance in *Arabidopsis* by biocontrol bacteria. MPMI 10(6):716–724
Vande Broek A, Dobbelaere S, van der Leyden J et al (2000) *Azospirillum*-plant root interactions: signalling and metabolic interactions. In: Triplett EW (ed) Prokaryotic nitrogen fixation: a model system for the analysis of a biological process. Horizon Scientific Press, Wymondham, pp 761–777
Veresoglou SD, Menexes G (2010) Impact of inoculation with *Azospirillum* spp. on growth properties and seed yield of wheat: a meta-analysis of studies in the ISI web of science from 1981 to 2008. Plant Soil 337:469–480
Vessey JK (2003) Plant growth promoting rhizobacteria as biofertilizers. Plant Soil 255:571–586
Vikram A (2007) Efficacy of phosphate solubilizing bacteria isolated from vertisols on growth and yield parameters of sorghum. Res J Microbiol 2:550–559
Wandersman C, Delepelaire P (2004) Bacterial iron sources: from siderophores to hemophores. Annu Rev Microbiol 58:611–647
Wani PA, Khan MS (2010) *Bacillus* species enhance growth parameters of chickpea (*Cicer arietinum* L.) in chromium stressed soils. Food Chem Toxicol 48:3262–3267
Welbaum GE, Sturz AV, Dong Z et al (2004) Managing soil microorganisms to improve productivity of agro-ecosystems. Crit Rev Plant Sci 23(2):175–193
Weller DM (2007) *Pseudomonas* spp. are ubiquitous bacteria in agricultural soils and have many traits that make them well suited as biocontrol agents of soilborne pathogens. Phytopathology 97(2):251–256
Xia Y, Embley TM, O'Donnell AG (1994) Phylogenetic analysis of *Azospirillum* by direct sequencing of PCR-amplified 16S rDNA. Syst Appl Microbiol 17:197–201
Young CC, Lin SY, Shen FT et al (2015) Molecular tools for identification and characterization of plant growth promoting rhizobacteria with emphasis in *Azospirillum* spp. In: Cassán et al (eds)

Handbook for *Azospirillum*: technical issues and protocols. Springer, International Publishing, Cham, pp 27–44

Youssef MMA, Eissa MFM (2014) Biofertilizers and their role in management of plant parasitic nematodes. J Biotechnol Pharm Res 5:1–6

Zaidi A, Khan MS, Ahemad M et al (2009) Plant growth promotion by phosphate solubilizing bacteria. Acta Microbiol Hung 56:263–284

Zhang Q, Zhou W, Liang G et al (2015) Effects of different organic manures on the biochemical and microbial characteristics of albic paddy soil in a short-term experiment. PLoS One 10(4):0124096

Zhengfei GA, Lansink O, Wossink A et al (2005) Damage abating inputs: a comparison of conventional and organic farming systems. Eur Rev Agric Econ 32(2):167–189

Pseudomonadaceae: From Biocontrol to Plant Growth Promotion

3

Roxane Roquigny, Amy Novinscak, Adrien Biessy, and Martin Filion

Abstract

Pseudomonas spp. are aerobic, Gram-negative bacteria that are ubiquitously found in soils. They are particularly well suited for plant root colonization and many strains display plant growth-promoting and/or biocontrol activity against various plant pathogens. Their ability to metabolize a wide array of nutrients, their rapidity and ease of growth and their natural abundance in variety of plant-soil environments make them promising organisms for the development of commercial biocontrol and biofertilizer products. In this chapter, we will discuss their diversity, genetics and ecology, while putting special emphasis on the mechanisms involved in biocontrol and/or plant growth promotion. Recent progress in genomics and transcriptomics, as well as future research on these organisms will also be discussed.

3.1 Introduction

The rhizosphere is the narrow zone of soil, influenced by a plant's root system (Rainey 1999). This zone is rich in nutrients when compared with the bulk soil due to the accumulation of a variety of plant exudates, such as amino acids and sugars, providing a rich source of energy and nutrients for bacteria (Gray and Smith 2005). This situation is reflected by the number of bacteria, commonly referred as rhizobacteria (Schroth and Hancock 1982), that are in the rhizosphere, generally 10–100 times higher than that in the bulk soil (Weller and Thomashow 1994). It has been determined that only 1–2 % of bacteria are able to promote plant growth in the rhizosphere (Antoun and Kloepper 2001) and these bacteria are known as plant growth promoting

R. Roquigny • A. Novinscak • A. Biessy • M. Filion (✉)
Université de Moncton, Moncton E1A 3E9, NB, Canada
e-mail: martin.filion@umoncton.ca

S. Mehnaz (ed.), *Rhizotrophs: Plant Growth Promotion to Bioremediation*, Microorganisms for Sustainability 2, DOI 10.1007/978-981-10-4862-3_3

rhizobacteria (PGPR). The metabolites produced by PGPR can either directly improve plant growth or indirectly improve plant growth by minimizing the effects of soil-borne plant pathogens, a concept known as biocontrol. In some cases, biocontrol can also be observed alone without plant growth promotion, while both mechanisms often operate together. Some methods of direct plant promotion include production of the phytohormone auxin (Patten and Glick 2002), decrease of plant ethylene levels (Glick 2012) or increase in iron availability through the effects of siderophores (Cézard et al. 2015). Biocontrol mechanisms can include competition, antibiosis (Podile and Kishore 2006) and induced systemic resistance (ISR; Bakker et al. 2007). Bacteria belonging to diverse genera have been identified as PGPR, of which *Pseudomonas* spp. and *Bacillus* spp. are predominant (Podile and Kishore 2006).

3.1.1 The Pseudomonadaceae

Pseudomonas spp. belong to the Pseudomonadaceae, which is a large bacterial family. Created in 1917 by Winslow and colleagues, it belongs to the class of Gammaproteobacteria (Winslow et al. 1917). These organisms are free-living bacteria that are commonly found in water and soil environments. Pseudomonadaceae encompasses four bacterial genera: *Pseudomonas*, *Xanthomonas*, *Gluconobacter* and *Zooglea*. The *Pseudomonas* genus was defined earlier than its family by Migula in 1894 (Migula 1894). At that time, distinction between genera was achieved using bacterial morphological properties. *Pseudomonas* spp. were defined as non-sporulating rod-shape cells which are usually motile. Taxonomy based solely on phenotypical traits was then replaced, due to advances in sequencing technologies, by a classification of *Pseudomonas* species into five "RNA homology" groups (Palleroni et al. 1973). The rRNA group 1 is the largest and encompasses the so called "fluorescent pseudomonads", which will be the focus of this book chapter.

3.1.2 Fluorescent Pseudomonads

Fluorescent pseudomonads are a functional group that comprises *Pseudomonas* species that produce a greenish fluorescent compound, known as pyoverdine, which is a siderophore (Cézard et al. 2015). Pyoverdines are secreted by fluorescent pseudomonads to capture and deliver iron to the cells. Microbial siderophores can also enhance iron uptake by plants that are able to recognize the bacterial ferric-siderophore complex (Masalha et al. 2000; Katiyar and Goel 2004). Over 100 pyoverdines have been identified from different species and strains of *Pseudomonas* (Meyer et al. 2008), representing about 20 % of the microbial siderophores characterized to date (Boukhalfa and Crumbliss 2002).

At the taxonomical level, the fluorescent pseudomonads include phytopathogenic cytochrome *c* oxidase positive species (*Pseudomonas cichorii*, *Pseudomonas marginalis* and *Pseudomonas tolaasii*), non-phytopathogenic, non-necrogenic strains (*Pseudomonas fluorescens*, *Pseudomonas putida*, *Pseudomonas chlororaphis*, and

Pseudomonas aeruginosa), and phytopathogenic necrogenic fluorescent *Pseudomonas* species without cytochrome *c* oxidase (*Pseudomonas syringae* and *Pseudomonas viridiflava*) (Choudhary et al. 2009). Various phenotypic methods have been used to cluster and identify bacteria according to several features such as morphology, pigmentation, and nutritional requirements. These methods have shown that *P. fluorescens* and *P. putida* were heterogeneous which led to *P. putida* being subdivided into biotypes A and B (biovars A and B). *P. fluorescens* was also subdivided into seven biotypes. Biotypes A to D and F were named biovars I to V, while biotype D became *P. chlororaphis* and biotype E became *P. chlororaphis* subsp. *aureofaciens* (Palleroni 1984).

3.1.3 Genomics and Genome Plasticity

During the last decade, several plant-associated fluorescent pseudomonad genomes have been sequenced including *P. putida* (Nelson et al. 2002), *P. fluorescens* (Paulsen et al. 2005) and *P. chlororaphis* (Shen et al. 2012). Recent breakthroughs in high-throughput sequencing technologies have been accompanied by an overwhelming increase in the number of *Pseudomonas* spp. genomes publicly available, which has enabled large scale comparative genomic studies. Such studies have highlighted the tremendous genomic diversity of plant-associated *Pseudomonas* spp. (Loper et al. 2012; Jun et al. 2016).

Plant-associated fluorescent pseudomonads possess a large genome which displays a mosaic structure with genes segregating between the "core genome" and the "accessory genome" (Silby et al. 2011). Genes from the core genome are conserved among the different strains which are thought to be responsible for essential cellular processes, whereas genes from the accessory genome, which are often unique to one or few strains, are responsible for the variability observed among strains. While strains from the fluorescent pseudomonad group share a relatively high number of genes, with 2,789 predicted protein-coding genes present in the genome of ten representative strains, corresponding to between 45 % and 52 % of the total number of genes (Loper et al. 2012), the core genome of the whole *Pseudomonas* genus was estimated to contain 1,224 protein-coding genes (Jun et al. 2016).

The genomic diversity and plasticity of plant-associated fluorescent pseudomonads come from their accessory genomes. Out of ten strains from the fluorescent pseudomonad group, 13,872 putative coding-protein genes were identified (Loper et al. 2012), of which 5,798 had no orthologs with other *Pseudomonas* spp. genomes. This diversity is often reflected in the wide range of secondary metabolites produced (Gross and Loper 2009) and it appears that plant-associated saprophytic *Pseudomonas* spp. genomes are a good place to search for operons involved in the production of new antimicrobial compounds (Gross et al. 2007; Van Der Voort et al. 2015).

Pseudomonas spp. genomes are in perpetual rearrangement as evidenced by the low synteny observed between closely-related strains (Silby et al. 2009; Wu et al. 2011; Loper et al. 2012). Mobile genetic elements, such as genomic islands, transposons or REP-elements are also abundant in the *Pseudomonas* spp. accessory genomes and often account for horizontal gene transfer acquired genomic material (Silby et al. 2011). The genomic diversity and plasticity of saprophytic plant-associated fluorescent pseudomonads is a key feature that empowers their valuable role in terms of plant growth promotion and biocontrol of plant pathogens.

3.2 Rhizosphere Competence of Plant-Beneficial Fluorescent Pseudomonads

Plant-beneficial fluorescent pseudomonads that possess genetic determinants for plant growth promotion and/or biocontrol of plant pathogens are not always effective when deployed in the field. Despite promising results under controlled condition, many authors have reported on the inconsistency of plant-growth promotion and biocontrol achieved by some *Pseudomonas* spp. strains in the field and have linked these results to an impaired rhizosphere colonization (Kloepper et al. 1980; Weller 1988). Rhizosphere colonization is a crucial step leading to disease-suppression given that: (i) an inverse correlation between population size of plant-beneficial *Pseudomonas* spp. and the disease incidence has been observed in several plant-pathogen systems (Bull et al. 1991; Raaijmakers et al. 1995); (ii) impaired rhizosphere colonization mutants, such as those obtained from *P. chlororaphis* subsp. *piscium* PCL1391, lost their disease suppression capability toward certain pathogens, such as *Fusarium oxysporum* f. sp. *radices-lycopersici* (Chin-A-Woeng et al. 2000); (iii) a linear relationship has been observed between population size of indigenous antibiotics producing *Pseudomonas* spp. and antibiotics accumulation in the rhizosphere (Raaijmakers et al. 1999; Mavrodi et al. 2012a), which is in line with biocontrol capabilities.

We define the rhizosphere competence (or rhizocompetence) of an introduced plant-beneficial *Pseudomonas* spp. strain as its aptitude to establish itself in the rhizosphere of a plant and to persist during several crop cycles while maintaining a high population level. It appeals to the capacity of an organism to forge a successful trophic relationship with the plant as well to its ability to compete with indigenous microorganisms coveting the very same ecological niches. Various approaches have been undertaken to identify traits involved in the rhizosphere competence of plant-beneficial *Pseudomonas* spp. including site-directed mutagenesis (Lugtenberg et al. 2001), rhizosphere-induced gene monitoring with promoter-trapping technology (IVET; Rainey 1999) and broader population-based approaches by assessing traits which distinguish fluorescent pseudomonads from the rhizosphere to those isolated from the bulk soil (Latour et al. 2003). Many traits indispensable for rhizosphere colonization have been characterized, such as flagella, chemotaxis, adhesion, etc. (Lugtenberg et al. 2001) often by monitoring the speed at which bacteria reach the root tip of seedlings grown under gnotobiotic conditions with total disregard for the key role of competition in rhizosphere colonization. Nonetheless, comparative studies carried out in agricultural soils have highlighted the superior root-colonizing ability of certain genotypes over others (Raaijmakers and Weller 2001; Ghirardi et al. 2012), and have successfully identified major competitive enhancing traits for rhizosphere colonization (Ghirardi et al. 2012).

3.2.1 Interactions Between Plant-Beneficial *Pseudomonas* spp. and Their Host

3.2.1.1 Rhizoplane Colonization by Plant-Beneficial *Pseudomonas* spp.

Some root exudates, such as malic acid and citric acid, act as chemoattractants for beneficial (and deleterious) bacteria. Hence, motility (De Weger et al. 1987) and especially flagella-driven chemotaxis towards exudate components (De Weert et al. 2002), is an important trait for the rhizosphere competence of *Pseudomonas* spp. In fact, the rhizosphere is not considered as a homogeneous environment but rather as a succession of favourable and less favourable ecological niches; chemotaxis allows *Pseudomonas* spp. to set a course to the most advantageous locations which are generally located at the junctions between epidermal root cells and sites of side roots appearance (Chin-A-Woeng et al. 1997). Adhesion to the root surface is an important mechanism in root colonization, and several determinants have been described. The hair-like structures pili and a root-adhesion outer membrane protein, homologous to OprF from the plant pathogen *P. syringae*, have been shown to be involved in the adhesion to the root surface of several plants by *Pseudomonas* spp. (Vesper 1987; De Mot et al. 1992). The plant root surface glycoprotein agglutinin has been implicated in the adhesion of *P. putida* to the root, an adhesion mediated by the coding-protein gene *aggA* (Anderson 1983; Buell and Anderson 1992). Transition from transient adhesion to irreversible attachment to root surfaces constitutes the first step to the formation of a microcolony (or biofilm), which will soon become a mature biofilm. Lap (large adhesion proteins) has been shown to be involved in this transition in *P. putida* (Hinsa et al. 2003). Biofilms are multicellular aggregates encased in a complex matrix mainly composed of extracellular polymeric substances (EPS), proteins and eDNA (extracellular DNA; Flemming and Wingender 2010). Biofilms enable plant-beneficial *Pseudomonas* spp. to resist harsh conditions including desiccation and high concentrations of toxic compounds (Danhorn and Fuqua 2007).

3.2.1.2 Antibiotic Production

Large populations of antibiotic-producing *Pseudomonas* spp. have been observed in several fields (Raaijmakers et al. 1997; Mazurier et al. 2009; Parejko et al. 2012) and have been frequently associated with disease-suppressiveness (Raaijmakers and Weller 1998; Weller et al. 2002; Mazurier et al. 2009). One may think that the capacity to produce broad-spectrum antibiotics, such as 2,4-diacetylphloroglucinol (DAPG) or phenazine derivatives might enhance the ecological competence throughout antagonism towards competitors. However, it remains to be demonstrated. Carroll et al. (1995) showed that the incapacity to produce DAPG did not reduce the rhizosphere competence of strain *P. fluorescens* F113 in the rhizosphere of sugarbeets. In contrast, phenazine defective mutants of strains *P. synxantha* 2–79 and *P. chlororaphis* subsp. *aureofaciens* 30–84 were not able to maintain high population levels in the rhizosphere of wheat in the presence of indigenous microorganisms, whereas a *P. chlororaphis subsp. aureofaciens* 30–84 phenazine defective mutant colonized to the same extent as its parent strain when inoculated in the rhizosphere of wheat grown in pasteurized soil (Mazzola et al. 1992). These results suggest the involvement of phenazine production in competitive rhizosphere colonization by plant-beneficial *Pseudomonas* spp., but do not demonstrate that antibiosis is the mechanism involved. It has been suggested that phenazines do not affect the immediate competitors of plant-beneficial *Pseudomonas* spp. (Mavrodi et al. 2006; Pierson and Pierson 2010) and could serve other purposes (Price-Whelan et al. 2006).

3.3 Fluorescent Pseudomonad Mechanisms Leading to Plant Growth Promotion

With regards to plant growth promotion, fluorescent pseudomonads are often divided in two groups, based on their mode of action. The first group, which will be covered in this section, consists of fluorescent pseudomonads that directly influence plant growth, seed emergence or improve crop yields and are often referred as bio-fertilizers (Glick et al. 1999). The second group is known as biocontrol fluorescent pseudomonads that are able to indirectly influence plant growth by reducing the negative pressure that plant pathogens put on the plant's growth and development.

3.3.1 Direct Plant Growth Promotion

Various mechanisms of direct plant growth promotion have been studied, such as the production of phytohormones, including auxins, cytokinins and gibberellins, the reduction of ethylene levels in plants through the action of ACC-deaminase enzyme and mechanisms to increase nutrient availability in the plant, such as increasing phosphorus uptake by solubilisation of inorganic phosphates, the production of iron-chelating siderophores to increase iron uptake and nitrogen fixation. Only few species of *Pseudomonas* have shown the ability to fix nitrogen, including *Pseudomonas stutzeri* (Krotzky and Werner 1987) and *Pseudomonas azotifigens* (Hatayama et al. 2005). As these species are not part of the fluorescent pseudomonad group, nitrogen fixation will not be further discussed.

Plant growth hormones (e.g. auxins, cytokinins and gibberellins) are synthesized in extremely low concentrations in plants and act as chemical messengers and growth and development regulators in plants (Martínez-Viveros et al. 2010). In addition to being synthesized by plants, these phytohormones are synthesized by a number of bacteria associated with plants and soil (Martinez-Toledo et al. 1988; Bottini et al. 1989). The production of phytohormones by *Pseudomonas* species is considered to be one of their main mechanisms of plant growth promotion (Egamberdieva 2005). To date, auxins are the most well studied phytohormones in rhizobacteria (Karadeniz et al. 2006; Spaepen et al. 2007). Bacterial production of phytohormones is interesting as there is currently no evidence for metabolic effects of phytohormones in bacteria (Persello-Cartiaux et al. 2003).

3.3.1.1 Indole-3-Acetic Acid

Auxin is a phytohormone produced by plants and involved in growth regulation. It was discovered that a majority of bacteria in the rhizosphere are able to produce the auxin indole-3-acetic acid (IAA) and auxins are able to influence plant growth in beneficial and deleterious ways (Patten and Glick 1996). Several pathways for IAA synthesis from L-tryptophan have been investigated, such as the indole-3-pyruvic acid pathway (Costacurta et al. 1994; Patten and Glick 1996), the indole-3-acetamide pathway (Patten and Glick 1996) and the side chain oxidase pathway (Oberhänsli et al. 1991; Patten and Glick 1996).

Many factors can influence the production of IAA levels including the IAA production pathway (Persello-Cartiaux et al. 2003) and the localization of the IAA synthesis genes, either in the bacterial chromosome or on a plasmid (Patten and Glick 1996). The impact of bacterial IAA on plants has been either beneficial or deleterious and its effect seems to depend on the level of IAA produced inherently by the plant (Dubeikovsky et al. 1993; Persello-Cartiaux et al. 2003). In cases where plants produce low levels of IAA, the addition of bacterial IAA can be beneficial on the plant roots. Beneficial effects of bacterial IAA have been shown to stimulate root hair formation and increase the number and length of lateral and primary roots (Davies 1995). When the plant is producing adequate levels of IAA, the addition of bacterial IAA can be detrimental on root length. At deleterious levels, IAA has been shown to be inhibitory to primary root growth (Davies 1995). Cucumber plants inoculated with a wild type IAA producing *P. protegens* CHA0 strain demonstrated enhanced growth, while inoculation with an IAA overproducing mutant stunted the cucumber growth (Beyeler et al. 1999). The IAA overproducing strain *P. fluorescens* BSP53a stimulated root development in black currant, but suppressed root development in sour cherry cuttings (Dubeikovsky et al. 1993). The authors suggest that their results indicate that the amount of IAA produced by black currant (*Ribes nigrum* L.) plants was suboptimal while the amount secreted by sour cherry (*Prunus cerasus* L.) cuttings was already optimal for the plant and the additional level of IAA produced by *P. fluorescens* BSP53a was inhibitory. The range of optimal IAA concentration for plants may be small as *P. putida* GR12-2, a low level producer of IAA, resulted in a two to three fold increase in the length of canola (*Brassica napus* L.) seedling roots (Glick 1995; Caron et al. 1996) while an over-producing mutant of *P. putida* GR12-2 (producing four times the amount of the wild type) significantly inhibited the growth of canola roots (Xie et al. 1996).

3.3.1.2 Cytokinins

Cytokinins are N^6-substituted aminopurines that are synthesized in plant roots and are translocated to the shoots through the xylem. They are involved in multiple functions and act as plant growth regulators and influence plant physiological and developmental processes such as cell division, seed germination, root development, accumulation of chlorophyll, leaf expansion, and delay of leaf and chloroplast senescence (Patrick 1987; Salisbury and Ross 1992; Arshad and Frankerberger 1993; Chernyad'ev 2009). Several natural cytokinins are known and include isopentenyladenine and compounds differing in the presence or absence and location of a hydroxyl group: zeatin, trans-zeatin, cis-zeatin and dihydrozeatin (Chernyad'ev 2009). At very low concentrations (as low as 10^{-8} M) cytokinin use in plant growth promotion can be efficient, environmentally safe and inexpensive (Chernyad'ev 2009). They have been shown to act in conjunction with auxins. In *in vitro* plant cell cultures, a high cytokinin/auxin ratio promoted shoot production while auxin alone initiated root growth and equimolar amounts of cytokinin and auxin caused undifferentiated callus cells to proliferate (Crozier et al. 2000).

Cytokinins may also be produced by rhizosphere microorganisms that live in close proximity to the root and these cytokinins also may influence plant growth and development (Nieto and Frankenberger 1990; Arshad and Frankerberger 1993; De Salamone et al. 2001). Inoculation of plants with bacteria producing cytokinins has been shown to stimulate shoot growth and reduce root/shoot ratio in plants suffering due to drought (Arkipova et al. 2007).

3.3.1.3 Gibberellins

Gibberellins (GAs) are a large group of important tetracyclic diterpenoid acids and are produced by plants and influence a range of developmental processes in plants including stem elongation, seed germination, seedling emergence, and flower and fruit growth (Davies 1995; Crozier et al. 2000; King and Evans 2003; Sponsel 2003). In most of these processes, gibberellins act in combination with other phytohormones and other regulatory factors, demonstrating highly integrated signaling pathways (Trewavas 2000).

Many GAs have been identified using modern analytical techniques and 136 GAs have been identified in plants, fungi and bacteria (Arshad and Frankerberger 1993; Bottini et al. 2004). Three β-hydroxylated, C19 gibberellins GA_1, GA_3 and GA_4, have all been reported as being directly involved in promotion of shoot elongation in plants (Crozier et al. 2000). Gibberellic acid (GA_3) is the main product of gibberellins in bacteria (Bruckner and Blecschmidt 1991). It is a terpenoid hormone involved in regulating plant growth and development (Karakoç and Aksöz 2006). Gibberellin production has been observed in various *Pseudomonas* spp. *P. putida* H-2-3 that produces bioactive GA_1 and GA_4 significantly increased the growth of a GA-deficient rice (*Oryza sativa* L.) cultivar *Waito-C* (Kang et al. 2014). This strain was also able to enhance plant growth as well as tolerance to drought and salt stresses in soybean (*Glycine max* (L.) Merr.) plants through various mechanisms, including GA production (Kang et al. 2014). Inoculation of GA-producing *Pseudomonas* sp. 54RB led to increased growth and yield in soybean plants (Afzal et al. 2010). GA_3 production is influenced by cultural conditions and these factors include pH, temperature and incubation time (Kahlon and Malhotra 1986).

3.3.1.4 1-Aminocyclopropane-1-Carboxylate (ACC) Deaminase

Ethylene is a gaseous phytohormone that acts at low concentrations in the regulation of all processes of plant growth, development and senescence (Shaharoona et al. 2006; Saleem et al. 2007). In addition to acting as a plant growth regulator, it has also been identified as a stress hormone. At high concentrations, ethylene can be harmful as it induces defoliation, premature senescence and cellular processes that inhibit stem and root growth (Li et al. 2005). In response to various environmental stressors, plants will synthesize 1-aminocyclopropane-1-carboxylate (ACC) which is the immediate precursor for ethylene (Chen et al. 2002; Glick et al. 2007). Some of the ACC produced by the plants is secreted into the rhizosphere and reabsorbed by the plant roots, where it will be converted to ethylene. Accumulation of ethylene in the roots leads to poor root growth and further stress. The ability to degrade ACC by bacteria in the rhizosphere helps in the re-establishment of a healthy root system that can surmount environmental stresses (Martinez-Viveros et al. 2010).

Many bacteria synthesize the enzyme ACC deaminase that will degrade ethylene to α-ketobutyrate and ammonia (Glick et al. 1998; Glick 2005). A significant amount of ACC might be excreted by the plants roots and taken up by soil microorganisms and hydrolyzed by ACC deaminase, decreasing the amount of ACC in the environment, preventing ethylene accumulation in plants and allows the bacteria to use ACC as a nitrogen source (Penrose and Glick 2003; Persello-Cartiaux et al. 2003; Glick 2005).

Pseudomonas strains demonstrating ACC deaminase activity have been isolated in soil (Govindasamy et al. 2008). Reed and Glick (2005) inoculated canola seeds with ACC-deaminase producing *Pseudomonas asplenii* and observed an increase in dry matter content of the root and aerial parts. Arshad et al. (2008) demonstrated that a strain of *Pseudomonas* sp. with ACC deaminase activity was able to partially eliminate the effect of drought stress on the growth of peas. Tomato plants pretreated with *P. fluorescens* and *P. migulae* (both displaying ACC deaminase activity) were healthier and demonstrated better growth under high salinity stress compared to plants pretreated with an ACC deaminase deficient mutant or without bacterial treatment (Ali et al. 2014). *P. fluorescens* strains transformed with ACC deaminase gene and its regulatory region increased length of canola plants (Wang et al. 2000).

3.3.1.5 Phosphate Solubilizing *Pseudomonas* spp.

Phosphorus (P) is an essential plant nutrient for growth and development with low availability in many agricultural soils (Martínez-Viveros et al. 2010). Many soils have a high total P content due to the application of P fertilizers over long periods of time (Dey 1988), however, a large portion of P is present in insoluble forms and is not available for plant nutrition (Mullen 2005).

Phosphate solubilizing bacteria constitute between 1 % and 50 % of the total population of cultivable bacteria in soil (Chabot et al. 1993; Khan et al. 2009). A considerably higher concentration of phosphate solubilizing bacteria is found in the rhizosphere as compared to bulk soil (Katznelson et al. 1962; Raghu and MacRae 1966). The ability of rhizosphere bacteria to solubilize insoluble P minerals has been attributed to their secretion of organic acids (e.g. gluconate, citrate, lactate, and succinate) and phosphatases (Gyaneshwar et al. 1999; Rodríguez and Fraga 1999) to convert the insoluble phosphate into soluble ions (Podile and Kishore 2006). These bacteria solubilize quantities in excess of their nutritional demands, thereby making it available for plants (Chen et al. 2006).

Increased plant growth and phosphate uptake have been reported in many crop species as a result of the inoculation of phosphate solubilizing *Pseudomonas* species, for example in rice (Gusain et al. 2015), in soybean (Fankem et al. 2015; Afzal et al. 2010), in pea (Oteino et al. 2015) and in wheat (Babana and Antoun 2006).

Additionally, Afzal et al. (2010) found increased nodulation in soybean plants that were co-inoculated with *Bradyrhizobium* strain TAL 377 and *Pseudomonas* sp. strain 54RB as compared to only *Bradyrhizobium* TAL 377. They suggest that the increase in nodulation could be due to *Pseudomonas*-induced phosphate solubilisation (as well as an increase in gibberellic acid), which increased root proliferation and stimulated plant growth (Afzal et al. 2010).

3.3.1.6 Siderophores

Hundreds of siderophores have been identified and reported for cultivable microorganisms, some of which are recognized and used by different microorganisms, while other are species-specific (Crowley 2006). These compounds are produced by various types of bacteria in response to iron deficiency, normally occurring in neutral to alkaline pH soils, due to low iron solubility at high pH (Sharma and Johri 2003). Many plants can use various bacterial siderophores as iron sources, although the total concentrations may be too low to significantly contribute to plant iron uptake.

Among most of the bacterial siderophores studied, those produced by *Pseudomonas* species are known for their high affinity to iron. The most abundant siderophore in *Pseudomonas* sp. is pyoverdine. Carrillo-Castañeda et al. (2002) reported positive effects on alfalfa (*Medicago sativa* L.) plantlet growth after the inoculation of siderophore producing *Pseudomonas* sp. grown in iron limited cultures. The inoculated alfalfa seeds increased their germination as well as the root and stem dry weight.

3.3.1.7 Pyrroloquinoline Quinone

Pyrroloquinoline quinone (PQQ) is the main cofactor in redox enzymes names quinoproteins and was first identified in 1979 as a cofactor in bacterial methanol dehydrogenase (Salisbury et al. 1979) and glucose dehydrogenase (Duine et al. 1979). The production of the PQQ molecule is encoded by the *pqq* operon which consists of six core genes, *pqqABCDEF* (Goldstein et al. 2003; Oteino et al. 2015). Additional genes in the PQQ operon (*pqqHIJKM*) have been identified in *P. fluorescens* B16 (Choi et al. 2008). PQQ has antioxidant properties and is involved in plant growth promotion through phosphate solubilisation. Glucose dehydrogenase uses PQQ as a redox cofactor for the oxidation of glucose to gluconic acid. This acid is then diffused into the areas surrounding the bacteria and helps in the acidic solubilisation of insoluble phosphates in soil (Duine et al. 1990; Stites et al. 2000; Misra et al. 2012). Other studies have shown that PQQ is also involved in the biocontrol ability in certain *P. fluorescens* strains (James and Gutterson 1986; de Werra et al. 2009).

In addition to its role in phosphate solubilisation, PQQ is suspected to be directly involved in plant growth promotion. A PQQ mutant of *P. fluorescens* B16 lost its growth promoting ability in tomato, cucumber, Arabidopsis (*Arabidopsis thaliana* (L.) Heynh.), and hot pepper (*Capsicum annuum* (L.)), which were restored when the PQQ genes were complemented in the B16 mutant (Choi et al. 2008). This group also directly applied synthetic PQQ to cucumber plants and saw an increase in the fresh weight of the plants (Choi et al. 2008). They also applied synthetic PQQ to germinating seedlings of Arabidopsis and hot pepper and observed increases in the fresh and dry weight of Arabidopsis and the size of the cotyledons of the hot peppers, indicating that PQQ is directly involved in plant growth promotion (Choi et al. 2008).

3.3.2 Plant-Beneficial Fluorescent Pseudomonads in Biocontrol

The fluorescent pseudomonads group contains numerous organisms that have the capacity to suppress diseases in several plant-pathogen systems and that can act as effective biological control agents (BCAs; Haas and Défago 2005). In this section, we will attempt to summarize the knowledge gathered on biocontrol of plant pathogens using fluorescent pseudomonads.

3.3.2.1 Plant-Pathogens and Systems Controlled by Fluorescent Pseudomonads

Since the publication of an important review by Weller in 2007, a large number of studies have focused on plant beneficial fluorescent pseudomonads and their antagonistic activity toward plant pathogens. The biocontrol capability of fluorescent pseudomonads is particularly interesting as they exhibit a wide activity and are able to target a broad spectrum of plant pathogens. Among these, the fungus *Gaeumannomyces graminis* var. *tritici*, responsible for the take-all disease of wheat, is the most studied and described plant pathogen system and serves as a model system for *Pseudomonas* spp./pathogen interactions (Kwak and Weller 2013). In this system, *P. protegens* CHA0 has been shown to control the disease through characterized antifungal activity determinants, such as the production of DAPG (Keel et al. 1992). When comparing the biocontrol activity of *P. protegens* CHA0 and a mutant unable to produce DAPG, Keel et al. (1992) found that the DAPG-mutant showed less inhibition of *G. graminis* var. *tritici in vitro*, and less suppression effect on take-all of wheat as compared to the wild-type. However, although various roles of DAPG are known, such as an inducer of plant resistance and a signal molecule that affects gene expression (Dubuis et al. 2007), the precise mode of action of DAPG in disease suppression is still a matter of debate.

An increasing number of papers have been published about new fluorescent pseudomonads demonstrating biocontrol activity, so we present a table listing studies that have occurred since Weller's review published in 2007 (Table 3.1). Most of these studies have focused on fungal diseases, whereas studies describing the biocontrol of bacterial and viral diseases using fluorescent pseudomonads are rare. In this context, in the last 10 years, *Rhizoctonia solani* has been the most investigated pathogen, while the majority of biocontrol stains described belong to the species *P. fluorescens*.

With the recent developments and costs reductions associated with next-generation genome sequencing, genome sequencing of fluorescent pseudomonads of biocontrol interest has been significantly increasing. Next-generation sequencing not only allows the comparison of different biocontrol strains and exploring their functional heterogeneity against their origins (Loper et al. 2012; Rong et al. 2012) but it is also increasingly used to identify genes of biocontrol interest, such as those involved in secondary metabolite biosynthesis (Massart et al. 2015; Roquigny et al. 2015).

Table 3.1 Plant pathogens controlled by plant-beneficial *Pseudomonas* spp. and the mode of action involved in biocontrol

Category of pathogen	Pathogens	Plant system	Biocontrol strain	Biocontrol mechanism	References
Fungi	*Alternaria tenuissima*	Cardoon	*Pseudomons* sp. PS2	Antibiosis (PCA, 2-OH-PHZ, IAA)[a]	Jošić et al. (2012a, b)
	Botrytis cinerea	Tobacco	*P. putida* B001	ISR	Park et al. (2011)
		Alfalfa	*P. fluorescens* UM270	Antibiosis (phenazines, DAPG, HCN, IAA), competition (siderophores), ISR (ACC deaminase, auxin)	Hernández-León et al. (2015)
	Fusarium oxysporum	Mungbean	*Pseudomonas* sp. NAFP-19, NAFP-31 and NAFP-32	Antibiosis, competition	Noreen et al. (2015)
	F. oxysporum f. sp. *radicids-cucumerinum*	Cucumber	*P. aeruginosa* P23	Antibiosis (DAPG), competition (siderophores)	Bradley and Punja (2010)
	F. oxysporum f. sp. *radices-lycopersici*	Tomato	*P. chlororaphis* M71	Antibiosis (potential phenazines and DAPG), competition (siderophores)	Puopolo et al. (2011)
	F. oxysporum	Cymbidium orchids	*Pseudomonas* sp. BRL-1	Competition (siderophores)	Sen et al. (2009)
	Fusarium solani	Mungbean	*Pseudomonas* sp. NAFP-19, NAFP-31 and NAFP-32	Antibiosis, competition	Noreen et al. (2015)
	Gaeumannomyces graminis var. *tritici*	Wheat	*P. fluorescens* JC14-07, HC9-07, and HC13-07	Antibiosis (PCA)	Yang et al. (2011)
		Wheat	*P. fluorescens* VUPf5	Antibiosis (HCN, PCA), competition (siderophores)	Lagzian et al. (2013)
		Barley	*Pseudomonas* sp. DSMZ 13134	Antibiosis, competition, ISR	Frölich et al. (2012)

	Macrophomina phaseolina	Sorghum	*P. fluorescens* SRI-156	Antibiosis (IAA), siderophores	Gopalakrishnan et al. (2011)
		Broadbean	*P. fluorescens* RF36	Antibiosis (IAA)	Devi et al. (2011)
		Mungbean	*Pseudomonas* sp. NAFP-19, NAFP-31 and NAFP-32	Antibiosis, competition	Noreen et al. (2015)
		Safflower	*P. fluorescens* CTPF31	Competition (siderophores)	Govindappa et al. (2011)
		Sunflower	*P. aeruginosa* PF23	Potential antibiosis (biosurfactants)	Tewari and Arora (2014)
	Mucor hiemalis f. sp. *hiemalis*	Cymbidium orchids	*Pseudomonas* sp. BRL-1	Competition (siderophores)	Sen et al. (2009)
	Phytophthora drechsleri	Cucumber	*P. fluorescens* strains	Antibiosis (DAPG, PCA, PLT)	Shirzad et al. (2012)
			P. fluorescens CV6	Antibiosis (HCN), competition (siderophores)	Maleki et al. (2010)
	Phytophthora infestans	Potato	*P. chlororaphis* R47	Antibiosis (HCN, phenazines, PRN, HPR), competition (siderophores)	Guyer et al. (2015)
	Polymyxa betae	Sugar beet	P. putida	Currently unknown mechanism	Aksoy and Kutluk Yilmaz (2008)
	Pyrenophora teres	Barley	*P. fluorescens* MKB156	ISR	Khan et al. (2010)
	Pythium irregulare	Soybean	*P. protogens* WayneR1	Antibiosis (DAPG, HCN, PLT)	McSpadden Gardener et al. (2007) and Rong et al. (2012)
	Pythium myriotylum	Cocoyam	*Pseudomonas* sp. CMR12a	Antibiosis (PCN, HCN, biosurfactants)	Perneel et al. (2007)
	Pythium ultimum	Alfalfa	*P. fluorescens* UP61.2, UP143.8 and UP148.3	Antibiosis (DAPG, HCN, PLT, PRN)	Quagliotto et al. (2009)
	Ralstonia solanacearum	Tomato	*P. brassicacearum* J12	Antibiosis (DAPG, HCN), siderophores	Zhou et al. (2012)
		Eggplant	*Pseudomonas* sp. EB67	Antibiosis (DAPG, IAA) siderophores	Ramesh et al. (2009)

(continued)

Table 3.1 (continued)

Category of pathogen	Pathogens	Plant system	Biocontrol strain	Biocontrol mechanism	References
	Rhizoctonia solani	Potato	*P. aeruginosa* RZ9	Currently unknown mechanism	Mrabet et al. (2015)
		Isabgol	*P. aeruginosa* SD12	Antibiosis (HCN, 1-OH-PHZ), competition (siderophores)	Patra (2012)
		Lettuce	*P. jessenji* RU47	ISR (siderophores)	Adesina et al. (2009)
		Sugar beet	*Pseudomonas poae* RE*1-1-14	Currently unknown mechanism	Müller et al. (2013)
		Bean	*P. fluorescens* UTPF5 (formerly *P. fluorescens* P-5)	Antibiosis (DAPG, HCN)	Afsharmanesh et al. (2010)
		Mungbean	*Pseudomonas* sp. NAFP-19, NAFP-31 and NAFP-32	Antibiosis, competition	Noreen et al. (2015)
		Broad bean	*P. fluorescens* RF36	Antibiosis (IAA)	Devi et al. (2011)
		French bean	*P. fluorescens* strains	Currently unknown mechanism	Negi et al. (2011)
	Rhizoctonia cerealis	Wheat	*P. chlororaphis* subsp. *aurantiaca* Pa40	Antibiosis (2-OH-PHZ, PRN, HCN), competition (siderophores), ISR	Jiao et al. (2013)
	Rhizoctonia oryzae	Wheat	*P. fluorescens* 29G9 and Wood3R, *P. chlororaphis* 48G9	Secondary metabolites (DAPG, PRN, CLP, PCA)	Mavrodi et al. (2012b)
	Sclerotinia sclerotiorum	Tomato	*Pseudomonas* sp. Psf5	Antibiosis, siderophores	Hammami et al. (2013)
	Sclerotium rolfsii	Tomato	*Pseudomonas* sp. PCI2	Antibiosis (IAA)	Pastor et al. (2010, 2012)

Bacteria	*Clavibacter michiganensis* subsp. *Michiganensis*	Tomato	*P. fluorescens* LBUM300	Antibiosis (DAPG, HCN)	Lanteigne et al. (2012)
	Pectobacterium carotovorum subsp. *carotovorum*	Orchids	*Pseudomonas* sp. BRL-1	Competition (IAA, siderophores)	Sen et al. (2009)
	P. carotovorum	Tobacco	*P. putida* B001	ISR	Park et al. (2011)
	Pseudomonas savastanoi	Olive	*P. fluorescens* PCIF7	Unknown antibiosis mechanism	Maldonado-González et al. (2013)
	Streptomyces scabies	Potato	*P. fluorescens* LBUM223	Antibiosis (PCA)	Arseneault et al. (2013, 2014)
Virus	Tobacco mosaic virus	Tobacco	*P. putida* B001	ISR	Park et al. (2011)

[a]*PCA* phenazine 1-carboxylic acid, *2-OH-PHZ* 2-hydroxyphenazine, *IAA* Indole acetic acid, *HCN* hydrogen cyanide, *DAPG* 2,4-diacetylphloroglucinol, *PLT* pyoluteorin, *PRN* pyrrolnitrin, *HPR* 2-hexyl-5-propyl-alkylresorcinol

3.3.2.2 Mechanisms Involved in Disease Suppression

Over the years, it has been demonstrated that fluorescent pseudomonads display numerous capabilities to suppress plant diseases due to various genetic and phenotypic characteristics. To date, several mechanisms of disease suppression have been detected in *Pseudomonas* spp. and the main ones are competition for iron (Berg 2009), plant induced systemic resistance (ISR; Bakker et al. 2007) and antibiosis (Raaijmakers et al. 2002). Once fluorescent pseudomonads are established in the plant rhizosphere, more than one mechanisms of biocontrol may be used in parallel.

3.3.2.2.1 Competition

Previously described as one of the primary ways for bacteria to establish themselves in the rhizosphere, competition is also one of the main mechanisms used by BCAs to compete with plant pathogens for space and nutrients, leading to reduced disease development (Haas and Défago 2005). One of the main nutrients that leads to competition is iron due to its limiting presence in soil (Loper and Buyer 1991). Fluorescent pseudomonads will compete for iron through the production of the siderophore pyoverdine. Siderophores have demonstrated biocontrol capacity under *in vitro* and *in vivo* conditions on pathogenic fungi or fungal-like organisms, including *Pythium* spp. and *Fusarium* spp. (Loper and Buyer 1991; León et al. 2009; Sen et al. 2009). Since the 1990s, it was suggested that siderophores production was dependent on a large range of biotic and abiotic factors (Loper and Buyer 1991; O'Sullivan and O'Gara 1992); notably, the nature and the concentration of nitrogen and carbon sources, the level of phosphate, and the soil's pH and temperature (O'Sullivan and O'Gara 1992). For example, in the case of *P. aeruginosa* PAO1, a high phosphate concentration inhibits pyoverdine production. Succinic acid and ammonium sulphate have been identified to be the best sources of carbon and nitrogen for pyoverdine production with an optimum carbon to nitrogen ratio of 4 to 1 (Barbhaiya and Rao 1985). Loper and Buyer (1991) concluded that among these factors, pH might be the most important for iron availability in soil. More recently, similar studies were performed and demonstrated the importance of minerals and carbon sources for siderophore biosynthesis and therefore for microbial competition (Duffy and Défago 1999; Guiñazú et al. 2010, 2013).

3.3.2.2.2 Induced Resistance

Some physical or chemical stresses have been shown to be responsible for an induced state of resistance in plants which protects against pathogenic infections (Pieterse et al. 2014). Resistance can be triggered either by pathogenic or non-pathogenic microorganisms, therefore two main types of resistance have been identified: systemic acquired resistance (SAR), which is usually triggered by pathogenic microorganisms infecting the plant but to a level that does not cause disease development, and induced systemic resistance (ISR) generally triggered by beneficial microorganisms. This distinction between the agents responsible for inducing either SAR or ISR is not always clear and it has been shown that a beneficial microorganism may induce SAR response (Van De Mortel et al. 2012), and a pathogenic microorganism an ISR response (Pieterse et al. 2014). Finally, in some cases, both SAR and ISR have been shown to be induced in parallel (Van Wees et al. 2000). The focus of this section will be on the typical ISR response generally induced by plant-beneficial *Pseudomonas* spp.

Before considering ISR, it is however important to briefly described SAR, which has been fully reviewed throughout the years (Spoel and Dong 2012; Dangl et al. 2013; Henry et al. 2013; Pieterse et al. 2013, 2014). SAR in plants is triggered after a local activation of immunity through physical or chemical recognition of a pathogen by the plant. Then, a complex network of signals is activated which may lead to a systemic defense response in the plant. Two major types of signals: PTI ((PAMP (pathogen-associated-molecular-pattern)-triggered immunity)) and ETI (effector-triggered immunity) (Thonart et al. 2012), may be involved. Briefly, PTI is considered as the first line of defense of the plant, which is often bypassed by the plant pathogen through suppression of PTI or preventing the pathogen's detection. ETI is instead considered as a manifestation of the second line of defense, often described as a gene-for-gene resistance leading most of the time to a programmed cell death in order to stop the infection. In other words, PTI and ETI are the local defense lines of the plant, which in turn may trigger SAR if the pathogen is capable of escaping these first and second lines of defense. SAR is mediated by the plant hormone salicylic acid (SA). An increase in the SA level throughout the plant is essential for the establishment of SAR (Van Loon and Bakker 2006). According to current knowledge, SAR seems to remain active for the lifetime of the plant in spite of a lessened induced state that can be observed over time (Van Loon and Bakker 2006).

On the other hand, ISR is a typical mode of action of plant-beneficial *Pseudomonas* spp. in disease suppression. During ISR, bacterial determinants like flagella and secondary metabolites such as DAPG, lipopolysaccharides, and siderophores are detected by the plant host, leading to the secretion of hormonal mediators (Bakker et al. 2013). ISR is mediated by jasmonic acid (JA) and ethylene (ET). Many good reviews have been published on the subject (Bakker et al. 2007, 2013; Pieterse et al. 2014). According to the review by Bakker et al. (2007), noteworthy determinants of *Pseudomonas* spp. are the production of siderophores such as pseudobactin (pyoverdine) and antibiotics such as DAPG. More recently, the capacity of *P. chlororaphis* subsp. *aurantiaca* Pa40 in eliciting an ISR response in wheat infected by *Rhizoctonia cerealis* was linked to the production of phenazines rather than other bacterial determinants (Jiao et al. 2013).

Following the accumulation of JA, ET (in ISR) and/or SA (in SAR) different metabolic cascades connected with the different pathways (ISR or SAR) result in systemic chemical changes in the plant such as the release of proteins or defense strengthening physical barriers of the plant (Jones and Dangl 2006; Dangl et al. 2013). For instance, a hypersensitive response, which is associated with a programmed cell death, can be observed as the hallmark of ETI (Spoel and Dong 2012). After the activation of plant-defense pathways, the plant cell wall is reinforced by the deposition of glucan polymers (Spoel and Dong 2012). Current trends are to study a plant's transcriptome to better understand ISR-SAR activation differences. One of the first studies was on *Arabidopsis thaliana* whose resistance was induced by beneficial *P. fluorescens* WCS417r (Verhagen et al. 2004). This study revealed that in *A. thaliana*, gene activity of the transcription factors implicated in the regulation of JA and ET-dependent defenses was upregulated in the root, but not in the leaves (Verhagen et al. 2004). Walters et al. (2013) reviewed controlling crop diseases using induced resistance and discussed notably that as induced resistance is a host response, its expression under field conditions is likely to be influenced by a number of factors, including the environment, genotype, crop nutrition and the extent to which plants are already induced.

3.3.2.2.3 Antibiosis

Antibiosis is described as the capacity to produce and secrete antibiotic compounds or other antimicrobial diffusible compounds leading to the inhibition of a pathogen's growth and, in most cases, to the reduction of the pathogen's population. Several secondary metabolites produced by fluorescent pseudomonads have been studied and their activity has been demonstrated by comparing the activity of wild-type strains to isogenic non-antibiotic producing mutant strains. These types of studies have shown that antibiosis is one of the most important mechanisms for biocontrol in fluorescent pseudomonads (Siddiqui 2005). Plant-beneficial *Pseudomonas* spp. are able to produce and secrete a wide range of antimicrobial compounds. The best example of this is *P. protegens* CHA0, which can synthetize more than ten compounds displaying antagonistic activity towards pathogens: DAPG, hydrogen cyanide (HCN), pyoluteorin (PLT), pyrrolnitrin (PRN), and multiple phenazine compounds (Haas et al. 1991; Haas and Défago 2005). For a description of the nature, the biosynthesis, and the function of antibiotic compounds produced by fluorescent pseudomonads, readers are referred to several excellent reviews on this topic (Keel et al. 1992; Raaijmakers et al. 2002; Haas and Défago 2005; Fernando et al. 2006; Mavrodi et al. 2006, 2010). As one of the most studied group of antibiotics involved in plant-beneficial *Pseudomonas* spp./pathogen interactions, we will describe in more detail the phenazine derivatives group, their production and their action in the field.

3.3.2.2.4 Phenazines

Phenazines play a vital role in the biocontrol of plant diseases (Tambong and Höfte 2001; Chin-A-Woeng et al. 2003; Mavrodi et al. 2006; D'aes et al. 2011; Hua and Höfte 2015) and may also contribute to biofilm formation and virulence (Price-Whelan et al. 2006; Pierson and Pierson 2010; Selin et al. 2010). The most common phenazine derivatives are pyocyanin, phenazine-1-carboxylic acid (PCA) and phenazine-1-carboxamide (PCN). These compounds appear to improve the stability of colonies by producing a biofilm that allows the bacteria to attach to roots or seeds of plants (Mavrodi et al. 2006). The genes responsible for the production of PCA are organized in an operon of seven genes: *phzABCDEFG* (Mavrodi et al. 2006, 2010). This operon is accompanied by genes involved in the regulation, transport, resistance and PCA conversion to other phenazine derivatives. The phenazine operon is well conserved as the loss of the ability to produce phenazines is usually associated with a reduced ability to survive in the environment (Mavrodi et al. 2013).

As previously indicated, fluorescent pseudomonads can produce a large range of antimicrobial secondary metabolites, however, the capacity to produce a greater number of antibiotics is not necessarily associated with a better biocontrol response (Perneel et al. 2007). The production of HCN, PCA, PCN, PLT and PRN by *Pseudomonas* sp. CMR5c would suggest that this strain could be a perfect biocontrol agent with a broad spectrum activity. However, despite the myriad of secondary metabolites being produced, *Pseudomonas* sp. CMR5c was not as effective as *Pseudomonas* sp. CMR12a, which produces phenazine derivatives, against *Pythium myriotylum* in cocoyam (Perneel et al. 2007). The authors concluded that in this system, phenazines are key factors in the biological control of cocoyam root rot rather than pyrrolnitrin and pyoluteorin.

Another important point to consider concerning phenazines implication in biocontrol is the quantity or the dose being produced by a given BCA. Pathogen destruction is linked to high levels of antibiotics (Haas and Keel 2003) and a decrease in the pathogen population may depend on the concentration of PCA produced by a BCA (Arseneault et al. 2014). This hypothesis is supported by the fact that the level and the timing of antibiotic biosynthetic gene expression depends on the bacterial population density. The higher a bacterial concentration is, the more antibiotic accumulation will occur in soil (Mavrodi et al. 2012a). In general, the scientific community agrees on the necessity to have a minimal threshold of BCA present in order for biocontrol to occur. For phenazine-producing plant-beneficial *Pseudomonas* spp., this level has been estimated between 10^4 and 10^6 CFU/g of root (Raaijmakers and Weller 1998; Haas and Défago 2005). The ability to quantify antibiotics directly in soils is increasingly interesting to scientists. For example, it was observed that for PCA, 100 μM localized produced amounts were sufficient for the inhibition of pathogens (Mendes et al. 2011). It has, however, been suggested that a sub-inhibitory concentration of antibiotics might, in some cases, suppress disease development through the alteration of the transcriptional activity of key pathogenesis genes in the pathogen (Davies et al. 2006; Raaijmakers and Mazzola 2012). Arseneault et al. (2013) have suggested that transcriptional changes in a pathogen leading to reduced virulence due to the exposure to sub-inhibitory concentration of antibiotics is a key factor in biocontrol and could be considered as an independent mechanism of antibiosis (Arseneault et al. 2013). More specifically, the reduction of potato common scab disease symptoms was not linked to a reduction in *Streptomyces scabies* following the inoculation of potato plants with PCA-producing *P. fluorescens* LBUM 223, but to a significant alteration of gene expression, notably genes involved in pathogenicity, suggesting a novel biocontrol mechanism (Arseneault et al. 2017, unpublished results).

3.3.3 Regulation of Biocontrol Mechanisms

As previously mentioned, the production of secondary metabolites is usually dependent on bacterial population density, a phenomenon known as quorum sensing (QS). Once the quorum is reached, bacteria are able to modify the expression of some operons involved in secondary metabolite biosynthesis by "sensing" the accumulation of small signaling molecules called autoinducers. For *Pseudomonas* spp., N-acyl-L-homoserine lactones (AHL) have been identified as key signal molecules and their synthesis and recognition generally involves the LuxI/LuxR-like protein family system (Lee et al. 2010). QS is involved in many cellular processes such as antibiotic synthesis and biofilm formation. In *P. aeruginosa*, this is the main mechanism for regulating the production of PCA (Pierson and Pierson 1996; Chin-A-Woeng et al. 2003). More recently, in *P. chlororaphis* subsp. *aurantiaca* StFRB508, multiple AHLs produced via two different quorum-sensing systems demonstrated the regulation of a same QS-regulated-function; PCA production (Morohoshi et al. 2013). Biosynthesis of phenazines seems to occur late during the growth phase of *Pseudomonas* spp. (Mavrodi et al. 2006) as it depends on the population density and certain environmental conditions such as temperature, pH, and the availability of certain nutrients (Chin-A-Woeng et al. 2003). Throughout the years, a lot of excellent review articles that have focused on quorum sensing have been published and we refer the reader to these for more details on the subject (Miller and Bassler 2001; Schauder and Bassler 2001; Compant et al. 2005; Williams 2007; Ng and Bassler 2009).

3.4 Conclusions and Future Perspectives

Fluorescent pseudomonads are proven to be an important group of plant growth promoters and biocontrol agents. They are able to utilize various mechanisms to increase plant growth, protect plants from disease, and to colonize and maintain significant populations in the rhizosphere of many different plants. However, many questions remain as to how to utilize our knowledge of root colonization, plant growth promotion and biocontrol activities of *Pseudomonas* species to use these strains in large-scale agricultural contexts. The advent of next-generation sequencing technologies will allow future research to investigate the accessory genomes of many *Pseudomonas* species to better understand their unique plant growth promotion and biocontrol activities. Next-generation sequencing will also allow researchers to focus on the rhizosphere as a whole and better understand the interactions of *Pseudomonas* species with the indigenous rhizosphere population and the plant through transcriptome and metagenome analyses. Better understanding of these complex interactions may gain insight to overcome inconsistent disease control, which remains a major impediment to widespread use and commercialization of plant growth promoting *Pseudomonas* species.

References

Adesina MF, Grosch R, Lembke A et al (2009) In vitro antagonists of *Rhizoctonia solani* tested on lettuce: rhizosphere competence, biocontrol efficiency and rhizosphere microbial community response. FEMS Microbiol Ecol 69:62–74

Afsharmanesh H, Ahmadzadeh M, Javan-Nikkhah M et al (2010) Characterization of the antagonistic activity of a new indigenous strain of *Pseudomonas fluorescens* isolated from onion rhizosphere. J Plant Pathol 92:187–194

Afzal A, Bano A, Fatima M (2010) Higher soybean yield by inoculating with N-fixing and P-solubilizing bacteria. Agron Sustain Dev 30:487–495

Aksoy H-M, Kutluk Yilmaz N-D (2008) Antagonistic effects of natural *Pseudomonas putida* biotypes on *Polymyxa betae* Keskin, the vector of beet necrotic yellow vein virus in sugar beet. J Plant Dis Prot 115:241–246

Ali S, Trevor CC, Glick BR (2014) Amelioration of high salinity stress damage by plant growth-promoting bacterial endophytes that contain ACC deaminase. Plant Physiol Biochem 80:160–167

Anderson AJ (1983) Isolation from root and shoot surfaces of agglutinins that show specificity for saprophytic pseudomonads. Can J Bot 61:3438–3443

Antoun H, Kloepper JW (2001) Plant growth promoting rhizobacteria. In: Brenner S, Miller JH (eds) Encyclopedia of genetics. Academic, New York, pp 1477–1480

Arkipova TN, Prinsen EA, Veselov SU et al (2007) Cytokinin producing bacteria enhance plant growth in drying soil. Plant Soil 292:305–315

Arseneault T, Goyer C, Filion M (2017) Phenazine production by *Pseudomonas* sp. LBUM223 contributes to the biological control of potato common scab. Phytopathology 103:995–1000

Arseneault T, Pieterse CMJ, Gerin-Ouellet M et al (2014) Long-term induction of defense gene expression in potato by *Pseudomonas* sp. LBUM223 and *Streptomyces scabies*. Phytopathology 104:926–932

Arshad M, Frankerberger WT Jr (1993) Microbial production of plant growth regulators. In: Metting FB Jr, Dekker M (eds) Soil microbial ecology. Applications in agricultural and environmental management. Marcel Dekker, New York, pp 307–348

Arshad M, Shaharoona B, Mahmood T (2008) Inoculation with *Pseudomonas* spp. containing ACC-deaminase partially eliminates the effects of drought stress on growth, yield, and ripening of pea (*Pisum sativum* L.) Pedosphere 18:611–620

Babana A, Antoun H (2006) Effect of Tilemsi phosphate rock-solubilizing microorganisms on phosphorus uptake and yield of field-grown wheat (*Triticum aestivum* L.) in Mali. Plant Soil 287:51–58

Bakker PAHM, Pieterse CMJ, Van Loon LC (2007) Induced systemic resistance by fluorescent *Pseudomonas* spp. Phytopathology 97:239–243

Bakker PAHM, Doornbos RF, Zamioudis C et al (2013) Induced systemic resistance and the rhizosphere microbiome. Plant Pathol J 29:136–143

Barbhaiya H, Rao K (1985) Production of pyoverdine, the fluorescent pigment of *Pseudomonas aeruginosa* PAO1. FEMS Microbiol Lett 27:233–235

Berg G (2009) Plant-microbe interactions promoting plant growth and health: perspectives for controlled use of microorganisms in agriculture. Appl Microbiol Biotechnol 84:11–18

Beyeler M, Keel C, Michaux P et al (1999) Enhanced production of indole-3-acetic acid by a genetically modified strain of *Pseudomonas fluorescens* CHA0 affects root growth of cucumber, but does not improve protection of the plant against *Pythium* root rot. FEMS Microbiol Ecol 28:225–233

Bottini R, Fulchieri M, Pearce Pharis DRP (1989) Identification of gibberellins A1, A3 and Iso-A3 in cultures of *Azosporillum lipoferum*. Plant Physiol 10:45–47

Bottini R, Cassán F, Piccoli P (2004) Gibberellin production by bacteria and its involvement in plant growth promotion and yield increase. Appl Microbiol Biotechnol 65:497–503

Boukhalfa H, Crumbliss AL (2002) Chemical aspects of siderophore mediated iron transport. Biometals 15:325–339

Bradley GG, Punja ZK (2010) Composts containing fluorescent pseudomonads suppress *Fusarium* root and stem rot development on greenhouse cucumber. Can J Microbiol 56:896–905

Bruckner B, Blecschmidt D (1991) The gibberellin fermentation. Crit Rev Biotechnol 11:163–192

Buell C, Anderson A (1992) Genetic analysis of the *agg*A locus involved in agglutination and adherence of *Pseudomonas putida*, a beneficial fluorescent pseudomonad. Mol Plant-Microbe Interact 5:154–162

Bull CT, Weller DM, Thomashow LS (1991) Relationship between root colonization and suppression of *Gaeumannomyces-graminis* var *tritici* by *Pseudomonas fluorescens* strain 2–79. Phytopathology 81:954–959

Caron M, Patten CL, Ghosh S et al (1996) Effects of the plant growth promoting rhizobacterium *Pseudomonas putida* GR 122 on the physiology of canola roots. Plant Growth Regul Soc Am Q 7:18–20

Carrillo-Castañeda G, Juárez Muños J, Peralta-Videa JR et al (2002) Alfalfa growth promotion by bacteria grown under iron limiting conditions. Adv Environ Res 6:391–399

Carroll H, Moenne-Loccoz Y, Dowling DN et al (1995) Mutational disruption of the biosynthesis genes coding for the antifungal metabolite 2,4 diacetylphloroglucinol does not influence the ecological fitness of *Pseudomonas fluorescens* F113 in the rhizosphere of sugarbeets. Appl Environ Microbiol 61:3002–3007

Cézard C, Farvacques N, Sonnet P (2015) Chemistry and biology of pyoverdines, *Pseudomonas* primary siderophores. Curr Med Chem 22:165–186

Chabot R, Antoun H, Cescas MP (1993) Stimulation de la croissance du maïs et de la laitue romaine par des microorganismes dissolvant le phosphate inorganique. Can J Microbiol 39:941–947

Chen H, Qualls RG, Miller GC (2002) Adaptive responses of *Lepidium latifolium* to soil flooding: biomass allocation, adventitious rooting, aerenchyma formation and ethylene production. Environ Exp Bot 48:119–128

Chen YP, Rekha PD, Arun AB et al (2006) Phosphate solubilizing bacteria from subtropical soil and their tricalcium phosphate solubilizing abilities. Appl Soil Ecol 34:33–41

Chernyad'ev II (2009) The protective action of cytokinins on the photosynthetic machinery and productivity of plants under stress. Appl Biochem Microbiol 45:351–362

Chin-A-Woeng TF, De Priester W, Van Der Bij AJ et al (1997) Description of the colonization of a gnotobiotic tomato rhizosphere by *Pseudomonas fluorescens* biocontrol strain VCS365, using scanning electron microscopy. Mol Plant-Microbe Interact 10:79–86

Chin-A-Woeng TF, Bloemberg GV, Mulders IHM et al (2000) Root colonization by phenazine-1-carboxamide producing bacterium *Pseudomonas chlororaphis* PCL 1391 is essential for biocontrol of tomato foot and root rot. Mol Plant-Microbe Interact 13:1340–1345

Chin-A-Woeng TF, Bloemberg GV, Lugtenberg BJJ (2003) Phenazines and their role in biocontrol by *Pseudomonas* bacteria. New Phytol 157:503–523

Choi O, Kim J, Kim J-G et al (2008) Pyrrolquinoline quinone is a plant growth promotion factor produced by *Pseudomonas fluorescens* B16. Plant Physiol 146:657–668

Choudhary DK, Prakash A, Wray V et al (2009) Insights of the fluorescent pseudomonads in plant growth regulation. Curr Sci 97:170–179

Compant S, Duffy B, Nowak J et al (2005) Use of plant growth-promoting bacteria for biocontrol of plant diseases: principles, mechanisms of action, and future prospects. Appl Environ Microbiol 71:4951–4959

Costacurta A, Keijers V, Vanderleyden J (1994) Molecular cloning and sequence analysis of an *Azospirillum brasilense* indole-3-pyruvate decarboxylase gene. Mol Gen Genet 243:463–472

Crowley DE (2006) Microbial siderophores in the plant rhizosphere. In: Barton LL, Abadía J (eds) Iron nutrition in plants and rhizospheric microorganisms. Springer, Dordrecht, pp 169–198

Crozier A, Kamiya Y, Bishop G et al (2000) Biosynthesis of hormones and elicitor molecules. In: Buchanan BB, Gruissem W, Jones RL (eds) Biochemistry and molecular biology of plants. Am Soc Plant Physiol, Rockville, pp 850–929

D'aes J, Hua GKH, De Maeyer K et al (2011) Biological control of *Rhizoctonia* root rot on bean by phenazine and cyclic lipopeptide producing *Pseudomonas* CMR12a. Phytopathology 101:996–1004

Dangl JL, Horvath DM, Staskawicz BJ (2013) Pivoting the plant immune system from dissection to deployment. Science 341:746–751

Danhorn T, Fuqua C (2007) Biofilm formation by plant-associated bacteria. Annu Rev Microbiol 61:401–422

Davies PJ (1995) The plant hormones: their nature, occurrence and functions. In: Davies PJ (ed) Plant hormones: physiology, biochemistry and molecular biology. Kluwer, Dordrecht, pp 1–12

Davies J, Spiegelman GB, Yim G (2006) The world of sub-inhibitory antibiotic concentrations. Curr Opin Microbiol 9:445–453

De Mot R, Proost P, Van Damme J et al (1992) Homology of the root adhesion of *Pseudomonas fluorescens* OE 28.3 with porin F of *P.neruginosa* and *P. syringae*. Mol Gen Genet 231:489–493
De Salamone IEG, Hynes RK, Nelson LM (2001) Cytokinin production by plant growth promoting rhizobacteria and selected mutants. Can J Microbiol 47:404–411
De Weert S, Vermeiren H, Mulder IH et al (2002) Flagella-driven chemotaxis towards exudate components is an important trait for tomato root colonization by *Pseudomonas fluorescens*. Mol Plant-Microbe Interact 15:1173–1180
De Weger LA, Van Der Vlugt CI, Wijfjes AH et al (1987) Flagella of a plant-growth-stimulating *Pseudomonas fluorescens* strain are required for colonization of potato roots. J Bacteriol 169:2769–2773
De Werra P, Péchy-Tarr M, Keel C et al (2009) Role of gluconic acid production in the regulation of biocontrol traits of *Pseudomonas fluorescens* CHA0. Appl Environ Microbiol 75:4162–4174
Devi SI, Talukdar N, Sharma KC (2011) Screening of rhizobacteria for their plant growth promotion ability and antagonism against damping off and root rot diseases of broad bean (*Vicia faba* L.) Indian J Microbiol 51:14–21
Dey KB (1988) Phosphate solubilizing organisms in improving fertility status. In: Sen SP, Palit P (eds) Biofertilizers: potentialities and problems. Plant Physiology Forum, Naya Prokash, Calcutta, pp 237–248
Dubeikovsky AN, Mordukhova EA, Kochetkov VV et al (1993) Growth promotion of black currant softwood cuttings by recombinant strain *Pseudomonas fluorescens* BSP53a synthesizing an increased amount of indole-3-acetic acid. Soil Biol Biochem 25:1277–1281
Dubuis C, Keel C, Haas D (2007) Dialogues of root-colonizing biocontrol pseudomonads. Eur J Plant Pathol 119:311–328
Duffy BK, Défago G (1999) Environmental factors modulating antibiotic and siderophore biosynthesis by *Pseudomonas fluorescens* biocontrol strains. Appl Environ Microbiol 65:2429–2438
Duine JA, Frank JJ, van Zeeland JK (1979) Glucose dehydrogenase from *Acinetobacter calcoaceticus*: a “quinoprotein”. FEBS Lett 108:443–446
Duine JA, van der Meer RA, Groen BW (1990) The cofactor pyrrolquinoline quinone. Annu Rev Nutr 10:297–318
Egamberdieva D (2005) Characterization of *Pseudomonas* species isolated from the rhizosphere of plants grown in serozem soil, semiarid region of Uzbekistan. Sci World J 5:501–509
Fankem H, Tchakounte GVT, Nkot LN et al (2015) Common bean (*Phaseolus vulgaris* L.) and soya bean (*Glycine max*) growth and nodulation as influenced by rock phosphate solubilizing bacteria under pot grown conditions. Int J Agric Pol Res 5:242–250
Fernando WD, Nakkeeran S, Zhang Y (2006) Biosynthesis of antibiotics by PGPR and its relation in biocontrol of plant diseases. In: Siddiqui Z (ed) PGPR: biocontrol and biofertilization. Springer, Dordrecht, pp 67–109
Flemming H-C, Wingender J (2010) The biofilm matrix. Nat Rev Microbiol 8:623–633
Frölich A, Buddrus-Schiemann K, Durner J et al (2012) Response of barley to root colonization by *Pseudomonas* sp. DSMZ 13134 under laboratory, greenhouse, and field conditions. J Plant Interact 7:1–9
Ghirardi S, Dessaint F, Mazurier S et al (2012) Identification of traits shared by rhizosphere-competent strains of fluorescent pseudomonads. Microb Ecol 64:725–737
Glick BR (1995) The enhancement of plant growth by free-living bacteria. Can J Microbiol 41:109–117
Glick BR (2005) Modulation of plant ethylene levels by the bacteria enzyme ACC deaminase. FEMS Microbiol Lett 251:1–7
Glick BR (2012) Plant growth-promoting bacteria: mechanisms and applications. Scientifica 2012:963401
Glick BR, Penrose D, Li J (1998) A model for the lowering of plant ethylene concentrations by plant growth-promoting bacteria. J Theor Biol 190:63–68
Glick BR, Patten CL, Holguin G et al (1999) Biochemical and genetic mechanisms used by plant growth promoting bacteria. Imperial College Press, London

Glick BR, Cheng Z, Czarny J et al (2007) Promotion of plant growth by ACC deaminase-producing soil bacteria. Eur J Plant Pathol 119:329–339
Goldstein A, Lester T, Brown J (2003) Research on the metabolic engineering of the direct oxidation pathway for extraction of phosphate from ore has generated preliminary evidence for PQQ biosynthesis in *Escherichia coli* as well as possible role for the highly conserved region of quinoprotein dehydrogenases. Biochem Biophys Acta 1647:266–271
Gopalakrishnan S, Humayun P, Kiran BK et al (2011) Evaluation of bacteria isolated from rice rhizosphere for biological control of charcoal rot of sorghum caused by *Macrophomina phaseolina* (Tassi) Goid. World J Microbiol Biotechnol 27:1313–1321
Govindappa M, Ravishankar RV, Lokesh S (2011) Screening of *Pseudomonas fluorescens* isolates for biological control of *Macrophomina phaseolina* root-rot of safflower. Afr J Agric Res 6:6256–6266
Govindasamy V, Senthilkumar M, Gaikwad K et al (2008) Isolation and characterization of ACC deaminase gene from two plant growth promoting rhizobacteria. Curr Microbiol 57:312–317
Gray EJ, Smith DL (2005) Intracellular and extracellular PGPR: commonalities and distinctions in the plant-bacterium signaling processes. Soil Biol Biochem 37:395–412
Gross H, Loper JE (2009) Genomics of secondary metabolite production by *Pseudomonas* spp. Nat Prod Rep 26:1408–1446
Gross H, Stockwell VO, Henkels MD et al (2007) The genomisotopic approach: a systematic method to isolate products of orphan biosynthetic gene clusters. Chem Biol 14:53–63
Guiñazú LB, Andrés JA, Del Papa MF et al (2010) Response of alfalfa (*Medicago sativa* L.) to single and mixed inoculation with phosphate-solubilizing bacteria and *Sinorhizobium meliloti*. Biol Fertil Soils 46:185–190
Guiñazú LB, Andrés JA, Rovera M et al (2013) Evaluation of rhizobacterial isolates from Argentina, Uruguay and Chile for plant growth promoting characteristics and antagonistic activity towards *Rhizoctonia* sp. and *Macrophomina* sp. *in vitro*. Eur J Soil Biol 54:69–77
Gusain YS, Kamal R, Mehta CM et al (2015) Phosphate solubilizing and indole-3-acetic acid producing bacteria from the soil of Garhwal Hymalaya aimed to improve the growth of rice. J Environ Biol 36:310–307
Guyer A, De Vrieze M, Bönisch D et al (2015) The anti-*Phytophthora* effect of selected potato-associated *Pseudomonas* strains: from the laboratory to the field. Front Microbiol. doi:10.3389/fmicb.2015.01309
Gyaneshwar P, Parekh LJ, Archana G et al (1999) Involvement of a phosphate-starvation inducible glucose dehydrogenase in soil phosphate solubilisation by *Enterobacter asburiae*. FEMS Microbiol Lett 171:223–229
Haas D, Défago G (2005) Biological control of soil-borne pathogens by fluorescent pseudomonads. Nat Rev Microbiol 3:307–319
Haas D, Keel C (2003) Regulation of antibiotic production in root-colonizing *Pseudomonas* spp. and relevance for biological control of plant disease. Annu Rev Phytopathol 41:117–153
Haas D, Keel C, Laville J et al (1991) Secondary metabolites of *Pseudomonas fluorescens* strain CHA0 involved in the suppression of root diseases. In: Hennecke H, Verma DPS (eds) Advances in molecular genetics of plant-microbe interactions. Springer, Dordrecht, pp 450–456
Hammami I, Hsouna AB, Hamdi N et al (2013) Isolation and characterization of rhizosphere bacteria for the biocontrol of the damping-off disease of tomatoes in Tunisia. C R Biol 336:557–564
Hatayama K, Wawai S, Shoun H et al (2005) *Pseudomonas azotifigens* sp. nov., a novel nitrogen-fixing bacterium isolated from a compost pile. Int J Syst Evol Microbiol 55:1539–1544
Henry E, Yadeta KA, Coaker G (2013) Recognition of bacterial plant pathogens: local, systemic and transgenerational immunity. New Phytol 199:908–915
Hernández-León R, Rojas-Solís D, Contreras-Pérez M et al (2015) Characterization of the antifungal and plant growth-promoting effects of diffusible and volatile organic compounds produced by *Pseudomonas fluorescens* strains. Biol Control 81:83–92
Hinsa SM, Espinosa-Urgel M, Ramos JL et al (2003) Transition from reversible to irreversible attachment during biofilm formation by *Pseudomonas fluorescens* WCS365 requires an ABC transporter and a large secreted protein. Mol Microbiol 49:905–918

Hua GKH, Höfte M (2015) The involvement of phenazines and cyclic lipopeptide sesselin in biocontrol of *Rhizoctonia* root rot on bean (*Phaseolus vulgaris*) by *Pseudomonas* sp. CMR12a is influenced by substrate composition. Plant Soil 388:243–253

James DW Jr, Gutterson NI (1986) Multiple antibiotics produced by *Pseudomonas fluorescens* HV37a and their differential regulation by glucose. Appl Environ Microbiol 52:1183–1189

Jiao Z, Wu N, Hale L et al (2013) Characterisation of *Pseudomonas chlororaphis* subsp. *aurantiaca* strain Pa40 with the ability to control wheat sharp eyespot disease. Ann Appl Biol 163:444–453

Jones JDG, Dangl JL (2006) The plant immune system. Nature 444:323–329

Jošić D, Pivić R, Miladinović M et al (2012a) Antifungal activity and genetic diversity of selected *Pseudomonas* spp. from maize rhizosphere in Vojvodina. Genetika 44:377–388

Jošić D, Protolipac K, Starović M et al (2012b) Phenazine producing *Pseudomonas* isolates decrease *Alternaria tenuissima* growth, pathogenicity and disease incidence on cardoon. Arch Biol Sci 64:1495–1503

Jun S-R, Wassenaar TM, Nookaew I et al (2016) Diversity of *Pseudomonas* genomes, including populus-associated isolates, as revealed by comparative genome analysis. Appl Environ Microbiol 82:375–383

Kahlon SS, Malhotra S (1986) Production of gibberellic acid by fungal mycelium immobilized in sodium alginate. Enzym Microb Technol 8:613–616

Kang S-M, Radhakrishnan R, Khan AL et al (2014) Gibberellin secreting rhizobacterium, *Pseudomonas putida* H-2-3 modulates the hormonal and stress physiology of soybean to improve the plant growth under saline and drought conditions. Plant Physiol Biochem 84:115–124

Karadeniz A, Topcuoğlu ŞF, İnan S (2006) Auxin, gibberellin, cytokinin and abcissic acid production in some bacteria. World J Microbiol Biotechnol 22:1061–1064

Karakoç S, Aksöz N (2006) Some optimal cultural parameters for gibberellic acid biosynthesis by *Pseudomonas* sp. Turk J Biol 30:81–85

Katiyar V, Goel R (2004) Siderophore-mediated plant growth promotion at low temperature by mutant of fluorescent pseudomonad. Plant Growth Regul 42:239–244

Katznelson H, Peterson EA, Rovatt JW (1962) Phosphate dissolving microorganisms on seed and in the root zone of plants. Can J Bot 40:1181–1186

Keel C, Schnider U, Maurhofer M et al (1992) Suppression of root diseases by *Pseudomonas fluorescens* CHA0- importance of the bacterial secondary metabolite 2,4-diacetylphloroglucinol. Mol Plant Microbe Interact 5:4–13

Khan MS, Zaidi A, Wani PA (2009) Role of phosphate-solubilizing microorganisms in sustainable agriculture – a review. Agron Sustain Dev 27:29–43

Khan MR, Brien EO, Carney BR et al (2010) A fluorescent pseudomonad shows potential for the control of net blotch disease of barley. Biol Control 54:41–45

King RW, Evans LT (2003) Gibberellins and flowering of grasses and cereals: prising open the lid of the “Florigen” black box. Annu Rev Plant Physiol Plant Mol Biol 54:307–328

Kloepper J, Schroth M, Miller T (1980) Effects of rhizosphere colonization by plant growth-promoting rhizobacteria on potato plant development and yield. Phytopathology 70:1078–1082

Krotzky A, Werner D (1987) Nitrogen fixation in *Pseudomonas stutzeri*. Arch Microbiol 147:48–57

Kwak Y-S, Weller DM (2013) Take-all of wheat and natural disease suppression: a review. Plant Pathol J 29:125–135

Lagzian A, Saberi Riseh R, Khodaygan P et al (2013) Introduced *Pseudomonas fluorescens* VUPf5 as an important biocontrol agent for controlling *Gaeumannomyces graminis* var. *tritici* the causal agent of take-all disease in wheat. Arch Phytopathol Plant Protect 46:2104–2116

Lanteigne C, Gadkar VJ, Wallon T et al (2012) Production of DAPG and HCN by *Pseudomonas* sp. LBUM300 contributes to the biological control of bacterial canker of tomato. Phytopathology 102:967–973

Latour X, Delorme S, Mirleau P et al (2003) Identification of traits implicated in the rhizosphere competence of fluorescent pseudomonads: description of a strategy based on population and model strain studies. Agronomie 23:397–405

Lee Y, Yeom J, Jim J et al (2010) Phenotypic and physiological alterations by heterologous acylhomoserine lactone synthase expression in *Pseudomonas putida*. Microbiology 156:3762–3772
León M, Yaryura P, Montecchia M et al (2009) Antifungal activity of selected indigenous *Pseudomonas* and *Bacillus* from the soybean rhizosphere. Int J Microbiol. doi:10.1155/2009/572049
Li Q, Saleh-Lakha S, Glick BR (2005) The effect of native and ACC deaminase-containing *Azospirillum brasilense* Cd 1843 on the rooting of carnation cuttings. Can J Microbiol 51:511–514
Loper JE, Buyer JS (1991) Siderophores in microbial interactions on plant surfaces. Mol Plant Microbe Interact 4:5–13
Loper JE, Hassan KA, Mavrodi DV et al (2012) Comparative genomics of plant-associated *Pseudomonas* spp.: insights into diversity and inheritance of traits involved in multitrophic interactions. PLoS Genet 8:e10002784
Lugtenberg JB, Dekkers L, Bloemberg GV (2001) Molecular determinants of rhizosphere colonization by *Pseudomonas*. Annu Rev Phytopathol 39:461–490
Maldonado-González MM, Prieto P, Ramos C et al (2013) From the root to the stem: interaction between the biocontrol root endophyte *Pseudomonas fluorescens* PICF7 and the pathogen *Pseudomonas savastanoi* NCPPB 3335 in olive knots. Microb Biotechnol 6:275–287
Maleki M, Mostafaee S, Mokhtarnejad L et al (2010) Characterization of *Pseudomonas fluorescens* strain CV6 isolated from cucumber rhizosphere in Varamin as a potential biocontrol agent. Aust J Crop Sci 4:676–683
Martinez-Toledo MV, Moreno RJ, Gonzalez-Lopez J (1988) Root exudates of *Zea mays* and production of auxins, gibberellins and cytokinins by *Azobacter chroococcum*. Plant Soil 110:149–152
Martínez-Viveros O, Jorquera MA, Crowley DE et al (2010) Mechanisms and practical considerations involved in plant growth promotion by rhizobacteria. J Soil Sci Plant Nutr 10:293–319
Masalha J, Kosegarten H, Elmaci Ö et al (2000) The central role of microbial activity for iron acquisition in maize and sunflower. Biol Fertil Soils 30:433–439
Massart S, Perazzolli M, Höfte M et al (2015) Impact of the omic technologies for understanding the modes of action of biological control agents against plant pathogens. Biocontrol 60:725–746
Mavrodi DV, Blankenfeldt W, Thomashow LS (2006) Phenazine compounds in fluorescent *Pseudomonas* spp. biosynthesis and regulation. Annu Rev Phytopathol 44:417–445
Mavrodi DV, Peever TL, Mavrodi OV et al (2010) Diversity and evolution of the phenazine biosynthesis pathway. Appl Environ Microbiol 76:866–879
Mavrodi DV, Mavrodi OV, Parejko JA et al (2012a) Accumulation of the antibiotic phenazine-1-carboxylic acid in the rhizosphere of dryland cereals. Appl Environ Microbiol 78:804–812
Mavrodi OV, Walter N, Elateek S et al (2012b) Suppression of *Rhizoctonia* and *Pythium* root rot of wheat by new strains of *Pseudomonas*. Biol Control 62:93–102
Mavrodi DV, Parejko JA, Mavrodi OV et al (2013) Recent insights into the diversity, frequency and ecological roles of phenazines in fluorescent *Pseudomonas* spp. Environ Microbiol 15:675–686
Mazurier S, Corberand T, Lemanceau P, Raaijmakers JM (2009) Phenazine antibiotics produced by fluorescent pseudomonads contribute to natural soil suppressiveness to *Fusarium* wilt. ISME J 3:977–991
Mazzola M, Cook RJ, Thomashow LS et al (1992) Contribution of phenazine antibiotic biosynthesis to the ecological competence of fluorescent pseudomonads in soil habitats. Appl Environ Microbiol 58:2616–2614
McSpadden Gardener BB, Gutierrez LJ, Joshi R et al (2007) Distribution and biocontrol potential of *phlD*$^{+}$ pseudomonads in corn and soybean fields. Phytopathology 95:715–724
Mendes R, Kruijt M, De Bruijn I et al (2011) Deciphering the rhizosphere microbiome for disease-suppressive bacteria. Science 332:1097–1100
Meyer JM, Gruffaz C, Raharinosy V et al (2008) Siderotyping of fluorescent *Pseudomonas*: molecular mass determination by mass spectrometry as a powerful pyoverdine siderotyping method. Biometals 21:259–271

Migula W (1894) Über ein neues System der Bakterien. Arb Bakteriologischen Inst Tech Hochschule Karlsruhe 1:235–238

Miller MB, Bassler BL (2001) Quorum sensing in bacteria. Annu Rev Microbiol 55:165–199

Misra HS, Rajpurohit YS, Khairnar NP (2012) Pyrroloquinoline-quinone and its versatile roles in biological processes. J Biosci 37:313–325

Morohoshi T, Wang WZ, Suto T et al (2013) Phenazine antibiotic production and antifungal activity are regulated by multiple quorum-sensing systems in *Pseudomonas chlororaphis* subsp. *aurantiaca* StFRB508. J Biosci Bioeng 116:580–584

Mrabet M, Elkahoui S, Tarhouni B et al (2015) Potato seed dressing with *Pseudomonas aeruginosa* strain RZ9 enhances yield and reduced black scurf. Phytopathol Mediterr 54:265–274

Mullen MD (2005) Phosphorus in soils: biological interactions. In: Hillel D, Rosenzweig C, Powlson D et al (eds) Encyclopedia of soils in the environment. Academic, Oxford, pp 210–215

Müller H, Zachow C, Alavi M et al (2013) Complete genome sequence of the sugar beet endophyte *Pseudomonas poae* RE* 1-1-14, a disease-suppressive bacterium. Genome Announc 1:e00020–e00013

Negi YK, Prabha D, Garg SK et al (2011) Genetic diversity among cold-tolerant fluorescent *Pseudomonas* isolates from Indian Hymalayas and their characterization for biocontrol and plant growth-promoting activities. J Plant Growth Regul 30:128–143

Nelson K, Weinel C, Paulsen I et al (2002) Complete genome sequence and comparative analysis of the metabolically versatile *Pseudomonas putida* KT2440. Environ Microbiol 4:799–808

Ng W-L, Bassler BL (2009) Bacterial quorum-sensing network architectures. Annu Rev Genet 43:197–222

Nieto KF, Frankenberger WT (1990) Microbial production of cytokinins. Soil Biochem 6:191–248

Noreen R, Ali SA, Hasan KA et al (2015) Evaluation of biocontrol potential of fluorescent *Pseudomonas* associated with root nodules of mungbean. Crop Prot 75:18–24

O'Sullivan DJ, O'Gara F (1992) Traits of fluorescent *Pseudomonas* spp. involved in suppression of plant root pathogens. Microbiol Rev 56:662–676

Oberhänsli T, Défago G, Haas D (1991) Indole-3-acetic acid (IAA) synthesis in the biocontrol strain CHA0 of *Pseudomonas fluorescens*: role of tryptophan side chain oxidase. J Gen Microbiol 137:2273–2279

Oteino N, Lally RD, Kiwanuka S (2015) Plant growth promotion induced by phosphate solubilizing endophytic *Pseudomonas* isolates. Front Microbiol. doi:10.3389/fmicb.2015.00745

Palleroni NJ (1984) Genus I *Pseudomonas*. In: Krieg NR, Holt JG (eds) Bergey's manual of determinative bacteriology. Williams and Wilkins Co, Baltimore, pp 141–168

Palleroni NJ, Kunisawa R, Contopoulou R et al (1973) Nucleic acid homologies in the genus *Pseudomonas*. Int J Syst Bacteriol 23:333–339

Parejko JA, Mavrodi DV, Mavrodi OV et al (2012) Population structure and diversity of phenazine-1-carboxylic acid producing fluorescent *Pseudomonas* spp. from dryland cereal fields of Central Washington State (USA). Microb Ecol 64:226–241

Park JY, Han SH, Lee JH et al (2011) Draft genome sequence of the biocontrol bacterium *Pseudomonas putida* B001, an oligotrophic bacterium that induces systemic resistance to plant diseases. J Bacteriol 193:6795–6796

Pastor N, Reynoso M, Tonelli M et al (2010) Potential biological control *Pseudomonas* sp. PCI2 against damping-off of tomato caused by *Sclerotium rolfsii*. J Plant Pathol 92:737–745

Pastor N, Carlier E, Andrés J et al (2012) Characterization of rhizosphere bacterial for control of phytopathogenic fungi of tomato. J Environ Manag 95:S332–S337

Patra DD (2012) Production, purification, and characterization of antifungal metabolite from *Pseudomonas aeruginosa* SD12, a new strain obtained from tannery waste polluted soil. J Microbiol Biotechnol 22:674–683

Patrick JW (1987) Are the hormones involved in assimilate transport? In: Hoad GV, Lenton JR, Jackson MB et al (eds) Hormone action in plant development: a critical appraisal. Butterworths Co. Ltd, Long Ashton, pp 178–188

Patten CL, Glick BR (1996) Bacterial biosynthesis of indole-3-acetic acid. Can J Microbiol 42:207–220
Patten CL, Glick BR (2002) Role of Pseudomonas putida indole acetic acid in development of the host plant root system. Appl Environ Microbiol 68:3795–3801
Paulsen IT, Press CM, Ravel J et al (2005) Complete genome sequence of the plant commensal *Pseudomonas fluorescens* Pf-5. Nat Biotechnol 23:873–878
Penrose DM, Glick BR (2003) Methods for isolating and characterizing ACC deaminase-containing plant growth promoting rhizobacteria. Physiol Plant 118:10–15
Perneel M, Heyrman J, Adiobo A et al (2007) Characterization of CMR5c and CMR12a, novel fluorescent *Pseudomonas* strains from the cocoyam rhizosphere with biocontrol activity. J Appl Microbiol 103:1007–1020
Persello-Cartiaux F, Nussaume L, Robaglia C (2003) Tales from the underground: molecular plant-rhizobacteria interactions. Plant Cell Environ 26:189–199
Pierson LS 3rd, Pierson EA (1996) Phenazine antibiotic production in *Pseudomonas aureofaciens*: role in rhizosphere ecology and pathogen suppression. FEMS Microbiol Lett 136:101–108
Pierson LS 3rd, Pierson EA (2010) Metabolism and function of phenazines in bacteria: impacts on the behavior of bacteria in the environment and biotechnological processes. Appl Microbiol Biotechnol 86:1659–1670
Pieterse CM, Poelman EH, Van Wees SC, Dicke M (2013) Induced plant responses to microbes and insects. Front Plant Sci. doi:10.3389/fpls.2013.00475
Pieterse CM, Zamioudis C, Berendsen RL et al (2014) Induced systemic resistance by beneficial microbes. Annu Rev Phytopathol 52:347–375
Podile AR, Kishore GK (2006) Plant growth-promoting rhizobacteria. In: Gnanamanickam SS (ed) Plant-associated bacteria. Springer, Dordrecht, pp 195–230
Price-Whelan A, Dietrich LE, Newman DK (2006) Rethinking "secondary" metabolism: physiological role for phenazine antibiotics. Nat Chem Biol 2:71–78
Puopolo G, Aida R, Pierson L et al (2011) Selection of a new *Pseudomonas chlororaphis* strain for the biological control of *Fusarium oxysporum* f. sp. *radices-lycopersici*. Phytopathol Mediterr 50:228–235
Quagliotto L, Azziz G, Bajsa N et al (2009) Three native *Pseudomonas fluorescens* strains tested under growth chamber and field conditions as biological agents against damping-off in alfalfa. Biol Control 51:42–50
Raaijmakers JM, Mazzola M (2012) Diversity and natural functions of antibiotics produced by beneficial and plant pathogenic bacteria. Annu Rev Phytopathol 50:403–424
Raaijmakers JM, Weller DM (1998) Natural plant protection by 2,4-diacetylphloroglucinol-producing *Pseudomonas* spp. in take-all decline soils. Mol Plant Microbe Interact 11:144–152
Raaijmakers JM, Weller DM (2001) Exploiting genotypic diversity of 2, 4-diacetyl phloroglucinol-producing *Pseudomonas* spp.: characterization of superior root-colonizing P. *fluorescens* strain Q8r1-96. Appl Environ Microbiol 67:2545–2554
Raaijmakers JM, Leeman M, Van Oorschot MM et al (1995) Dose-response relationships in biological control of *Fusarium* wilt of radish by *Pseudomonas* spp. Phytopathology 85:1075–1080
Raaijmakers JM, Weller DM, Thomashow LS (1997) Frequency of antibiotic-producing *Pseudomonas* spp. in natural environments. Appl Environ Microbiol 63:881–887
Raaijmakers JM, Bonsall RF, Weller DM (1999) Effect of population density of *Pseudomonas fluorescens* on production of 2,4-diacetyl phloroglucinol in the rhizosphere of wheat. Phytopathology 89:470–475
Raaijmakers JM, Vlami M, De Souza JT (2002) Antibiotic production by bacterial biocontrol agents. Anton Leeuw Int J G 81:537–547
Raghu K, MacRae IC (1966) Occurrence of phosphate-dissolving microorganisms in the rhizosphere of rice plants and in submerged soils. J Appl Bacteriol 29:582–586
Rainey PB (1999) Adaptation of *Pseudomonas fluorescens* to the plant rhizosphere. Environ Microbiol 1:243–257
Ramesh R, Joshi A, Ghanekar M (2009) Pseudomonads: major antagonistic endophytic bacteria to suppress bacterial wilt pathogen, *Ralstonia solanacearum* in the eggplant (*Solanum melongena* L.) World J Microbiol Biotechnol 25:47–55

Reed MLE, Glick BR (2005) Growth of canola (*Brassica napus*) in the presence of plant growth-promoting bacteria and either copper or polycyclic aromatic hydrocarbons. Can J Microbiol 51:1061–1969

Rodríguez H, Fraga R (1999) Phosphate solubilizing bacteria and their role in plant growth promotion. Biotechnol Adv 17:319–339

Rong X, Gurel FB, Meulia T et al (2012) Draft genome sequences of the *Pseudomonas fluorescens* biocontrol strains Wayne1R and Wood1R. J Bacteriol 194:724–725

Roquigny R, Arseneault T, Gadkar V et al (2015) Complete genome sequence of biocontrol strain *Pseudomonas fluorescens* LBUM223. Genome Announc 3:e00443–e00415

Saleem M, Arshad M, Hussain S et al (2007) Perspective of plant growth promoting rhizobacteria (PGPR) containing ACC deaminase in stress agriculture. J Ind Microbiol Biotechnol 34:635–648

Salisbury FB, Ross CW (1992) Plant physiology. Wadsworth Publishing Company, Belmont, pp 329–407

Salisbury SA, Forrest HS, Cruse WBT et al (1979) A novel coenzyme from bacterial primary alcohol dehydrogenases. Nature 280:843–844

Schauder S, Bassler BL (2001) The languages of bacteria. Genes Dev 15:1468–1480

Schroth MN, Hancock JG (1982) Disease-suppressive soil and root-colonizing bacteria. Science 216:1376–1381

Selin C, Habibian R, Poritsanos N et al (2010) Phenazines are not essential for *Pseudomonas chlororaphis* PA23 biocontrol of *Sclerotinia sclerotiorum*, but do play a role in biofilm formation. FEMS Microbiol Ecol 71:73–83

Sen S, Rai M, Acharya R et al (2009) Biological control of pathogens causing the Cymbidium pseudobulb rot complex using fluorescent *Pseudomonas* strain BRL-1. J Plant Pathol 91:751–755

Shaharoona B, Arshad M, Zahir ZA (2006) Effect of plant growth promoting rhizobacteria containing ACC-deaminase on maize (*Zea mays* L.) growth under axenic conditions and on nodulation in mung bean (*Vigna radiata* L.) Lett Appl Microbiol 42:155–159

Sharma A, Johri BN (2003) Growth promoting influence of siderophore-producing *Pseudomonas* strains GRP3A and PRS9 in maize (*Zea mays* L.) under iron limiting conditions. Microbiol Res 158:243–248

Shen X, Chen M, Hu H et al (2012) Genome sequence of *Pseudomonas chlororaphis* GP72, a root colonizing biocontrol strain. J Bacteriol 194:1269–1270

Shirzad A, Fallahzadeh-Mamaghani V, Pazhouhandeh M (2012) Antagonistic potential of fluorescent pseudomonads and control of crown and root rot of cucumber caused by *Phytophthora drechsleri*. Plant Pathol J 28:1–9

Siddiqui Z (2005) PGPR: prospective biocontrol agents of plant pathogens. Springer, Dordrecht

Silby MW, Cerdeño-Tárraga AM, Vernikos GS et al (2009) Genomic and genetic analyses of diversity and plant interactions of *Pseudomonas fluorescens*. Genome Biol 10:R51

Silby MW, Winstanley C, Godfrey SA et al (2011) *Pseudomonas* genomes: diverse and adaptable. FEMS Microbiol Rev 35:652–680

Spaepen S, Vanderleyden J, Remans R (2007) Indole-3-acetic acid in microbial and microorganism-plant signaling. FEMS Microbiol Rev 31:425–448

Spoel SH, Dong X (2012) How do plants achieve immunity? Defense without specialized immune cells. Nat Rev Immunol 12:89–100

Sponsel VM (2003) Gibberellins. In: Henry HL, Norman AW (eds) Encyclopedia of hormones, vol 2. Academic, Boston, pp 29–40

Stites TE, Mitchell AE, Rucker RB (2000) Physiological importance of quinoenzymes and the *O*-quinone family of cofactors. J Nutr 130:719–727

Tambong JT, Höfte M (2001) Phenazines are involved in biocontrol of *Pythium myriotylum* on cocoyam by *Pseudmonas aeruginosa* PNA1. Eur J Plant Pathol 107:511–521

Tewari S, Arora NK (2014) Multifunctional exopolysaccharides from *Pseudomonas aeruginosa* PF23 involved in plant growth stimulation, biocontrol and stress amelioration in sunflower under saline conditions. Curr Microbiol 69:484–494

Thonart P, Ongena M, Henry G (2012) PAMPs, MAMPs, DAMPs and others: an update on the diversity of plant immunity elicitors. Biotechnol Agron Soc Environ 16:257–268

Trewavas A (2000) Signal perception and transduction. In: Buchanan BB, Gruissem W, Jones RL (eds) Biochemistry and molecular biology of plants. Amer Soc Plant Physiol, Rockville, pp 930–987

Van De Mortel JE, De Vos RC, Dekkers E et al (2012) Metabolic and transcriptomic changes induced in *Arabidopsis* by the rhizobacterium *Pseudomonas fluorescens* SS101. Plant Physiol 160:2173–2188

Van Der Voort M, Meijer HJ, Schmidt Y et al (2015) Genome mining and metabolic profiling of the rhizosphere bacterium *Pseudomonas* sp. SH-C52 for antimicrobial compounds. Front Microbiol. doi:10.3389/fmicb.2015.00693

Van Loon L, Bakker PAHM (2006) Induced systemic resistance as a mechanism of disease suppression by rhizobacteria. In: Siddiqui ZA (ed) PGPR: biocontrol and biofertilization. Springer, Dordrecht, pp 39–66

Van Wees SC, De Swart EA, Van Pelt JA et al (2000) Enhancement of induced disease resistance by simultaneous activation of salicylate- and jasmonate-dependent defense pathways in *Arabidopsis thaliana*. Proc Natl Acad Sci 97:8711–8716

Verhagen BW, Glazebrook J, Zhu T et al (2004) The transcriptome of rhizobacteria-induced systemic resistance in *Arabidopsis*. Mol Plant Microbe Interact 17:895–908

Vesper SJ (1987) Production of pili (fimbriae) by *Pseudomonas fluorescens* and correlation with attachment to corn roots. Appl Environ Microbiol 53:1397–1405

Walters DR, Ratsep J, Havis ND (2013) Controlling crop diseases using induced resistance: challenges for the future. J Exp Bot 64:1263–1280

Wang C, Knill E, Glick BR et al (2000) Effect of transferring 1-amino cyclopropane-1-carboxylic acid (ACC) deaminase genes into *Pseudomonas fluorescens* strain CHA0 and its *gacA* derivative CHA96 on their growth-promoting and disease-suppressive capacities. Can J Microbiol 46:898–907

Weller DM (1988) Biological control of soilborne plant pathogens in the rhizosphere with bacteria. Annu Rev Phytopathol 26:379–407

Weller DM (2007) *Pseudomonas* biocontrol agents of soilborne pathogens: looking back over 30 years. Phytopathology 97:250–256

Weller DM, Thomashow LS (1994) Current challenges in introducing beneficial microorganisms into the rhizosphere. In: O'Gara F, Dowling DN, Boesten B (eds) Molecular ecology of rhizosphere microorganisms: biotechnology and the release of GMOs. VCH Verlagsgesellschaft, Weinheim, pp 1–18

Weller DM, Raaijmakers JM, Gardener BB et al (2002) Microbial populations responsible for specific soil suppressiveness to plant pathogens. Annu Rev Phytopathol 40:309–348

Williams P (2007) Quorum sensing, communication and cross-kingdom signalling in the bacterial world. Microbiology 153:3923–3938

Winslow CEA, Broadhurst J, Buchanan RE et al (1917) The families and genera of the bacteria. Preliminary report of the society of American Bacteriologists on characterization and classification of bacterial types. J Bacteriol 2:505–566

Wu XA, Monchy S, Taghavi S et al (2011) Comparative genomics and functional analysis of nich-specific adaptation in *Pseudomonas putida*. FEMS Microbiol Rev 35:299–323

Xie H, Pasternak JJ, Glick BR (1996) Isolation and characterization of mutants of the plant growth-promoting rhizobacterium *Pseudomonas putida* GR12-2 that overproduce indole acetic acid. Curr Microbiol 32:67–71

Yang M-M, Mavrodi DV, Mavrodi OV et al (2011) Biological control of take-all by fluorescent *Pseudomonas* spp. from Chinese wheat fields. Phytopathology 101:1481–1491

Zhou T, Chen D, Li C et al (2012) Isolation and characterization of *Pseudomonas brassicacearum* J12 as an antagonist against *Rasltonia solanacearum* and identification of its antimicrobial components. Microbiol Res 167:388–394

Rhizobial Bioformulations: Past, Present and Future

4

Naveen Kumar Arora, Maya Verma, and Jitendra Mishra

Abstract

The nitrogen fixing bacterial group known as rhizobia are very important and are used as biological fertilizers for two main purposes; one is to fulfil the nutritional requirements of increasingly populated world and other to overcome the problems arising due to chemical fertilizers. Rhizobial bioformulations are in the market since more than a century and can be the solution for deficiency of nitrogen in our food and soils. Rhizobia maintain the soil fertility along with higher crop yields due to the capability of biological nitrogen fixation (BNF). Currently, various types of rhizobial biofertilizers are commercially available in the market all over the world for agricultural purposes. These can be solid carrier based formulations (organic and inorganic), liquid formulations (with and without additives), synthetic polymer based formulations or metabolite based formulations, but there still is a great room for improvement. However, over the years there have been subtle changes in the rhizobial inoculants in terms of production and application.

4.1 Introduction

Nitrogen (N) is one of the most important nutrients for plant growth. About 1–5 % of total plant dry matter consists of N. It is an essential constituent of proteins, nucleic acids, chlorophyll, co-enzymes, phytohormones and secondary metabolites (Hawkesford et al. 2012). Due to its immense cellular need, N is required in large quantities. Although atmosphere contains 80% of dinitrogen (N_2) (Abd-Alla et al.

N.K. Arora (✉) • M. Verma • J. Mishra
Department of Environmental Microbiology, BBA University,
Lucknow 226025, Uttar Pradesh, India
e-mail: nkarora_net@rediffmail.com

S. Mehnaz (ed.), *Rhizotrophs: Plant Growth Promotion to Bioremediation*,
Microorganisms for Sustainability 2, DOI 10.1007/978-981-10-4862-3_4

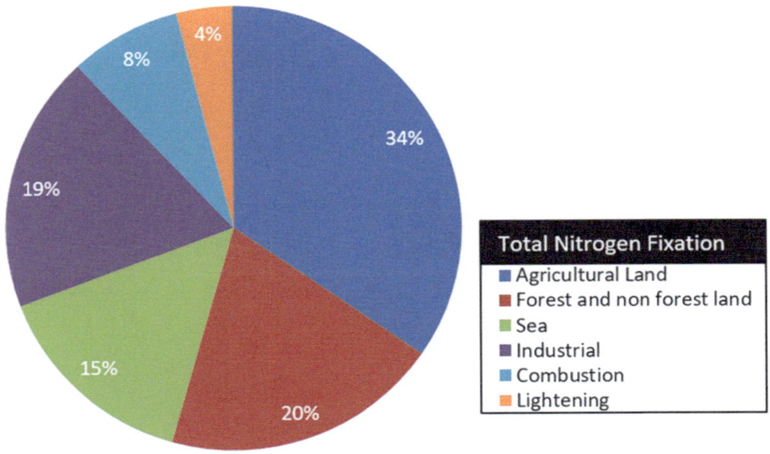

Fig. 4.1 Nitrogen fixation on earth (Modified from Bezdicek and Kennedy 1998)

2014a), it is relatively inert, and most organisms are unable to utilize it (Allito et al. 2015). Generally, the most reduced form of N that is ammonia (NH_3) or most oxidized form, nitrate (NO^{-}_3) are required to fulfil needs of plants. The fixed form of N is generated by conversion of N_2 to NH_3, a process also known as nitrogen fixation. Nitrogen fixation may occur by biological or non-biological means (Fig. 4.1). Non-biological fixation includes geochemical fixation by lightning (10% of the total N_2 fixation) (Bezdicek and Kennedy 1998) and industrial fixation by Haber-Bosch process (15% of total N_2 fixation) (Bezdicek and Kennedy 1998). However, BNF is carried out by some prokaryotes, including a small but diverse group of bacteria and archea, commonly referred as diazotrophs (Zehr et al. 2003; Kneip et al. 2007) (Fig. 4.2). They encode enzyme complex nitrogenase, that catalyses the conversion of N_2 gas to NH_3 (Santi et al. 2013). Amongst all of the fixed form of N_2, BNF is of immense importance. It is estimated that over half of the fixed N_2 is supplied biologically and has a profound agronomic, economic, and ecological impact (Smil 2001). Although BNF is a boon to agro-ecosystems, but N fertilizers are also used for meeting the total N requirement for agricultural production throughout the world. That is why a substantial increase has been noticed in demand for chemical N fertilizers as in comparsion to a century ago (Galloway et al. 2008). The increased use of chemical N fertilizers is problematic because it is susceptible to loss by leaching or denitrification (Luce et al. 2011) and regarded economically and environmentally undesirable. Moreover, manufacturing N fertilizers requires six times more energy than other fertilizers such as that of phosphorus or potassium (Da Silva et al. 1978).

There are various types of associations/interactions occurring between diazotrophs and their host plants (Santi et al. 2013). Amongst them most efficient are diazotrophic bacteria known as rhizobia, involved in the formation of root nodules in legumes (Santi et al. 2013). Currently, more than 98 species belonging to 14

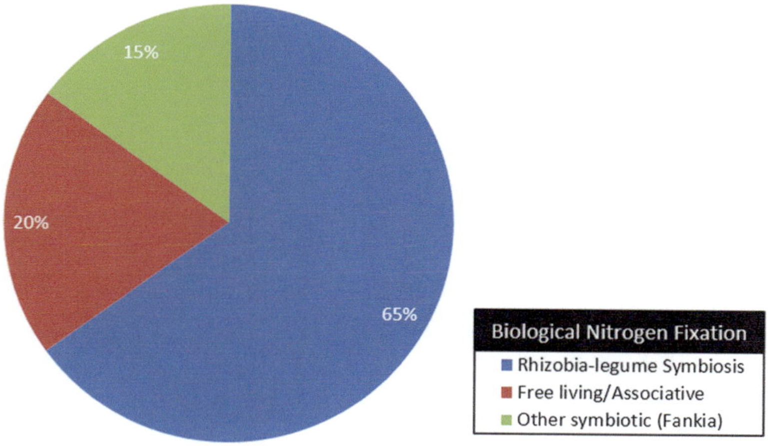

Fig. 4.2 Biological nitrogen fixation (Modified from Bouizgarne et al. 2015)

genera of α, β and ϒ proteobacteria have been described as rhizobia (Berrada and Fikri-Benbrahim 2014). Till date 12,000 nodulated legume species are known, and each has its own root nodulating partner(s) (Maróti and Kondorosi 2014). Legumes are second largest group of food and feed crops. These represent the third largest family of angiosperms and cover 12–15% of all available arable land which contributes more than 25% in the world's primary crop production (247 million tons of grain legumes annually; Ferguson et al. 2010). Rhizobia-legume symbiosis provides approximately 40 million tons of nitrogen into agricultural systems each year (Herridge et al. 2008) and plays a crucial role in increasing productivity and quality of crops specially protein content (Krapp et al. 2011). Besides this, these have been recognized as potential candidates to replace mineral N-fertilizers (Tairo and Ndakidemi 2013). It is estimated that the value of total nitrogen fixed by BNF process is equal to US $ 160–180 billion (Rajwar et al. 2013). Mostly the agricultural legumes have been studied for their symbiotic partner and others including wild and with little economic value are neglected (Ogasawara et al. 2003).

In the era of intensive agriculture, rising costs of N fertilizers and their adverse effects on environment have posed a threat to agroecosystems. Using rhizobial inoculants in the form of biofertilizers in place of N fertilzers has been considered as cheap and sustainable alternative (Arora et al. 2001; Kennedy et al. 2004; Mia et al. 2007). Use of biofertilizers is becoming more popular at the global level. The market trends also indicate that this could further increase in near future. The global biofertilizers market was estimated at US $ 535.8 million in 2014 and projected to reach up to US $ 1.88 billion by 2020 (Markets and Markets 2015). Amongst all types of biofertilizers, nitrogen fixers contribute maximum (75%) in agriculture (Grand View Research 2015) (Fig. 4.3).

Rhizobial biofertilizers have been applied to crops since more than a century. However, in last few years, research on field application of rhizobia has addressed

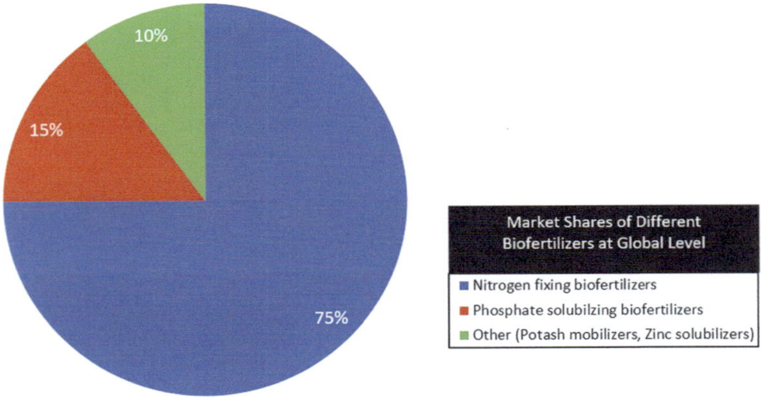

Fig. 4.3 Market share of different types of biofertilizers at global level (Grand View Research 2015)

the task of identifying the essential conditions required for its survival, usability and quality in the formulation products for enhancing crop production. Generally, the term bioformulation entails application of microorganism(s) as partial or complete substitute for chemical fertilizers/pesticides (Arora et al. 2010) but in a broad sense it is essential to define role of an active ingredient, a carrier material and an additive in preparation of bioformulation (Mishra and Arora 2016). Various types of bioformulations having different rhizobial species as active ingredients are being used, and indeed, have a profound effect on crop N requirement. However, newer techqinues for identification of different carriers, additives and delivery systems have provided robustness to convential bioformulations and show potential to subside the current need of mineral N fertilizers. In the present review, the journey of development and advances in rhizobial inoculants since their inception in the market, are discussed.

4.2 History

Legumes have been used as a food source since ancient times and also known as soil improvers. However, real role of rhizobia was identified much later. German botanist Leonhard Fuchsius, published the first drawings of nodulated legumes in 1542 (Fuchsius 1542). Malpighi (1679) also observed nodules on the bean roots (*Phaseolus vulgaris* and *Vicia faba*). BNF by legumes was first time proposed by Boussingault (1838) when he was doing crop rotation experiments with legumes and found increased N content causes superior nutritive quality in legumes and benefits to the soil. Lachmann (1858) during the microscopic study of nodules found that the nodules contain vibrio like particles. Further, these particles were also described as bacteria like and Woronin (1866) validated that root nodules in legumes were formed by a specific group of bacteria. Another milestone in the field of rhizobia-legume symbiosis was the discovery of the ability of root nodules to fix gaseous N which was demonstrated by German scientists Hellriegel and Wilfarth, in 1886 and two years later, they published their observations (Hellriegel and Wilfarth 1888). In the same year Dutch microbiologist Beijerinck, first time isolated a bacterium from root nodules and named it as *Bacillus radicicola* (Beijerinck 1888) and later Frank renamed it *Rhizobium leguminosarum* (Frank 1889).

The first commercial N biofertilizer of rhizobia, 'Nitragin' was patented by Nobbe and Hiltner (1896). Famous agricultural chemist Guthrie stated about rhizobial inoculants "one of the most valuable contributions ever made by science to practical agriculture" (Guthrie 1896). In the nineteenth century, Löhis and Hansen (1921) classified the rhizobia into two groups according to their growth patterns as slow growers and fast growers. Baldwin and Fred (1929) proposed the cross inoculation between some leguminous host plants and rhizobia. This cross inoculation concept was suggested for taxonomic characterization of rhizobia, based on cross inoculation groups (Eckhardt et al. 1931; Fred et al. 1932).

Before the discovery of rhizobia, the inoculation of seed or soil was done in the crop by "soil transfer method", in which soil from legume grown field to field or field to seed were applied before planting (Fred et al. 1932). However, artificial inoculation techniques by using pure cultures on agar slants and broths also began (Nobbe and Hiltner 1896; Fred et al. 1932). Inoculant production techniques started to change from those of the early 1900s, and use of solid carrier based formulations were started for legume inoculants, which were developed to enhance the shelf life and field efficacy of inoculants. In this context, soil and peat formulations were used (Fred et al. 1932). Peat was the most important carrier for long term storage of inoculants because of some beneficial properties, e.g., high water holding capacity, chemical and physical evenness, non-toxic and environment friendly nature (Ferreira and Castro 2005). Although peat was being used as the most common type of carrier for rhizobia based inoculants (Bezdicek et al. 1978; Khavazi et al. 2007; Albareda et al. 2008) but some constraints were also reported, such as it may contain inhibitory factors for the microbes (Brockwell 1985) and lack of availability in many parts of the world. Hence, the interest in using other carrier materials was also on. The carrier materials such as lignite (Kandasamy and Prasad 1971), filter mud (Philpotts 1976), coal-bentonite mixture (Deschodt and Strijdom 1976), cellulose (Pugashetti et al. 1971), coal (Crawford and Berryhill 1983), bagasse, wheat straw, compost of coir dust and soil, charcoal, manure, compost, powdered coconut shells, ground teak leaves also started to find their way in rhizobial inoculants (Tilak and Subba Rao 1978). Solid carrier based inoculants were also developed as granular inoculants for direct application in soil (Brockwell et al. 1980), and key benefits of this technique were easy storage, handling, and application (Smith 1992).

It is reported that liquid based formulations for rhizobia were developed as alternatives and were effective (Van Schreven et al. 1953; Singleton et al. 2002). Smith et al. (1981) reported that nodule numbers increased when applied with liquid rhizobial inoculants. The freeze-dried inoculants based on lyophilization techniques were first time identified as commercially beneficial in 1958 (Brockwell 1982). Bonish (1979) applied diluted soil samples to inoculate clover seedlings growing in lab conditions. At that time gel based microbial inoculants were also developed as alternatives to powdered carrier-based inoculants by entrapping rhizobia in polymer gels such as polyacrylamide-entrapped *Rhizobium* (PER) (Dommergues et al. 1979), alginate-entrapped *Rhizobium* (AER), xanthan-entrapped *Rhizobium* (XER); which gave satisfactory results in wet conditions (Jung et al. 1982). Kremer and Peterson (1982) reported that freeze-dried rhizobia suspended in dried oil were resistant against high soil temperatures. Brockwell et al. (1988) found whole soil inoculation technique (WSIT) suitable for clover production. In this method, rhizobial populations of soil were used to inoculate plants for assessing the N_2 fixing capability of that soil. Thies et al. (1991) developed simple functions to predict the need for inoculation based on numbers of rhizobia in the soil and soil nitrate levels.

Use of flavonoids (genistein/daidzein) for enhancing soybean yield was patented by Smith and Zhang (1999) which was commercialized as SoyaSignal™ (Leibovitch et al. 2001). Ballard and Charman (2000) used the Brockwell technique (WSIT) to evaluate the symbiotic N_2-fixing potential of soil samples. Other formulation types also appeared in the market, such as the vermiculite-based Gold Coat TM *Rhizobium* inoculant (Paau et al. 1991), liquid seed-applied soybean inoculant, Cell-Tech® (Smith 1995), liquid in-furrow inoculant LIFT (Smith 1995) and air-dried clay powder for alfalfa, Nitragin® Gold (Smith 1995). It can be said that rhizobial species have been successfully marketed globally and inoculants are produced and used in many countries in all continents of the world (Nelson 2004).

Hussain et al. (1995) introduced precursor based inoculum technology for growth enhancement of lentil crop in which indole acetic acid (IAA) precursor (tryptophan) was added with inoculum. The commercialisation of a genetically engineered strain of *Sinorhizobium meliloti* was approved in 1997 (EPA 1997). After few years, modifications in liquid inoculants were proposed; Singleton et al. (2002) developed additives and cell protectants based liquid formulations for improved growth performance. The additives promote cell survival in storage and after application to seed or soil (Singleton et al. 2002). Commonly used additives for rhizobial inoculants were polyvinyl pyrrolidone (PVP), carboxymethyl cellulose (CMC), gum arabic, sodium alginate and glycerol. PVP is a synthetic vinyl polymer that improves survival of rhizobia by protecting it from desiccation and also from harmful seed coat exudates (Singleton et al. 2002). The CMC has important rheological property and increases the gel viscosity of carriers to make it more suitable for viability of rhizobial cells (Rohr 2007). Gum arabic is a complex carbohydrate extracted from Acacia and commonly used as adhesive to protect the rhizobia against desiccation (Wani et al. 2007). Sodium alginate is non-toxic compound and used to enhance the survival of inoculant because it has limited heat transfer property and high water activity (Jung et al. 1982). Glycerol is used as additive because it protects rhizobial cells from desiccation by slowing the drying rate (Manikandan et al. 2010).

Besides above mentioned carriers for rhizobial inoculants, waste water sludge was also used as a carrier and it was firstly reported by Ben Rebah et al. (2002a). There are various other types of formulations with different carriers developed for rhizobial inoculants and some of them have been patented, e.g., the patent no. 521.850 (Belgian) for *Rhizobium* which uses diatomaceous earth and colloidal silica; the British patent no. 1.777.077 for the use of bentonite for *Rhizobium*. For more advancement in formulation techniques, genetic modification of rhizobia is also being done to increase the efficiency of N_2 fixation such as genes that regulate the Hup system (recycles the hydrogen released during nitrogen fixation) have been identified and transferred (Brito et al. 2002).

From the beginning of twentieth century, extensive research has been carried out for development of state of the art rhizobial bioformulations and advents of newer techniques have provided inputs in this direction. There are various modifications related to rhizobial inoculants that have been done since their inception and the major events in the history are compiled in Fig. 4.4. The rhizobial inoculants are being used from long time in diverse types of formulations and detail of some formulations are given in Table 4.1.

4.3 Present Scenario

The application of rhizobial bioformulations is one of the cheapest and eco-friendly approaches for improving production of leguminous plants and fixation of atmospheric nitrogen (Thakare and Rasal 2000). It has been estimated that 2000 tons of rhizobial inoculants of worth US$ 50 million are produced worldwide every year (Ben Rebah et al. 2007) and this quantity is sufficient to inoculate 20 million hectares of legumes (Herridge et al. 2002). Currently, there are various types of rhizobial bioformulations available in the market for agricultural purposes and all are categorized mainly into two major groups - solid formulations and liquid formulations (Burges and Jones 1998). Peat is still most common carrier material in rhizobial inoculant production (Kaljeet et al. 2011) and other carrier materials such as coal, bagasse, coir dust, perlite are also used (Albareda et al. 2008). In a recent study, Ruíz-Valdiviezo et al. (2015) worked on the granular peat based and perlite based bioformulations. The additive based solid inoculants are also used such as sawdust based formulations amended with CMC (Aeron et al. 2012).

Liquid formulations are used for legume inoculation as more suitable technique for mechanical sowing in large areas (Fernandes-Júnior et al. 2009). Liquid formulations typically contain aqueous, oil or polymer based products and these formulations may have the desired strain and its nutrients, which are more tolerant to adverse conditions (Brahmaprakash and Sahu 2012). One of the methods for liquid formulations is water-in-oil emulsions (Vandergheynst et al. 2007), which is beneficial for desiccation sensitive organisms as it slows down water evaporation. Currently, additive based liquid formulations are in greater use and demand (Rivera et al. 2014; Ruíz-Valdiviezo et al. 2015).

Polymer gel based formulation techniques and synthetic polymer based techniques are also in focus (Fernandes-Júnior et al. 2009). Synthetic formulations based on a mixture of polymers have been continuously investigated (Bashan 1998; John et al. 2011). Alvarez et al. (2010) used silica gel as efficient formulation technique for rhizobial inoculants. Denardin and Freire (2000) reported that blends of natural or synthetic polymers are able to maintain viability of rhizobial cells for over 6 months. For agricultural and environmental uses, these polymers include alginate,

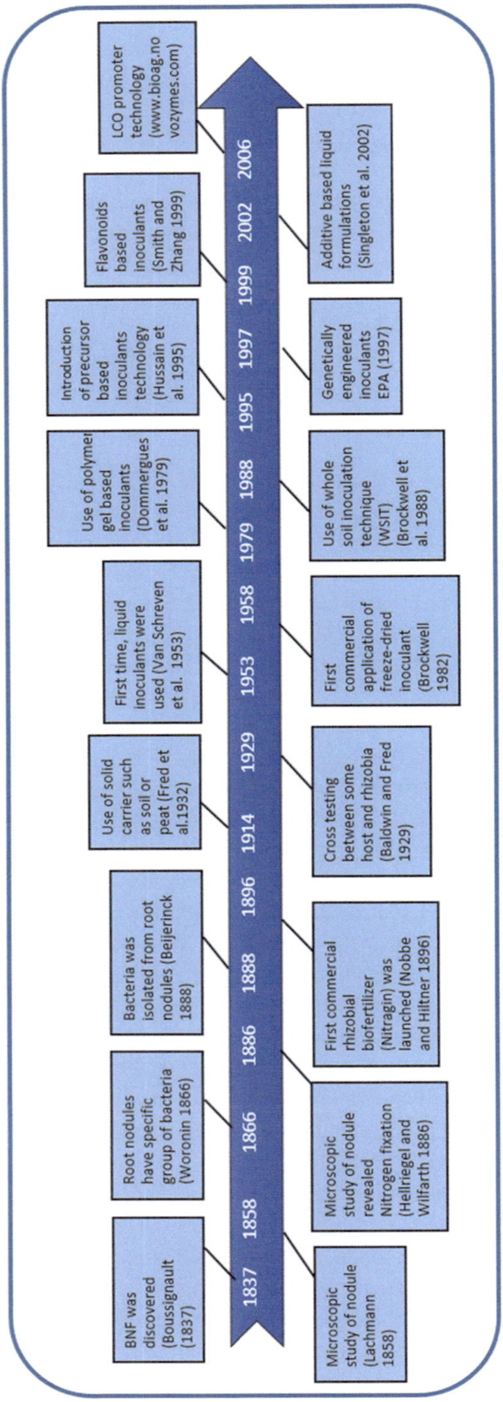

Fig. 4.4 Timeline of major discoveries in the development of rhizobial bioformulations

Table 4.1 Overview of some of the rhizobial bioformulations around the globe

Formulation type	Additives	Microorganism	Plant species	References
Liquid (culture media or water)	Glycerol, PVP, trehalose, FeEDTA	*Bradyrhizobium japonicum*	Soybean	Singleton et al. (2002)
	PVP; FeEDTA	Several rhizobia	Soybean	Albareda et al. (2008)
		B. japonicum		
	Unknown (commercial)	*B. japonicum*	Soybean	Maurice et al. (2001)
	Gum Arabic	*Bradyrhizobium* sp.; *Rhizobium* sp.	*Acacia mangium*, Greengram, *Leucaena leucocephala*	Diouf et al. (2003), Gamal-Eldin and Elbanna (2011), and Wani et al. (2007)
Organic inoculants (peat)	None or with undisclosed additives	*B. japonicum*;	Chickpea, faba beans, maize, pea, soybean	Clayton et al. (2004), Hamaoui et al. (2001),Hungria et al. (2010), Hynes et al. (2001), Khalid et al. (2004), and Revellin et al. (2000)
		Rhizobium sp.;		
	Applications as:	*Rhizobium leguminosarum*		
	Seed coating and pellets	*bv. viceae*		
	Vermiculite	Rhizobia	*Calliandra calothyrsus*	Kokalis-Burelle et al. (2003) and Odee et al. (2002)
	Pyrophyllite (hydrous aluminum silicate)	*Trichoderma virens* and *Burkholderia cepacia*	Bell pepper	Meyer et al. (2001)
	Arabic gum	Several *Rhizobium* and *Bradyrhizobium*	Bean, *Lupinus*, *Hedysarum*	Albareda et al. (2009) and Temprano et al. (2002)
Coir dust/coco peat vermicompost/ earthworm compost	Lignite	*Bacillus megaterium* and *R. leguminosarum*	Soybean	Raja Sekar and Karmegam (2010)
Sawdust	Composted by inoculation with *Cephalosporium* sp. and *Azospirillum Brasilense*	*B. japonicum*, *B. Arachis*, *R. meliloti*, *R. lotus*	Groundnuts, lucerne and a grass mixture of bird's foot trefoil and ryegrass, soybean	Kostov and Lynch (1998)

(continued)

Table 4.1 (continued)

Formulation type	Additives	Microorganism	Plant species	References
Sawdust	None	*R. leguminosarum* and *Pseudomonas fluorescens*	*Trifolium repense*	Arora et al. (2008)
Fibers from brewer's spent barley grain grape bagasse, cork compost	Gum Arabic, CMC	Several rhizobia; *B. japonicum*	Soybean	Albareda et al. (2008)
Wastewater sludge	Acid, alkaline and oxidative pre-treatments	*S. meliloti, R. leguminosarum* bv *viciae*, *B. japonicum* and *B. elkanii*	Not tested	Ben Rebah et al. (2002a, b)
Clay soil	Elemental sulfur	*Rhizobium* sp. and *Thiobacillus* sp.	Groundnut	Anandham et al. (2007)
Loess soil	None	Phosphate solubilizing bacteria (PSB) and *Rhizobium* sp.	None	Li et al. (2011)
Clay minerals, perlite	Gum Arabic, CMC	Several rhizobia; *B. japonicum;* *B. megaterium*	Soybean	Albareda et al. (2008)
Local soils	None	Rhizobia	*Calliandra calothyrsus*, rice	Hashem (2001) and Odee et al. (2002)
Perlite	Gum Arabic	*Rhizobium* and *Bradyrhizobium*	Bean, *Lupinus*, *Hedysarum*, Soybean	Temprano et al. (2002)
Alginate	None	*Rhizobium* spp.	*Leucaena* *Leucocephala*	Forestier et al. (2001)
CMC/corn starch	MgO	Rhizobia *Azospirillum amazonense*, *B. tropica*	Cowpea	Fernandes-Júnior et al. (2009)

Modified from Bashan et al. (2014)

agar, λ and κ carrageenan, pectin, chitosan, bean gum, and proprietary polymers (Bashan et al. 2014). Granular vermicompost, produced from essential oil bearing crop, scented geranium (*Pelargonium graveolens*), is used as efficient carrier for rhizobia (Ben Rebah et al. 2007). Some natural compounds such as cow urine and calliterpenone added to rhizobial bioformulations increased the numbers of rhizobial cells by tenfold when supplemented at 12.5–25.0 µl/ml (Kalra et al. 2010).

In case of rhizobial inoculants, mono-inoculation, co-inoculation or multistrain inoculation are also being used (Arora et al. 2014; Malusá and Vassilev 2014; Verma et al. 2014). The application of rhizobia in combination with other plant growth promoting rhizobacteria (PGPR) are used as suitable alternatives to promote plant growth both under normal and stress conditions, for example, plant growth promoting bacteria (PGPB) and rhizobia enhance nodulation, nitrogen fixation symbiotically (Rodrigues et al. 2013; Arora et al. 2014; Chibeba et al. 2015) and increase the grain yield by involving diverse mechanisms (Hungria et al. 2015). It is reported that Rubiya (2006) developed the "Multigeneric diazotrophic co-flocs" (*Azospirillum, Azotobacter* and *Rhizobium*) and reported good improvement in rice yield. The biofilm based inoculants containing a fungal-rhizobia consortium were also applied significantly for increased N_2 fixation in legumes (Jayasinghearachchi and Seneviratne 2004). The soil-made inoculants are also used by mixing clay soil inoculant with powdered elemental sulphur and inoculation of sulphur oxidizing bacteria (*Thiobacillus* sp.) with rhizobia synergistically promoted the yield and oil content of groundnut in sulphur-deficient soils (Anandham et al. 2007). Khare and Arora (2011) reported that on applying pyocyanin-producing pseudomonads together with rhizobia, an enhancement in nodulation ability is observed, which causes better growth and productivity of groundnut even in the presence of fungal phytopathogens. The consortium based liquid bioformulation technique is now being used as an important way for sustainable agriculture (Pindi and Satyanarayana 2013). Some of the recent trends include arbuscular mycorrhizal fungi (AMF) and rhizobia co-inoculation, which enhance the growth and yield of crops due to higher nutrient uptake (Abd-Alla et al. 2014b). AMF causes the enhancement in nitrogen uptake process by rhizobia and their association with legumes (Tajini et al. 2011). Meng et al. (2015) stated that AMF and rhizobia simplify the nitrogen uptake process and transfer in soybean/maize inter-cropping system. This AMF and rhizobia co-inoculation has great potential for stressed soils. Zhu et al. (2016) reported that AMF and nitrogen fixing bacteria enhance alfalfa yield under saline conditions.

There is also an emerging formulation technique in which the addition of microbial/plant associated secondary metabolites to bioformulations increases agricultural productivity by improving the inoculants efficiency (Morel et al. 2015). In the current market, metabolite based formulations for rhizobial inoculants are highly focused and the additions of flavonoids and phytohormones are being used for rhizobial inoculants. Plant inoculation with rhizobial cells, previously induced with flavonoids during growth, significantly alleviates the effects of adverse conditions (Nápoles et al. 2009; Muñoz et al. 2014). The addition of flavonoids to inoculated crops enhances the nitrogen fixation (Dashti et al. 2000), improves the rhizobial competitiveness and nodulation (Pan and Smith 2000). Although flavonoids are expensive, these act at very low concentrations and produced industrially for sustainable agriculture (Mishra and Arora 2016; Morel et al. 2016).

The phytohormones produced by microbes or plants show positive effects on plant growth (Hedin and McCarty 1994; Dazzo et al. 2000; Chandra et al. 2007; Tank and Saraf 2010; Kudoyarova et al. 2015). The inoculations of plants with phytohormones producing rhizobia have positive effects on plant development and seed priming with phytohormones (IAA, gibberellins, abscisic acid and ethylene) increase the germination rate and finally crop yields (Roberto et al. 2012; Bianco et al. 2014; Kudoyarova et al. 2014; Qiu et al. 2014). Addition of phytohormones to bioformulations increases plant development and yield in comparison to bioformulations alone (Morel et al. 2016). Although various types of rhizobial bioformulations are commercially available in the current market having diverse types of additives, adjuvants and metabolites, it needs more exploration for future use.

4.4 Limitations: With the Current System

The demand of bioinoculants, mainly rhizobial inoculants, is high and these are produced commercially at global level (Brockwell and Bottomley 1995). Although rhizobial bioformulations have very significant value in agriculture, these also have some limitations. It is obvious that on application of inoculants in field, the introduced microbes face a very hostile environment and sometimes their population decreases, which leads to failure (Bashan and Levanony 1988; Arora et al. 2001). In some cases, applied rhizobial inoculants are unable to increase sufficient crop yield because of competitions faced by indigenous rhizosphere microflora of plants (Merwe et al. 1974; Olufajo and Adu 1993; Mazid and Khan 2014). Generally, predatory organisms, protozoans and bacteriophages are already present in the soil (Somasegaran and Hoben 1994). The environmental conditions also affect the inoculant efficacy and adverse abiotic stresses (hot, dry and saline conditions) can cause rapid decrease in rhizobial populations (Mazid et al. 2011; Deshwal et al. 2013). Other factors such as bacterial survival on the seed are mainly affected by three factors: desiccation, the toxic nature of seed coat exudates and high temperatures (Deaker et al. 2004). Bacterial survival on seed directly affects the total legume/crop yield (Brockwell and Bottomley 1995). Sterilization techniques of inoculant carrier (Strijdom and van Rensburg 1981) and compatibility towards crop also affect the applicability of inoculants (Lupwayi et al. 2006). Shelf life of inoculants is a very major factor for their efficacy which mainly depends on several factors (production technology, carrier and packaging material used, transport activity) to sustain the quality of inoculants (Arora et al. 2010). Often import and proper storage of inoculants are also problematic because in absence of proper care, viability of inoculants decreases with loss in their beneficial properties (Kaljeet et al. 2011). The storage of bioformulations needs special facilities and skills, which most producers, shopkeepers, and farmers do not possess (Arora et al. 2010). The use of genetically improved rhizobia as inoculants has some legislative constraints because it requires permission from environmental protection agencies to release into the environment and the second problem is less understanding of microbial ecology (Geetha and Joshi 2013).

All inoculant producers claim that their products promote crop productivity but actually most of the products are available in the market without robust scientific data to favour their efficacy (Herrmann et al. 2015). In this regard, Brockwell et al. (1995) stated that most of the inoculants (90%) have no practical impact on the yield of target crop. Similarly, Olsen et al. (1996) reported that commercially available rhizobial inoculants lack proper population of rhizobial cells. Recently, Herrmann et al. (2015) analysed various inoculants and reported that more than 50% of the inoculants have high levels of contamination. It is also reported that contaminants have detrimental effects on the quality of rhizobial inoculants (Sparrow and Ham 1983; Rodriguez-Navarro et al. 1991) and 25% contaminants of the commercial inoculants can be opportunistic human pathogens (Olsen et al. 1996; Gomez et al. 1997).

The quality of inoculants is judged by their efficacy on application in field and if their quality and field performance is poor, then the product become unsuccessful in the market. Hence a lot of inoculants produced globally are of poor quality and the reason behind this issue is the lack of efficient quality control programmes (Somasegaran 1991; Brockwell and Bottomley 1995; Catroux et al. 2001). This drawback can lead to a negative impact on the future of the inoculant industry. In this context, various researchers suggested the need of an appropriate regulatory quality control program at international level for successful production and use of inoculants by end users (farmers) (Olsen et al. 1994; Herrmann et al. 2015; Arora et al. 2016). Thus, there is a requirement of strict regulations for rhizobial bioformulations to overcome above mentioned problems related with its worldwide productions and applications. All these factors have to be taken into accord to develop future rhizobial inoculants.

4.5 Future Prospects

The future of rhizobial formulations is directed to overcome or improve the lacunae associated with the present systems. As there are various types of rhizobial formulations available in the market, but each product has some limitations especially regarding their use, efficacy, survival or market availability. Hence, it is challenging task to develop a state of art formulation technology that fulfils all the required traits and make it available to farmers with global acceptancy. In this context, novel scientific approaches and information has also led to some good results and promise. For example, genotypic diversity of rhizobia can be assessed accurately by polymerase chain reaction (PCR) fingerprinting techniques such as enterobacterial repetitive intergenic consensus PCR (ERIC- PCR; Pongslip 2012), repetitive element palindromic PCR (rep-PCR; Menna et al. 2009) and enterobacterial repetitive sequences (BOX- PCR; Granada et al. 2014). These techniques help in the characterization of rhizobial strain which controls quality of inoculants (Pongslip 2012). Similarly, two primers random amplification polymorphic DNA (TP-RAPD) and amplified fragment length polymorphism (AFLP) are highly discriminating fingerprinting techniques and differentiate at species or below species level (Gzyl et al. 2005). However, emphases should also be given on techniques for increasing population density and survival of rhizobial strains in inoculants. Damasceno et al. (2013) devised electrospinning technique of rhizobia immobilization in nanofibers. They showed that encapsulation of rhizobial cell with polyvinyl alcohol (PVA) nanofibers enhances protection from dehydration and also minimized the effects of toxic chemicals. Nanoparticles made of inorganic or organic materials may enhance the quality of carrier-based microbial inoculants (Malusa et al. 2012). Sivasakthivelan and Saranraj (2013) also stated that survival of cells is mandatory for better commercialization of rhizobial inoculants in the global market. Suman et al. (2016) used the hydrogel based inoculants and experimentally proved that this method has low-cost and long shelf-life and also increased plant development in drought-prone environments. There has been a conflict in choosing a carrier for rhizobia inoculant development, as the carrier choice differs greatly in different inoculants. It is clear that same basic carriers are still in use. Various workers addressed that to increase the inoculant quality and efficiency, and to reduce costs and environmental impacts, alternative carrier materials have to be explored (Ben Rebah et al. 2007; Albareda et al. 2008). Polymeric inoculants (Bashan et al. 2014) and alginate beads (Sivakumar et al. 2014) have already been tested and need more exploration for their future use in inoculants.

Recently, Arora and Mishra (2016) provided their view on using metabolites and additives in bioformulations. Metabolites such as EPS can be used as carriers and also can play role in protection of cells and help in the nodulation process. The rhizobia nodulation genes *nod* ABC are involved in the synthesis of lipo-chitin oligosaccharides (LCOs) metabolites which are synthesized in the response of root exudates or flavonoids (Abdel-Lateif et al. 2012). LCOs increase plant growth development on applying alone or in combination with rhizobia and have given better results in field conditions (Miransari and Smith 2009; Marks et al. 2013). LCOs also act as growth regulators in a wide variety of plants, including non-legumes (Zhang and Smith 2002; Prithiviraj et al. 2003). Novozyme, first started to manufacture bioformulations containing LCOs and this technology is known as LCO promoter technology. These days, single formulations containing LCOs (Ratchet® and Torque®) or formulations with bacteria (OptimizeII®, Signum® and DynaStartMax®) are being marketed (Bioag.novozymes.com 2015).

Precursor in inoculant technology is also emerging as promising technique for efficient rhizobial bioformulations development (Naveed et al. 2015). Qureshi et al. (2012) found that the co-inoculation of *Rhizobium* and *Bacillus* sp. in the presence of a phytohormone precursor L-tryptophan (L-TRP) improved the pod and straw yield. Similarly, Qureshi et al. (2013) showed that interaction of L-TRP and rhizobial species increased the fresh fodder and dry matter yield in comparison to their separate application. Micronutrients have important role in nitrogen fixation and in this regard, Arora et al. (2009) recommended that the supplementation of Mo and Fe (up to certain concentrations) in soils along with the rhizobial formulations enhance the symbiotic nitrogen fixation process. Mmbaga et al. (2014) reported that the rhizobial inoculants supplemented with exogenous nutrients (phosphorus and potassium) improved photosynthesis, nutrient uptake, nodulation, growth and yield of the crop. According to Marks et al. (2013) and Morel et al. (2015), the bioformulations based on a mixture of various compounds, e.g., phytohormones, LCOs, flavonoids may be very useful for higher plant growth, and further research is required in this field.

The discovery of novel stress tolerating rhizobial species is also thought to be imperative in developing bioformulations that will survive in stress conditions (high temperature, drought, salinity) (Arora et al. 2012; Laranjo et al. 2014; Rao 2014). Currently, various types of rhizobial species have been discovered but only few of them are used as inoculants (Table 4.2). Stress tolerant bacteria have lots of scope for bioformulations production and also have important role in reclamation of waste lands (Arora et al. 2000, 2006). In a study, Karuppasamy et al. (2011) showed that

Table 4.2 Classification of rhizobia and their inoculants used in global market

Genus	Species	Inoculant crop	References
Rhizobium	*leguminosarum*	*Lactuca sativa and Daucus carota, Pea*	Flores-Félix et al. (2013) and Clayton et al. (2004)
	galegae	*Galega orientalis*	Vassileva and Ignatov (2002)
	tropici	*Zea mays*	Marks et al. (2015)
	endophyticum	*P. vulgaris*	López-López et al. (2010)
	phaseoli	*Vigna radiate*	Zahir et al. (2010)
	fabae	*V. faba*	Tian et al. (2008)
	etli	*P. vulgaris*	Soares et al. (2006)
	undicola	*Neptunia natans*	de Lajudie et al. (1998)
	gallicum	*P. vulgaris*	Sassi-Aydi et al. (2012)
	giardinii	*P. vulgaris*	Amarger et al. (1997)
	hainanensis	NA	NA
	huautlense	*Sesbania herbacea*	Wang and Martínez-Romero (2000)
	mongolense	*Medicago ruthenica*	Van Berkum et al. (1998)
	yanglingense	NA	NA
	larrymoorei	NA	NA
	indigoferae	NA	NA
	sullae	*Hedysarum coronarium*	Fitouri et al. (2012)

(continued)

Table 4.2 (continued)

Genus	Species	Inoculant crop	References
	loessense	NA	NA
	cellulosilyticum	*P. vulgaris*	Diez-Mendez et al. (2015)
	miluonense	*Lespedeza chinensis*	Gu et al. (2007)
	multihospitium	NA	NA
	oryzae	*Glycine max*	Waswa (2013)
	pisi	NA	NA
	mesosinicum	NA	NA
	alamii	*Helianthus annuus*	Alami et al. (2000)
	alkalisoli	NA	NA
	tibeticum	*Trigonella foenumgraecum*	Abd-Alla et al. (2014c)
	tubonense	NA	NA
	halophytocola	NA	NA
	radiobacter	*Graminaceous crops*	Humphry et al. (2007)
	rhizogenes	NA	NA
	rubi	NA	NA
	vitis	NA	NA
	nepotum	NA	NA
Ensifer	*meliloti*	*Medicago truncatula, Mucuna pruriens*	Olah et al. (2005) and Aeron et al. (2012)
	fredii	*G. max*	Albareda et al. (2008)
	sahelense	NA	NA
	terangae	NA	NA
	medicae	NA	NA
	arboris	NA	NA
	kostiense	NA	NA
	xingianens	NA	NA
	adhaerens	NA	NA
	kummerowiae	NA	NA
	americanum	*P. vulgaris*	Mnasri et al. (2012)
	mexicanus	*P. vulgaris*	Lloret et al. (2007)
	numidicus	NA	NA
Shinella	*kummerowiae*	NA	NA

(continued)

Table 4.2 (continued)

Genus	Species	Inoculant crop	References
Mesorhizobium	*loti*	*Lotus corniculatus*	Karaś et al. (2015)
	huakuii	NA	NA
	cicero	*Cicer arietinum*	Rokhzadi et al. (2008)
	tianshanense	NA	NA
	mediterraneum	*Hordeum vulgare, Cicer arietinum*	Peix et al. (2001) and Dudeja et al. (2011)
	plurifarium	NA	NA
	amorphae	NA	NA
	chacoense	NA	NA
	septentrionale	NA	NA
	temperatum	*G. max*	Waswa (2013)
	thiogangeticum	NA	NA
	albiziae	*Albizia kalkora, G. max*	Wang et al. (2007) and Waswa (2013)
	caraganae	NA	NA
	gobiense	NA	NA
	tarimense	NA	NA
	australicum	NA	NA
	opportunistum	NA	NA
	metallidurans	NA	NA
	alhagi	NA	NA
	camelthorni	NA	NA
	abyssinicae	NA	NA
	muleiense	NA	NA
	hawassense	NA	NA
	qingshengii	NA	NA
	robiniae	NA	NA
	shonense	NA	NA
	shangrilense	NA	NA
	silamurunense	NA	NA
	tamadayense	NA	NA
Phyllobacterium	*trifolii*	*Fragaria ananassa*	Flores-Felix et al. (2015)
Methylobacterm	*nodulans*	*Crotalaria perrottetii*	Jourand et al. (2004)

(continued)

Table 4.2 (continued)

Genus	Species	Inoculant crop	References
Microvirga	*lupine*	NA	NA
	lotononidis	*Leobordea sp.*	Ardley et al. (2013)
	zambiensis	NA	NA
Ochrobactrum sp.	*cytisi*	*Cucumis sativus*	Xu et al. (2015)
	lupine	*Lupinus albus*	Trujillo et al. (2005)
Azorhizobium	*caulinodans*	*Leucaena leucocephala*	Waelkens et al. (1995)
	dobereinereae	NA	NA
	oxalatiphilum	NA	NA
Devosia	*neptuniae*	*Neptunia natans*	Rivas et al. (2003)
Bradyrhizobium	*japonicum*	*G. max*	Zerpa et al. (2013)
	elkanii	*Vigna unguiculata*	Soares et al. (2006)
	iaoningensese	NA	NA
	yuanmingense	*G. max*	Soe and Yamakawa (2013)
	betae	NA	NA
	canariense	NA	NA
	iriomotense	NA	NA
	jicamae	NA	NA
	lablabi	NA	NA
	huanghuaihaiense	NA	NA
	cytisi	NA	NA
	daqingense	NA	NA
	denitrificans	NA	NA
	oligotrophicum	NA	NA
	pachyrhizi	NA	NA
Burkholderia	*caribensis*	*Amaranthus cruentus* and *A. hypochondriacus*	Parra-Cota et al. (2014)
	cepacia	*P. vulgaris*	Peix et al. (2001)
	tuberum	*Macroptilium atropurpureum*	Annette et al. (2013)
	phymatum	*P. vulgaris*	Talbi et al. (2013)

(continued)

Table 4.2 (continued)

Genus	Species	Inoculant crop	References
	nodosa	NA	NA
	sabiae	NA	NA
	mimosarum	NA	NA
	rhizoxinica	NA	NA
	diazotrophica	NA	NA
	endofungorum	NA	NA
	heleia	NA	NA
	symbiotica	*Mimosa cordistipula*	Sheu et al. (2012)
	ambifaria	*Zea mays, A. cruentus and A. hypochondriacus*	Ciccillo et al. (2002) and Parra-Cota et al. (2014)
	vietnamiensis	*Oryza sativa*	Choudhury and Kennedy (2004)
Cupriavidus	*taiwanensis*	*Rhynchosia ferulifolia*	Garu et al. (2009)
Pseudomonas	NA	NA	Zhao et al. (2013)

Modified from Berrada and Fikri-Benbrahim (2014)
NA Not Available

the growth of tree legumes *Samanea saman* could be improved by application of stress tolerant rhizobia. Ahmad et al. (2013) also stated that halo-tolerant, auxin producing *Rhizobium* strains improve osmotic stress tolerance in mung bean. Recently, benefits of using exopolysaccharides (EPS) in bioformulation is documented (Tewari and Arora 2014; Thenmozhi and Dinakar 2014). EPS protects inoculated rhizobial cells from stress factors such as salinity, dessication and pH (Ophir and Gutnick 1994; Qurashi and Sabri 2012). Use of EPS as efficient alternative carriers for inoculant production is also reported (Rodrigues et al. 2015). Maheshwari et al. (2012) suggested that the combined application of microbial inoculants and fertilizer worked as better choice for farmers to reduce the risk and expenses of chemical fertilizers.

Use of omics-based approaches (genomics and proteomics) can also be very useful in enhancing our understanding of rhizobia-legume symbiosis (Ramalingam et al. 2015). Omics based techniques including genomics, proteomics and metabolomics can go a long way in designing state of the art bioformulation for a particular soil and crop. Discovery of very similar signalling pathway in cereals as used by legumes to fix nitrogen has opened the door of non-legume fixation (de Bruijn 2016) and according to research by Rogers and Oldroyd (2014), in near future cereal crops capable of fixing nitrogen will also be available by the application of synthetic biological approaches.

4.6 Conclusion

Nitrogen is one of the key and limiting nutrients for agricultural ecosystems. The most important process to bring the atmospheric nitrogen in use for living creatures is BNF. Amongst microbes possessing the capacity to carry out BNF, rhizobia are the most important. Rhizobial inoculants are being used since long time around the globe, but still there is room for improvement so as to enhance the efficacy, productivity and credibility amongst the farmers. Liquid formulations are now picking up the pace in comparison to solid carrier based ones. Different types of additives and metabolites are also now being incorporated into the formulations to enhance their shelf life, performance and efficacy. Gel based encapsulated inoculants are also being used and with emergence of biotechnology, genetically improved inoculants and co-inoculants of rhizobia with other PGPRs/AMFs are also being worked upon in present time. Use of biotechnological tools and improvement in regulations can go a long way in designing a rhizobial bioformulation which will be more reliable and effective. To design a tailor made state of the art rhizobial formulation, it is very important to further our knowledge on plant-microbe interactions by using the latest tools and techniques. Also it would be better if the government and non-government organizations help in spreading the knowledge of usefulness of such bioformulations amongst the end-users. This is particularly required in developing nations. In future, we must see an even greater role and share of rhizobial inoculants in the market for sustainable supply of nitrogen to the future generations.

References

Abd-Alla MH, Issa AA, Ohyama T (2014a) Impact of harsh environmental conditions on nodule formation and dinitrogen fixation of legumes. In: Ohyama T (ed) Advances in biology and ecology of nitrogen fixation. InTech Open, Rijeka, p 282

Abd-Alla MH, El-Enany AE, Nafady NA et al (2014b) Synergistic interaction of *Rhizobium leguminosarum* bv. *viciae* and arbuscular mycorrhizal fungi as a plant growth promoting biofertilizers for faba bean (*Vicia faba* L.) in alkaline soil. Microbiol Res 169:49–58

Abd-Alla MH, Bashandy SR, Bagy MK et al (2014c) *Rhizobium tibeticum* activated with a mixture of flavonoids alleviates nickel toxicity in symbiosis with fenugreek (*Trigonella foenumgraecum* L.) Ecotoxicology 23(5):946–959

Abdel-Lateif K, Bogusz D, Hocher V (2012) The role of flavonoids in the establishment of plant roots endosymbioses with arbuscular mycorrhiza fungi, rhizobia and Frankia bacteria. Plant Signal Behav 7(6):636–641

Aeron A, Khare E, Arora NK et al (2012) Practical use of CMC-amended rhizobial inoculant for *Mucuna pruriens* cultivation to enhance the growth and protection against *Macrophomina phaseolina*. J Gen Appl Microbiol 58(2):121–127

Ahmad M, Zahir ZA, Nazli F et al (2013) Effectiveness of halo-tolerant, auxin producing *Pseudomonas* and *Rhizobium* strains to improve osmotic stress tolerance in mung bean (*Vigna radiata* L.) Braz J Microbiol 44(4):1341–1348

Alami Y, Achouak W, Marol C et al (2000) Rhizosphere soil aggregation and plant growth promotion of sunflowers by an exopolysaccharide-producing *Rhizobium* sp. strain isolated from sunflower roots. Appl Environ Microbiol 66:3393–3398

Albareda M, Rodríguez-Navarro DN, Camacho M et al (2008) Alternatives to peat as a carrier for rhizobia inoculants: solid and liquid formulations. Soil Biol Biochem 40:2771–2779

Albareda M, Rodriguez-Navarro DN, Temprano FJ (2009) Use of *Sinorhizobium* (*Ensifer*) *fredii* for soybean inoculants in south Spain. Eur J Agron 30:205–211

Allito BB, Ewusi-Mensah N, Alemneh AA (2015) Rhizobia strain and host-legume interaction effects on nitrogen fixation and yield of grain legume: a review. Mol Soil Biol 6(2):1–6

Alvarez GS, Pieckenstain FL, Desimone MF et al (2010) Evaluation of sol-gel silica matrices as inoculant carriers for *Mesorhizobium* spp. cells. In: Mendez-Vilas A (ed) Current research, technology and education topics in applied microbiology and microbial biotechnology. Formatex, pp 160–167

Amarger N, Macheret V, Laguerre G (1997) *Rhizobium gallicum* sp. nov., and *Rhizobium giardinii* sp. nov., from *Phaseolus vulgaris* nodules. Int J Syst Evol Microbiol 47(4):996–1006

Anandham R, Sridar R, Nalayini P et al (2007) Potential for plant growth promotion in groundnut (*Arachis hypogaea* L.) cv. ALR-2 by coinoculation of sulfur-oxidizing bacteria and *Rhizobium*. Microbiol Res 162:139–153

Annette A, Andrew L, Michelle L et al (2013) Nodulation and effective nitrogen fixation of *Macroptilium atropurpureum* (siratro) by *Burkholderia tuberum*, a nodulating and plant growth promoting beta-proteobacterium, are influenced by environmental factors. Plant Soil 369(1/2):543

Ardley JK, Reeve WG, O'Hara GW et al (2013) Nodule morphology, symbiotic specificity and association with unusual rhizobia are distinguishing features of the genus *Listia* within the southern African crotalarioid clade *Lotononis s.l.* Ann Bot 112(1):1–15

Arora NK, Mishra J (2016) Prospecting the roles of metabolites and additives in future bioformulations for sustainable agriculture. Appl Soil Ecol. doi.org/10.1016/j.apsoil.2016.05.020

Arora NK, Kumar V, Maheshwari DK (2000) Isolation of both fast and slow growing rhizobia effectively nodulating a medicinal legume, *Mucuna pruriens*. Symbiosis 29:121–137

Arora NK, Kumar V, Maheshwari DK (2001) Constraints, development and future of the inoculants with special reference to rhizobial inoculants. In: Maheshwari DK, Dubey RC (eds) Innovative approaches in microbiology. Singh and Singh, Dehradun, pp 241–245

Arora NK, Singhal V, Maheshwari DK (2006) Salinity-induced accumulation of poly-β-hydroxybutyrate in rhizobia indicating its role in cell protection. World J Microbiol Biotechnol 22:603–606

Arora NK, Khare E, Naraian R et al (2008) Sawdust as a superior carrier for production of multipurpose bioinoculants using plant growth promoting rhizobial and pseudomonad strain and their impact on productivity of *Trifolium repense*. Curr Sci 95(1):90–94

Arora NK, Khare E, Singh S et al (2009) Effect of Al and heavy metals on enzymes of nitrogen metabolism of fast and slow growing rhizobia under explanta conditions. World J Microbiol Biotechnol 26(5):811–816

Arora NK, Maheshwari DK, Khare E (2010) Plant growth promoting rhizobacteria: constraints in bioformulation, commercialization and future strategies. In: Maheshwari DK (ed) Bacteria and plant health. Springer, Berlin, pp 97–116

Arora NK, Tewari S, Singh S et al (2012) PGPR for protection of plant health under saline conditions. In: Maheshwari DK (ed) Bacteria in agrobiology: stress management. Springer Publication, Dordrecht, pp 239–258

Arora NK, Tiwari S, Singh R (2014) Comparative study of different carriers inoculated with nodule forming and free living plant growth promoting bacteria suitable for sustainable agriculture. J Pharm Chem Biol Sci 2(2):143–149

Arora NK, Verma M, Prakash J et al (2016) Regulation of biopesticides: global concerns and policies. In: Arora NK, Mehnaz S, Balestrini R (eds) Bioformulations: for sustainable agriculture. Springer, New Delhi, pp 283–299

Baldwin IL, Fred EB (1929) Nomenclature of the root nodule bacteria of the Leguminosae. J Bacteriol 17:141–150

Ballard RA, Charman N (2000) Nodulation and growth of pasture legumes with naturalised soil rhizobia 1. Annual *Medicago* spp. Aust J Exp Agric 40:939–948

Bashan Y (1998) Inoculants of plant growth-promoting bacteria for use in agriculture. Biotechnol Adv 16(4):729–770

Bashan Y, Levanony H (1988) Adsorption of the rhizosphere bacterium *Azospirillum brasilense* Cd to soil, sand and peat particles. J Gen Microbiol 134:1811–1820

Bashan Y, de Bashan LE, Prabhu SR et al (2014) Advances in plant growth-promoting bacterial inoculant technology: formulations and practical perspectives (1998–2013). Plant Soil 378:1–33

Beijerinck MW (1888) Cultur des Bacillus radicicola aus den Kno¨llchen. Bot Ztg 46:740–750

Ben Rebah F, Tyagi RD, Prevost D (2002a) Wastewater sludge as a substrate for growth and carrier for rhizobia: the effect of storage conditions on survival of *Sinorhizobium meliloti*. Bioresour Technol 83:145–151

Ben Rebah F, Tyagi RD, Prevost D et al (2002b) Wastewater sludge as a new medium for rhizobial growth. Water Qual Res J Can 37:353–370

Ben Rebah F, Prévost D, Yezza A et al (2007) Agro-industrial waste materials and wastewater sludge for rhizobial inoculant production: a review. Bioresour Technol 98:3535–3546

Berrada H, Fikri-Benbrahim K (2014) Taxonomy of the rhizobia: current perspectives. Br Microbiol Res J 4(6):616

Bezdicek DF, Kennedy AC (1998) In: Lynch JM, Hobbie JE (eds) Microorganisms in action. Blackwell, Oxford, p 243

Bezdicek DF, Evans DW, Abeda B et al (1978) Evaluation of peat and granular inoculum for soybean yield and N fixation under irrigation. Agron J 70:865–868

Bianco C, Senatore B, Arbucci S et al (2014) Modulation of endogenous indole-3-acetic acid biosynthesis in bacteroids within *Medicago sativa* nodules. Appl Environ Microbiol 80(14):4286–4293

Bioag.novozymes.com (2015) Novozymes BioAg Group. At http://www.bioag.novozymes.com

Bonish PM (1979) Clover rhizobia in soils: assessment of effectiveness using the plant infection count method. N Z J Agric Res 22:89–93

Bouizgarne B, Oufdou K, Ouhdouch Y (2015) Actinorhizal and rhizobial-legume symbioses for alleviation of abiotic stresses. In: Arora NK (ed) Plant microbes symbiosis: applied facets. Springer, New Delhi, pp 273–295

Boussingault J (1838) Recherches chimiques sur la vegetation enterprises dans le but d'examiner si les plantes prennent de I'atmosphere. Ann Chim Phys 67:1–54

Brahmaprakash GP, Sahu PK (2012) Biofertilizers for sustainability. J Indian Inst Sci 92(1):37–62

Brito B, Palacios JM, Imperial J et al (2002) Engineering the *Rhizobium leguminosarum* bv. *viciae* hydrogenase system for expression in free-living microaerobic cells and increased symbiotic hydrogenase activity. Appl Environ Microbiol 68(5):2461–2467

Brockwell J (1982) Inoculation methods for field experimenters and farmers. In: Vincent JM (ed) Nitrogen fixation in legumes. Academic Press, Australia, pp 211–227

Brockwell J (1985) Environmental interactions influencing innovative practices in legume inoculation. In: Shibles R (ed) Proceedings of the world soybean conference III. Westview Press, Boulder, pp 943–950

Brockwell J, Bottomley PJ (1995) Recent advances in inoculant technology and prospects for the future. Soil Biol Biochem 27:683–697

Brockwell J, Gault RR, Chase DL et al (1980) An appraisal of practical alternatives to legume seed inoculation: field experiments on seedbed inoculation with solid and liquid inoculants. Aust J Agric Res 31:47–60

Brockwell J, Holliday RA, Pilka A (1988) Evaluation of root-nodule bacteria introduced into field environment of the symbiotic nitrogen-fixing potential of soils by direct microbiological means. Plant Soil 108:163–170

Brockwell J, Bottomley PJ, Thies JE (1995) Manipulation of rhizobia microflora for improving legume productivity and soil fertility: a critical assessment. Plant Soil 174:143–180

Burges HD, Jones KA (1998) Trends in formulation of microorganisms and future research requirements. In: Burges HD (ed) Formulation of microbial biopesticides, beneficial microorganisms, nematodes and seed treatment. Kluwer Academic Publication, Dordrecht, pp 311–332

Catroux G, Hartmann A, Revellin C (2001) Trends in rhizobial inoculant production and use. Plant Soil 230:21–30

Chandra S, Choure K, Dubey RC et al (2007) Rhizosphere competent *Mesorhizobium loti* MP6 induces root hair curling, inhibits *Sclerotinia sclerotiorum* and enhances growth of Indian mustard (*Brassica campestris*). Braz J Microbiol 38:128–130

Chibeba AM, Guimarães MF, Brito OR et al (2015) Co-inoculation of soybean with *Bradyrhizobium* and *Azospirillum* promotes early nodulation. Am J Plant Sci 6:1641–1649

Choudhury ATMA, Kennedy IR (2004) Prospects and potentials for systems of biological nitrogen fixation in sustainable rice production. Biol Fertil Soils 39(4):219–227

Ciccillo F, Fiore A, Bevivino A et al (2002) Effects of two different application methods of *Burkholderia ambifaria* MCI 7 on plant growth and rhizospheric bacterial diversity. Environ Microbiol 4:238–245

Clayton GW, Rice WA, Lupwayi NZ et al (2004) Inoculant formulation and fertilizer nitrogen effects on field pea: crop yield and seed quality. Can J Plant Sci 84:89–96

Crawford SL, Berryhill DL (1983) Survival of *Rhizobium phaseoli* in coal-based legume inoculants applied to seeds. Appl Environ Microbiol 45:703–705

Da Silva JG, Serra GE, Moreira JR et al (1978) Energy balance for ethyl alcohol production for crops. Science 210:903–906

Damasceno R, Roggia I, Pereira C et al (2013) Rhizobia survival in seeds coated with polyvinyl alcohol (PVA) electrospun nanofibres. Can J Microbiol 59(11):716–719

Dashti N, Prithiviraj B, Zhou X et al (2000) Combined effects of plant growth promoting rhizobacteria and genistein on nitrogen fixation in soybean at suboptimal root zone temperatures. J Plant Nutr 23:593–604

Dazzo FB, Yanni YG, Rizk R et al (2000) Progress in multinational collaborative studies on the beneficial association between *Rhizobium leguminosarum* bv. *trifolii* and rice. In: Ladha JK, Reddy PM et al (eds) The quest for nitrogen fixation in rice. IRRI, Los Banos, pp 167–189

Deaker R, Roughley RJ, Kennedy IR (2004) Legume seed inoculation technology-a review. Soil Biol Biochem 36:1275–1288

de Bruijn FJ (2016) "Biological Nitrogen Fixation" book summary. Adv Microbiol 6:407–411

de Lajudie P, Laurent-Fulele E, Willems A et al (1998) *Allorhizobium undicola* gen. nov., sp. nov., nitrogen-fixing bacteria that efficiently nodulate *Neptunia natans* in Senegal. Int J Syst Bacteriol 48:1277–1290

Denardin ND, Freire JRJ (2000) Assessment of polymers for the formulation of legume inoculants. World J Microbiol Biotechnol 16:215–217

Deschodt CC, Strijdom WB (1976) Suitability of a coal bentonite base as carrier of rhizobia in inoculants. Phytophylactica 8:1–6

Deshwal VK, Singh SB, Kumar P et al (2013) Rhizobia unique plant growth promoting rhizobacteria: a review. Int J Life Sci 2(2):74–86

Diez-Mendez A, Menéndez E, García-Fraile P et al (2015) *Rhizobium cellulosilyticum* as a co-inoculant enhances *Phaseolus vulgaris* grain yield under greenhouse conditions. Symbiosis 67(1):135–141

Diouf D, Forestier S, Neyra M et al (2003) Optimisation of inoculation of *Leucaena leucocephala* and *Acacia mangium* with *Rhizobium* under greenhouse conditions. Ann For Sci 60:379–384

Dommergues YR, Diem HG, Divies D (1979) Polyacrylamide-entrapped *Rhizobium* as an inoculant for legumes. Appl Environ Microbiol 37:779–781

Dudeja SS, Singh NP, Sharma P et al (2011) Biofertilizer technology and pulse production. In: Singh A, Parmar N, Kuhad RC (eds) Bioaugmentation, biostimulation and biocontrol. Springer-Verlag, Berlin/Heidelberg, pp 43–63

Eckhardt MM, Baldwin IR, Fred EB (1931) Studies on the root-nodule bacteria of *Lupinus*. J Bacteriol 21:273–285

EPA (1997) Environmental protection act, fact sheet: commercialization of *Sinorhizobium* (*Rhizobium*) meliloti, RMBPC-2. At http://www.epa.gov/biotech_rule/pubs/factdft6.html

Ferguson BJ, Indrasumunar A, Hayashi S et al (2010) Molecular analysis of legume nodule development and auto regulation. J Integr Plant Biol 3:61–76

Fernandes-Júnior PI, Rohr TG, de Oliveira PJ et al (2009) Polymers as carriers for rhizobial inoculant formulations. Pesq Agrop Bras 44(9):1184–1190
Ferreira EM, Castro IV (2005) Residues of the cork industry as carriers for the production of legume inoculants. Silva Lusit 13:159–167
Fitouri SD, Trabelsi D, Saïdi S et al (2012) Diversity of rhizobia nodulating sulla (*Hedysarum coronarium* L.) and selection of inoculant strains for semi-arid Tunisia. Ann Microbiol 62:77–84
Flores-Félix JD, Menéndez E, Rivera LP et al (2013) Use of *Rhizobium leguminosarum* as a potential biofertilizer for *Lactuca sativa* and *Daucus carota* crops. J Plant Nutr Soil Sci 176:876–882
Flores-Felix JD, Silva LR, Rivera LP et al (2015) Plants probiotics as a tool to produce highly functional fruits: the case of *Phyllobacterium* and vitamin C in strawberries. PLoS One 10(4):e0122281
Forestier S, Alvarado G, Badjel SB et al (2001) Effect of *Rhizobium* inoculation methodologies on nodulation and growth of *Leucaena leucocephala*. World J Microbiol Biotechnol 17:359–362
Frank B (1889) Ueber die Pilzsymbiose der Leguminosen. Ber Deut Bot Ges 7:332–346
Fred EB, Baldwin IL, McCoy E (1932) Root nodule bacteria and leguminous plants, University of Wisconsin Studies in Science No.5. University of Wisconsin, Madison
Fuchsius L (1542) De historia stirpium commentarii isignes. Michael Isingrin, Basel
Galloway JN, Townsend AR, Erisman JW et al (2008) Transformation of the nitrogen cycle: recent trends, questions and potential solutions. Science 320:889–892
Gamal-Eldin H, Elbanna K (2011) Field evidence for the potential of *Rhodobacter capsulatus* as biofertilizer for flooded rice. Curr Microbiol 62:391–395
Garu G, Yates RJ, Deiana P et al (2009) Novel strains of nodulating *Burkholderia* have a role in nitrogen fixation with papilionoid herbaceous legumes adapted to acid, infertile soils. Soil Biol Biochem 41:125–134
Geetha SJ, Joshi SJ (2013) Engineering rhizobial bioinoculants: a strategy to improve iron nutrition. Sci World J 2013:1–15
Gomez M, Silva N, Hartmen A et al (1997) Evaluation of commercial soybean inoculants from Argentina. World J Microbiol Biotechnol 13:167–173
Granada CE, Strochein M, Vargas LK et al (2014) Genetic diversity and symbiotic compatibility among rhizobial strains and *Desmodium incanum* and *Lotus* spp. plants. Genet Mol Biol 37(2):396–405
Grand View Research (2015) Biofertilizers market analysis by product (nitrogen fixing, phosphate solubilizing), by application (seed treatment, soil treatment) and segment forecasts to 2022. Available at http://www.grandviewresearch.com/
Gu CT, Wang ET, Sui XH et al (2007) Diversity and geographical distribution of rhizobia associated with *Lespedeza* spp. in temperate and subtropical regions of China. Arch Microbiol 188:355–365
Guthrie FB (1896) Inoculation of soil for leguminous crops. Agric Gaz NSW 7:690–694
Gzyl A, Augustynowicz E, Mosiej E et al (2005) Amplified fragment length polymorphism (AFLP) versus randomly amplified polymorphic DNA (RAPD) as new tools for inter-and intra-species differentiation within Bordetella. J Med Microbiol 54(4):333–346
Hamaoui B, Abbadi JM, Burdman S et al (2001) Effects of inoculation with *Azospirillum brasilense* on chickpeas (*Cicer arietinum*) and faba beans (*Vicia faba*) under different growth conditions. Agronomy 21:553–560
Hashem MA (2001) Problems and prospects of cyanobacterial biofertilizer for rice cultivation. Aust J Plant Physiol 28:881–888
Hawkesford M, Horst W, Kichey T et al (2012) Functions of macronutrients. In: Marschner P (ed) Marschner's mineral nutrition of higher plants, 3rd edn. Academic, London, pp 135–189
Hedin PA, McCarty JC (1994) Effects of several commercial plant growth regulator formulations on yield and allelochemicals of cotton (*Gossypium hirsutum* L.) Agric Food Chem 42:1355–1357
Hellriegel H, Wilfarth H (1888) Untersuchungen u¨ ber die Stickstoffnahrung der Gramineon und Leguminosen. Beilageheft zu der Ztschr. Ver. Ru¨ benzucker-Industrie Deutschen Reichs
Herridge D, Gemell G, Hartley E (2002) Legume inoculants and quality control. ACIAR Proc 109:105–115

Herridge DF, Peoples MB, Boddey RM (2008) Global inputs of biological nitrogen fixation in agricultural systems. Plant Soil 311:1–18

Herrmann L, Atieno M, Brau L et al (2015) Microbial quality of commercial inoculants to increase BNF and nutrient use efficiency. In: de Bruijn FJ (ed) Biological nitrogen fixation volume 2. Wiley, Hoboken, pp 1031–1040

Humphry DR, Andrews M, Santos SR et al (2007) Phylogenetic assignment and mechanism of action of a crop growth promoting *Rhizobium radiobacter* strain used as a biofertiliser on graminaceous crops in Russia. Antonie Van Leeuwenhoek 91(2):105–113

Hungria M, Campo RJ, Souza EM et al (2010) Inoculation with selected strains of *Azospirillum brasilense* and A. *lipoferum* improves yields of maize and wheat in Brazil. Plant Soil 331:413–425

Hungria M, Nogueira MA, Araujo RS (2015) Soybean seed co-inoculation with *Bradyrhizobium* spp and *Azospirillum brasilense*: a new biotechnological tool to improve yield and sustainability. Am J Plant Sci 6:811–817

Hussain I, Arshad M, Zahir ZA et al (1995) Substrate dependent microbial production of auxins and their influence on growth and nodulation of lentil. Pak J Agric Sci 32:149–152

Hynes RK, Jans DC, Bremer E et al (2001) *Rhizobium* population dynamics in the pea rhizosphere of rhizobial inoculant strain applied in different formulations. Can J Microbiol 47:595–600

Jayasinghearachchi HS, Seneviratne G (2004) A Bradyrhizobial-*Penicillium* spp. biofilm with nitrogenase activity improves N_2 fixing symbiosis of soybean. Biol Fertil Soils 40:432–434

John RP, Tyagi RD, Brar SK et al (2011) Bioencapsulation of microbial cells for targeted agricultural delivery. Crit Rev Biotechnol 31:211–226

Jourand P, Giraud E, Béna G et al (2004) *Methylobacterium nodulans* sp. nov., for a group of aerobic, facultatively methylotrophic, legume root-nodule-forming and nitrogen-fixing bacteria. Int J Syst Evol Microbiol 54(6):2269–2273

Jung G, Mugnier J, Diem HG et al (1982) Polymer entrapped *Rhizobium* as an inoculant for legumes. Plant Soil 65:219–231

Kaljeet S, Keyeo F, Amir HG (2011) Influence of carrier materials and storage temperature on survivability of rhizobial inoculant. Asian J Plant Sci 10:331–337

Kalra A, Chandra M, Awasthi A et al (2010) Natural compounds enhancing growth and survival of rhizobial inoculants in vermicompost-based formulations. Biol Fertil Soils 46:521–524

Kandasamy R, Prasad NN (1971) Lignite as a carrier of rhizobia. Curr Sci 40:496

Karaś MA, Turska-Szewczuk A, Trapska D et al (2015) Growth and survival of *Mesorhizobium loti* inside *Acanthamoeba* enhanced its ability to develop more nodules on *Lotus corniculatus*. Microb Ecol 70(2):566–575

Karuppasamy K, Nagaraj S, Kathiresan K (2011) Stress tolerant *Rhizobium* enhances the growth of *Samanea saman* (JECQ) Merr. Afr J Basic Appl Sci 3(6):278–284

Kennedy IR, Choudhury ATMA, Kecskés ML (2004) Non-symbiotic bacterial diazotrophs in crop-farming systems: can their potential for plant growth promotion be better exploited? Soil Biol Biochem 36:1229–1244

Khalid A, Arshad M, Zahir ZA (2004) Screening plant growth promoting rhizobacteria for improving growth and yield of wheat. J Appl Microbiol 96:473–480

Khare E, Arora NK (2011) Dual activity of pyocyanin from *Pseudomonas aeruginosa*- antibiotic against phytopathogen and signal molecule for biofilm development by rhizobia. Can J Microbiol 57(9):708–713

Khavazi K, Rejali F, Seguin P et al (2007) Effects of carrier, sterilization method and incubation on survival of *Bradyrhizobium japonicum* in soybean (*Glycine max* L.) inoculants. Enzym Microb Technol 41:780–784

Kneip C, Lockhart P, Voß C, Maier UG (2007) Nitrogen fixation in eukaryotes–new models for symbiosis. BMC Evol Biol 7(1):55

Kokalis-Burelle N, Vavrina CS, Reddy MS et al (2003) Amendment of muskmelon and watermelon transplant media with plant growth-promoting rhizobacteria: effect on seedling quality, disease and nematode resistance. Hortic Technol 13:476–482

Kostov O, Lynch JM (1998) Composted sawdust as a carrier for *Bradyrhizobium, Rhizobium* and *Azospirillum* in crop inoculation. World J Microbiol Biotechnol 14:389–397
Krapp A, Berthomé R, Orsel M et al (2011) *Arabidopsis* roots and shoots show distinct temporal adaptation patterns toward nitrogen starvation. Plant Physiol 157:1255–1282
Kremer RJ, Peterson HL (1982) Effect of inoculant carrier on survival of *Rhizobium* on inoculated seed. Soil Sci 134:117–124
Kudoyarova GR, Korobova AV, Akhiyarova GR et al (2014) Accumulation of cytokinins in roots and their export to the shoots of durum wheat plants treated with the protonophore carbonyl cyanide m-chlorophenylhydrazone (CCCP). J Exp Bot 65:2287–2294
Kudoyarova GR, Arkhipova TN, Melent'ev AI (2015) Role of bacterial phytohormones in plant growth regulation and their development. In: Bacterial metabolites in sustainable agroecosystem. Springer, International Publishing, pp 69–86
Lachmann J (1858) Über knollen an den wurzeln der leguminosen [About bulbous roots of the legume]. Landwirthschaftliche Mitteilungen. Z Koniglichen
Laranjo M, Alexandre A, Oliveira S (2014) Legume growth-promoting rhizobia: an overview on the *Mesorhizobium* genus. Microbiol Res 169(1):2–17
Leibovitch S, Migner P, Zhang F et al (2001) Evaluation of the effect of SoyaSignal technology on soybean yield (*Glycine max* (L.) Merr.) under field conditions over 6 years in eastern Canada and the northern United States. J Agron Crop Sci 187:281–292
Li J, Zhang S, Shi S et al (2011) Four materials as carriers for phosphate dissolving *Rhizobium* sp. inoculants. Adv Mater Res 156–157:919–928
Lloret L, Ormeño-Orrillo E, Rincón R et al (2007) *Ensifer mexicanus* sp. nov. a new species nodulating *Acacia angustissima* (Mill.) Kuntze in Mexico. Syst Appl Microbiol 30:280–290
Löhis F, Hansen R (1921) Nodulating bacteria of leguminous plant. J Agric Res 20:543–556
López-López A, Rogel MA, Ormeno-Orrillo E et al (2010) *Phaseolus vulgaris* seed-borne endophytic community with novel bacterial species such as *Rhizobium endophyticum* sp. nov. Syst Appl Microbiol 33(6):322–327
Luce MS, Whalen JK, Ziadi N et al (2011) Nitrogen dynamics and indicesto predict soil nitrogen supply in humid temperate soils. Adv Agron 112:55
Lupwayi NZ, Clayton GW, Rice WA (2006) Rhizobial inoculants for legume crops. J Crop Improv 15:289–321
Maheshwari DK, Dubey RC, Aeron A et al (2012) Integrated approach for disease management and growth enhancement of *Sesamum indicum* L. utilizing *Azotobacter chroococcum* TRA2 and chemical fertilizer. World J Microbiol Biotechnol 28(10):3015–3024
Malpighi H (1679) Anatome plantarum. J Martyn Lond 99
Malusá E, Vassilev N (2014) A contribution to set a legal framework for biofertilisers. Appl Microbiol Biotechnol 98(15):6599–6607
Malusa E, Sas-Paszt L, Ciesielska J (2012) Technologies for beneficial microorganisms inocula used as biofertilizers. Sci World J 1–12
Manikandan R, Saravanakumar D, Rajendran L et al (2010) Standardization of liquid formulation of *Pseudomonas fluorescens* Pf1 for its efficacy against *Fusarium* wilt of tomato. Biol Control 54:83–89
Markets and Markets (2015) Biofertilizers market analysis by product (nitrogen fixing, phosphate solubilizing), by application (seed treatment, soil treatment) and segment forecasts to 2022. At http://www.marketsandmarkets.com/Market-Reports/compound-biofertilizers-customizedfertilizers-market-856.html
Marks BB, Megías M, Nogueira MA et al (2013) Biotechnological potential of rhizobial metabolites to enhance the performance of *Bradyrhizobium* spp. and *Azospirillum brasilense* inoculants with soybean and maize. AMB Express 3:21
Marks BB, Megías M, Ollero FJ et al (2015) Maize growth promotion by inoculation with *Azospirillum brasilense* and metabolites of *Rhizobium tropici* enriched on lipo-chitooligosaccharides (LCOs). AMB Express 5:71

Maróti G, Kondorosi É (2014) Nitrogen-fixing *Rhizobium*-legume symbiosis are polyploidy and host peptide-governed symbiont differentiation general principles of endosymbiosis? Front Microbiol 5:326

Maurice S, Beauclair P, Giraud JJ et al (2001) Survival and change in physiological state of *Bradyrhizobium japonicum* in soybean (*Glycine max* L. Merril) liquid inoculants after long-term storage. World J Microbiol Biotechnol 17:635–643

Mazid M, Khan TA (2014) Future of bio-fertilizers in Indian agriculture: an overview. Int J Agric Food Res 3(3):10–23

Mazid M, Khan TA, Mohammad F (2011) Potential of NO and H_2O_2 as signalling molecules in tolerance to abiotic stress in plants. J Indust Res Technol 1(1):56–68

Meng L, Zhang A, Wang F et al (2015) Arbuscular mycorrhizal fungi and *Rhizobium* facilitate nitrogen uptake and transfer in soybean/maize intercropping system. Front Plant Sci 6:339

Menna P, Pereira AA, Bangel EV et al (2009) Rep-PCR of tropical rhizobia for strain fingerprinting, biodiversity appraisal and as a taxonomic and phylogenetic tool. Symbiosis 48(1–3):120–130

Merwe SP, Strijdom BW, Uys CJ (1974) Groundnut response to seed inoculation under extensive agriculture practices in South African soils. Phytophylactica 6:295–302

Meyer SLF, Roberts DP, Chitwood DJ et al (2001) Application of *Burkholderia cepacia* and *Trichoderma virens*, alone and in combination, against *Meloidogyne incognita* on bell pepper. Nematropica 31:75–86

Mia MAB, Shamsuddin ZH, Zakaria W et al (2007) Associative nitrogen fixation by *Azospirillum* and *Bacillus* spp. in bananas. Infomusa 16(1–2):11–15

Miransari M, Smith D (2009) Rhizobial lipo-chitooligosaccharides and gibberellins enhance barley (*Hordeum vulgare* L.) seed germination. Biotechnology 8:270–275

Mishra J, Arora NK (2016) Bioformulations for plant growth promotion and combating phytopathogens: a sustainable approach. In: Arora NK, Mehnaz S, Balestrini R (eds) Bioformulations: for sustainable agriculture. Springer, New Delhi, pp 3–33

Mmbaga GW, Mtei KM, Ndakidemi PA (2014) Extrapolations on the use of *Rhizobium* inoculants supplemented with phosphorus (P) and potassium (K) on growth and nutrition of legumes. Agric Sci 5(12):1207

Mnasri B, Saïdi S, Chihaoui SA et al (2012) *Sinorhizobium americanum* symbiovar *mediterranense* is a predominant symbiont that nodulates and fixes nitrogen with common bean (*Phaseolus vulgaris* L.) in a Northern Tunisian field. Syst Appl Microbiol 35(4):263–269

Morel MA, Cagide C, Minteguiaga MA et al (2015) The pattern of secreted molecules during the co-inoculation of alfalfa plants with *Sinorhizobium meliloti* and *Delftia* sp. strain JD2: an interaction that improves plant yield. Mol Plant-Microbe Interact 28:134–142

Morel MA, Cagide C, Castro-Sowinski S (2016) The contribution of secondary metabolites in the success of bioformulations. In: Arora NK, Mehnaz S, Balestrini R (eds) Bioformulations: for sustainable agriculture. Springer, New Delhi, pp 235–250

Muñoz N, Soria-Díaz ME, Manyani H et al (2014) Structure and biological activities of lipo-chitooligosaccharide nodulation signals produced by *Bradyrhizobium japonicum* USDA 138 under saline and osmotic stress. Biol Fertil Soils 50:207–215

Nápoles MC, Guevara E, Montero F et al (2009) Role of *Bradyrhizobium japonicum* induced by genistein on soybean stressed by water deficit. Span J Agric Res 7:665–671

Naveed M, Mehboob I, Hussain MB et al (2015) Perspectives of rhizobial inoculation for sustainable crop production. In: Arora NK (ed) Plant microbes symbiosis: applied facets. Springer, New Delhi, pp 209–239

Nelson LM (2004) Plant growth promoting rhizobacteria (PGPR): prospects for new inoculants. Crop Manag 10:301–305

Nobbe F, Hiltner L (1896) Inoculation of the soil for cultivating leguminous plants. U.S. Patent 570 813

Odee DW, Indieka SA, Lesueur D (2002) Evaluation of inoculation procedures for *Calliandra calothyrsus* Meisn. grown in tree nurseries. Biol Fertil Soils 36:124–128

Ogasawara M, Suzuki T, Mutoh I et al (2003) *Sinorhizobium indiaense* sp. nov., and *Sinorhizobium abri* sp. nov., isolated from tropical legumes, *Sesbania rostrata* and *Abrus precatorius*, respectively. Symbiosis 34:53–68

Olah B, Briere C, Becard G et al (2005) Nod factors and a diffusible factor from arbuscular mycorrhizal fungi stimulate lateral root formation in *Medicago truncatula* via the DMI1/DMI2 signalling pathway. Plant J 44:195–207

Olsen PE, Rice WA, Bordeleau LM et al (1994) Analysis and regulation of legume inoculants in Canada: the need for an increase in standards. Plant Soil 161:127–134

Olsen PE, Rice WA, Bordeleau LM et al (1996) Levels and identities of nonrhizobial microorganisms found in commercial legume inoculant made with nonsterile peat carrier. Can J Microbiol 42:72–75

Olufajo OO, Adu JK (1993) Nodulation of soyabeans grown under field conditions and inoculated with *Bradyrhizobium japonicum* strains. In: Mulongoy K, Gueye M, Spencer DSC (eds) Biological nitrogen fixation and sustainability of tropical agriculture. Wiley, New York, pp 147–154

Ophir T, Gutnick DL (1994) A role for exopolysaccharides in the protection of microorganisms from desiccation. Appl Environ Microbiol 60(2):740–745

Paau AS, Graham LL, Bennett M (1991) Progress in formulation research for PGPR and biocontrol inoculants. In: Keel C, Koller B, Défago G (eds) Plant growth-promoting rhizobacteria-progress and prospects. IOBC/WPRS Bulletin, Zurich, pp 399–403

Pan B, Smith DL (2000) Pre-incubation of *Bradyrhizobium japonicum* cells with genistein reduces the inhibitory effects of mineral nitrogen on soybean nodulation and nitrogen fixation under field conditions. Plant Soil 223:235–242

Parra-Cota FI, Peña-Cabriales JJ, de los Santos-Villalobos S et al (2014) *Burkholderia ambifaria* and *B. caribensis* promote growth and increase yield in grain amaranth (*Amaranthus cruentus* and *A. hypochondriacus*) by improving plant nitrogen uptake. PLoS One 9(2):88–94

Peix A, Rivas-Boyero AA, Mateos PF et al (2001) Growth promotion of chickpea and barley by a phosphate solubilizing strain of *Mesorhizobium mediterraneum* under growth chamber conditions. Soil Biol Biochem 33:103–110

Philpotts H (1976) Filter mud as a carrier for *Rhizobium* inoculants. J Appl Bacteriol 41:277–281

Pindi PK, Satyanarayana SDV (2013) Liquid microbial consortium-a potential tool for sustainable soil health. J Biofertil Biopestic 3:124

Pongslip N (2012) Phenotypic and genotypic diversity of rhizobia. Bentham Science Publishers, Thailand, p 201

Prithiviraj B, Zhou X, Souleimanov A et al (2003) A host specific bacteria-to-plant signal molecule (Nod factor) enhances germination and early growth of diverse crop plants. Planta 216:437–445

Pugashetti BK, Gopalgowda HS, Patil RB (1971) Cellulose powder as legume inoculant base. Curr Sci 40:494–495

Qiu M, Li S, Zhou X et al (2014) De-coupling of root-microbiome associations followed by antagonist inoculation improves rhizosphere soil suppressiveness. Biol Fertil Soils 50(2):217–224

Qurashi AW, Sabri AN (2012) Bacterial exopolysaccharide and biofilm formation stimulate chickpea growth and soil aggregation under salt stress. Braz J Microbiol 43(3):1183–1191

Qureshi MA, Ahmad ZA, Akhtar N et al (2012) Role of phosphate solubilizing bacteria (PSB) in enhancing P-availability and promoting cotton growth. J Anim Plant Sci 22:204–210

Qureshi MA, Shahzad H, Imran Z et al (2013) Potential of *Rhizobium* species to enhance growth and fodder yield of maize in the presence and absence of L-tryptophan. J Anim Plant Sci 23(5):1448–1454

Raja Sekar K, Karmegam N (2010) Earthworm casts as an alternate carrier material for biofertilizers: assessment of endurance and viability of *Azotobacter chroococcum*, *Bacillus megaterium* and *Rhizobium leguminosarum*. Sci Hortic 124:286–289

Rajwar A, Sahgal M, Johri BN (2013) Legume-rhizobia symbiosis and interactions in agroecosystems. In: Arora NK (ed) Plant microbe symbiosis-fundamentals and advances. Springer, New Delhi, pp 233–265

Ramalingam A, Kudapa H, Pazhamala LT et al (2015) Proteomics and metabolomics: two emerging areas for legume improvement. Front Plant Sci 6:1116

Rao DLN (2014) Recent advances in biological nitrogen fixation in agricultural systems. Proc Indian Natl Sci Acad 80(2):359–378

Revellin C, Meunier G, Giraud JJ et al (2000) Changes in the physiological and agricultural characteristics of peat-based *Bradyrhizobium japonicum* inoculants after long-term storage. Appl Microbiol Biotechnol 54:206–211

Rivas R, Willems A, Subba-Rao NS et al (2003) Description of *Devosia neptuniae* sp. nov., that nodulates and fixes nitrogen in symbiosis with *Neptunia natans*, an aquatic legume from India. Syst Appl Microbiol 26(1):47–53

Rivera D, Obando M, Barbosa H et al (2014) Evaluation of polymers for the liquid rhizobial formulation and their influence in the *Rhizobium*-Cowpea interaction. Univ Sci 19(3):265–275

Roberto SR, Marinho de Assis A, Yamamoto LY et al (2012) Application timing and concentration of abscisic acid improve color of "Benitak" table grape. Sci Hortic 142:44–48

Rodrigues AC, Silveira JAG, Bonifacio A et al (2013) Metabolism of nitrogen and carbon: optimization of biological nitrogen fixation and cowpea development. Soil Biol Biochem 67:226–234

Rodrigues AC, Vendruscolo CT, Moreira AS et al (2015) *Rhizobium tropici* exopolysaccharides as carriers improve the symbiosis of cowpea-*Bradyrhizobium-Paenibacillus*. Afr J Microbiol Res 9(37):2037–2050

Rodriguez-Navarro DN, Temprano F, Orive R (1991) Survival of *Rhizobium* sp. (*Hedysarum coronarium* L.) on peat-based inoculants and inoculated seeds. Soil Biol Biochem 23:375–379

Rogers C, Oldroyd GE (2014) Synthetic biology approaches to engineering the nitrogen symbiosis in cereals. J Exp Bot 65(8):1939–1946

Rohr TG (2007) Rheological study of the mixture carboxymethylcellulose/starch and its use as a vehicle for the bacterial inoculation. Dissertation, Rural Federal University of Rio de Janeiro

Rokhzadi A, Asgharzadeh A, Darvish F et al (2008) Influence of plant growth promoting rhizobacteria on dry matter accumulation and yield of chickpea (*Cicer arietinum* L.) under field conditions. Am Eurasian J Agric Environ Sci 3(2):253–257

Rubiya E (2006) Coaggregated diazotrophic cultures- a novel delivery system of bioinoculation for lowland rice (*Oryza sativa* L.). Dissertation, Aannamalai University

Ruíz-Valdiviezo VM, Canseco LMCV, Suárez LAC et al (2015) Symbiotic potential and survival of native rhizobia kept on different carriers. Braz J Microbiol 46(3):735–742

Santi C, Bogusz D, Franche C (2013) Biological nitrogen fixation in non-legume plants. Ann Bot 111(5):743–767

Sassi-Aydi S, Aydi S, Abdelly C (2012) Inoculation with the native *Rhizobium gallicum* 8a3 improves osmotic stress tolerance in common bean drought-sensitive cultivar. Acta Agric Scand B-S P 62(2):179–187

Sheu SY, Chou JH, Bontemps C et al (2012) *Burkholderia symbiotica* sp nov., isolated from root nodules of *Mimosa* spp. native to north-east Brazil source. Int J Syst Evol Microbiol 62:2272–2278

Singleton P, Keyser H, Sande E (2002) Development and evaluation of liquid inoculants. In: Herridge D (ed) Inoculants and nitrogen fixation of legumes in Vietnam, vol 109e. ACIAR Proceedings, Canberra, pp 52–66

Sivakumar PK, Parthasarthi R, Lakshmipriya VP (2014) Encapsulation of plant growth promoting inoculant in bacterial alginate beads enriched with humic acid. Int J Curr Microbiol Appl Sci 3:415–422

Sivasakthivelan P, Saranraj P (2013) *Azospirillum* and its formulations: a review. Int J Microbiol Res 4(3):275–287

Smil V (2001) Enriching the earth: Fritz Haber, Carl Bosch and the transformation of world food production. The MIT Press, Cambridge, MA, p 338

Smith R (1992) Legume inoculant formulation and application. Can J Microbiol 38:485–492

Smith RS (1995) Inoculant formulations and applications to meet changing needs. In: Tikhonovich IA, Provorov NA, Romanov VI, Newton WE (eds) Nitrogen fixation: fundamentals and applications. Kluwer Academic Publishers, Dordrecht, pp 653–657

Smith D, Zhang F (1999) Composition for enhancing grain yield and protein yield of legumes grown under environmental conditions that inhibit or delay nodulation thereof, US Patent No. 5.922.316. US Patent and Trademark Office, Washington, DC

Smith RS, Ellis MA, Smith RE (1981) Effect of *Rhizobium japonicum* inoculant rates on soybean nodulation in a tropical soil. Agron J 73:505–508

Soares ALL, Ferreira PAA, Pereira JPAR et al (2006) Selected rhizobia of agronomic efficiency and diversity of native nodulating populations pardons (MG).II – bean. R Bras Ci Solo 30:803–811

Soe KM, Yamakawa T (2013) Low-density coinoculation of Myanmar *Bradyrhizobium yuanmingense* MAS34 and *Streptomyces griseoflavus* P4 to enhance symbiosis and seed yield in soybean varieties. Am J Plant Sci 4(09):1879

Somasegaran P (1991) Inoculant production with emphasis on choice of carriers, methods of production and reliability testing/quality assurance guidelines. In: Thompson JA (ed) Report on the expert consultation on legume inoculant production and quality control. FAO, Rome, pp 87–106

Somasegaran P, Hoben HJ (1994) Handbook for rhizobia. Springer, New York, p 450

Sparrow SD, Ham GE (1983) Survival of *Rhizobium phaseoli* in six carrier materials. Agron J 75:181–184

Strijdom BW, van Rensburg HJ (1981) Effect of steam sterilization and gamma irradiation of peat on quality of *Rhizobium* inoculants. Appl Environ Microbiol 41(6):1344–1347

Suman A, Verma P, Yadav AN et al (2016) Development of hydrogel based bio-inoculant formulations and their impact on plant biometric parameters of wheat (*Triticum aestivum* L.) Int J Curr Microbiol App Sci 5(3):890–901

Tairo EV, Ndakidemi PA (2013) Possible benefits of rhizobial inoculation and phosphorus supplementation on nutrition, growth and economic sustainability in grain legumes. Am J Res Commun 1(12):532–556

Tajini F, Trabelsi M, Drevon J (2011) Co-inoculation with *Glomus intraradices* and *Rhizobium tropici* CIAT899 increases P use efficiency for N_2 fixation in the common bean (*Phaseolus vulgaris* L.) under P deficiency in hydroaeroponic culture. Symbiosis 53:123–129

Talbi C, Argandoña M, Salvador M et al (2013) *Burkholderia phymatum* improves salt tolerance of symbiotic nitrogen fixation in *Phaseolus vulgaris*. Plant Soil 367:673–685

Tank N, Saraf M (2010) Salinity-resistant plant growth promoting rhizobacteria ameliorates sodium chloride stress on tomato plants. J Plant Interact 5:51–58

Temprano FJ, Albareda M, Camacho M et al (2002) Survival of several *Rhizobium/Bradyrhizobium* strains on different inoculant formulations and inoculated seeds. Int Microbiol 5:81–86

Tewari S, Arora NK (2014) Talc based exopolysaccharides formulation enhancing growth and production of *Hellianthus annuus* under saline conditions. Cell Mol Biol 60(5):73–81

Thakare CS, Rasal PH (2000) Evaluation of efficient *Rhizobium* strains for groundnut. Legum Res 23(4):259–261

Thenmozhi P, Dinakar S (2014) Exopolysaccharides (EPS) mediated induction of systemic resistance (ISR) in *Bacillus-Fusarium oxysporum* f. sp. *lycopersici* pathosystem in tomato (var. PKM-1). Int J Curr Microbiol App Sci 3(9):839–846

Thies JE, Singleton PW, Bohlool BB (1991) Modeling symbiotic performance of introduced rhizobia in the field by uses of indices of population size and nitrogen status of the soil. Appl Environ Microbiol 57:29–37

Tian CF, Wang ET, Wu LJ et al (2008) *Rhizobium fabae* sp. nov., a bacterium that nodulates *Vicia faba*. Int J Syst Evol Microbiol 58(12):2871–2875

Tilak KVBR, Subba Rao NS (1978) Carriers for legume (*Rhizobium*) inoculants. Fert News 23:25–28

Trujillo ME, Willems A, Abril A et al (2005) Nodulation of *Lupinus albus* by strains of *Ochrobactrum lupini* sp. nov. Appl Environ Microbiol 71:1318–1327

Van Berkum P, Beyene D, Baa G et al (1998) *Rhizobium mongolense* sp. nov., is one of three rhizobial genotypes identified which nodulate and form nitrogen-fixing symbioses with *Medicago ruthenica* [(L.) Ledebour]. Int J Syst Bacteriol 48:13–22

Van Schreven DA, Harmsen GW, Lindenbergh DT et al (1953) Experiments on the cultivation of *Rhizobium* in liquid media for use on zindderzee polders. Antonie van Leeuwenhoek 19:300–308

Vandergheynst JS, Scher HB, Gou HY et al (2007) Water in oil emulsions that improve the storage and delivery of the biolarvacide *Lagenidium giganteum*. BioControl 52:207–229

Vassileva V, Ignatov G (2002) Relationship between bacteroid poly-β-hydroxybutyrate accumulation and nodule functioning in the *Galega orientalis-Rhizobium galegae* symbiosis under diamine treatment. Physiol Plant 114(1):27–32

Verma JP, Yadav J, Tiwari KN et al (2014) Evaluation of plant growth promoting activities of microbial strains and their effect on growth and yield of chickpea (*Cicer arietinum* L.) in India. Soil Biol Biochem 70:33–37

Waelkens F, Voets T, Vlassak K et al (1995) The *nodS* gene of *Rhizobium tropici* strain CIAT899 is necessary for nodulation on *Phaseolus vulgaris* and on *Leucaena leucocephala*. Mol Plant-Microbe Interact 8:147–154

Wang ET, Martínez-Romero E (2000) *Sesbania herbacea–Rhizobium huautlense* nodulation in flooded soils and comparative characterization of *S. herbacea*-nodulating rhizobia in different environments. Microb Ecol 40(1):25–32

Wani PA, Khan MS, Zaidi A (2007) Effect of metal tolerant plant growth promoting *Bradyrhizobium* sp. (*Vigna*) on growth, symbiosis, seed yield and metal uptake by greengram plants. Chemosphere 70:36–45

Waswa MN (2013) Identifying elite rhizobia for commercial soybean (*Glycine max*) inoculants. Dissertation, Agricultural Technology, University of Nairobi

Woronin MS (1866) U¨ ber die bei der Schwarzerle (*Alnus glutinosa*) und bei der gewo¨hnlichen Gartenlupine (*Lupinus mutabilis*) auftretenden Wurzelanschwellungen. Me'moires de l'Academie Impe'riale des Sciences de St. Pe'tersbourg, VII Series, vol. X

Xu S, Liu Y, Wang J et al (2015) Isolation and potential of *Ochrobactrum* sp. NW-3 to increase the growth of cucumber. Int J Agric Policy Res 3(9):341–350

Zahir ZA, Kashif Shah M, Naveed M et al (2010) Substrate-dependent auxin production by *Rhizobium phaseoli* improves the growth and yield of *Vigna radiata* L. under salt stress conditions. J Microbiol Biotechnol 20(9):1288–1294

Zehr JP, Jenkins BD, Short SM et al (2003) Nitrogenase gene diversity and microbial community structure: a cross-system comparison. Environ Microbiol 5:539–554

Zerpa M, Mayz J, Méndez J (2013) Effects of *Bradyrhizobium japonicum* inoculants on soybean (*Glycine max* (L.) Merr.) growth and nodulation. Ann Biol Res 4(7):193–199

Zhang F, Smith DL (2002) Inter-organismal signalling in suboptimal environments: the legume-rhizobia symbiosis. Adv Agron 76:125–161

Zhao LF, Xu YJ, Ma ZQ et al (2013) Colonization and plant growth promoting characterization of endophytic *Pseudomonas chlororaphis* strain Zong1 isolated from *Sophora alopecuroides* root nodules. Braz J Microbiol 44(2):623–631

Zhu R, Tang F, Liu J et al (2016) Co-inoculation of arbuscular mycorrhizae and nitrogen fixing bacteria enhance alfalfa yield under saline conditions. Pak J Bot 48(2):763–769

Rhizotrophs in Saline Agriculture

5

Faizan Ullah and Asghari Bano

Abstract

The rhizo-deposition provides energy and nutritional inputs to soil with selection of large and distinct community of metabolically active soil microbiota that carries many biochemical transformations. Positive effects of *Rhizobium*, *Pseudomonas*, *Bacillus*, and *Azospirillum* on the mitigation of salt stress in inoculated plants have been documented. However, information is scarce regarding the mode of action of the beneficial microbes in improving salt tolerance to host plants. This chapter deals with the salt tolerance potential of rhizobacteria and their mechanism in planta. It has been shown that cooperative microbial activities can be exploited as a low input biotechnology. Addition of osmoprotectants increases tolerance of the microbes to salt. Difference in the utilization of C/N sources also makes the difference in the salt tolerance of rhizobia. Field experiments should be conducted with plant growth-promoting rhizobacteria (PGPR) isolated from stressed areas. Furthermore, efficiency of growth regulators previously used to ameliorate salt stress should be monitored in combination with PGPR, which may be useful as future strategy to mitigate salt stress for agriculture productivity and environmental sustainability. The mechanism of salt tolerance in PGPR appears similar to that of growth regulators applied exogenously to plants.

5.1 Introduction

Salt stress is an important environmental stress significantly affecting plant growth as well as deteriorating soil health and productivity. The FAO Land and Plant Nutrition Management Service estimated greater than 6 % of the land globally

F. Ullah
Department of Botany, University of Science and Technology, Bannu, KP, Pakistan

A. Bano (✉)
Department of Biosciences, Quaid Campus, University of Wah, Wah Cantt, Pakistan
e-mail: banoasghari@gmail.com

S. Mehnaz (ed.), *Rhizotrophs: Plant Growth Promotion to Bioremediation*, Microorganisms for Sustainability 2, DOI 10.1007/978-981-10-4862-3_5

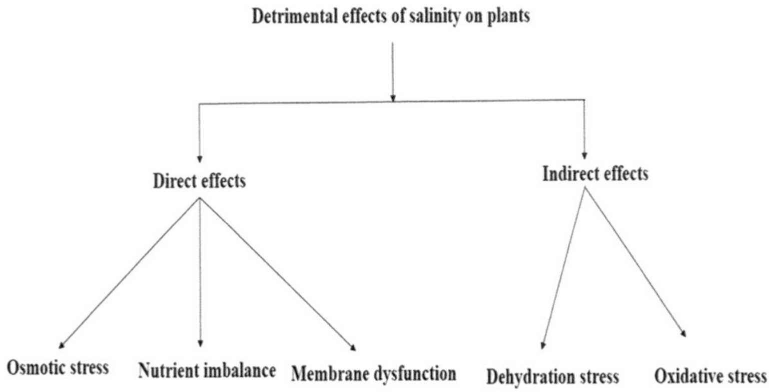

Fig. 5.1 Mechanism of the effects of osmotic stress on higher plants

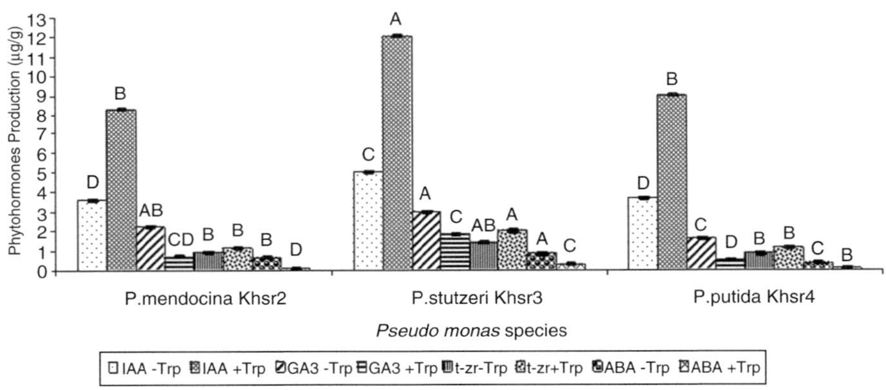

Fig. 5.2 Production of phytohormones (µg/ml) by Pseudomonas sp. in culture media supplemented with tryptophan @ µg/ml and without tryptophan. –Trp = without tryptophan, +Trp = with tryptophan. Adapted from Naz and Bano (2010)

affected by either salinity or sodicity (FAO 2008). The first response of salt stress is osmotic adjustment. Mechanism of the effects of osmotic stress on higher plants is given in Fig. 5.1. Plant survival under salt stress depends on maintaining a positive turgor, which is indispensable for the expansion growth of cell and for the stomatal conductance (Jaleel et al. 2007). The positive role of agrochemicals/growth regulators on the physiology of salt tolerant crops has been documented (Pospisilova et al. 2005; Gurmani et al. 2009).

Thrall et al. (2009) demonstrated that using plant microbe interaction to revegetate and restore the stressed soil is important and that the salt tolerance potential was higher in rhizobial population from saline soil. The genetic profile was also different from those isolated from non-saline area. Chakraborty et al. (2011) reported highly salt-tolerant bacteria from the rhizosphere of *Cynodon dactylon*, a facultative halophyte which can tolerate 10 % NaCl (Fig. 5.2).

The prerequisite for the effective usage of PGPR is the knowledge of how microbe community varies along environmental stress gradient and that how such variations are related to symbiotic effectiveness. Many reviews have been published on these fundamental biological understanding about plant microbe interactions (Muneer et al. 2012; Paul 2012). This article aims to provide salient features about osmotic adaptation in PGPR including *Rhizobium*, *Frankia*, and EM, also emphasizing the role of PGPR in crosstalk between biotic and abiotic stresses of plants, which may be useful as future strategy to mitigate salt stress for agriculture productivity and environmental sustainability.

5.2 Plant Growth Promotion by Rhizobacteria

Agricultural manipulation of free living and symbiotic rhizobacteria has gained considerable attention in modern agriculture (Akbari et al. 2007; Figueiredo et al. 2010). These beneficial bacteria are valuable component of the ecosystem and exhibit significant impact on biogeochemical cycle and induce systemic resistance in plants against pathogens (Walsh et al. 2001) and seedling establishment (Wei et al. 1996). Literature is documented on the use of rhizobacteria for improving growth, yield, and quality of economically important agricultural crops (Hayat et al. 2010; Ahemad and Khan 2011).

The most successful plant bacteria relationships in this context are (i) mutualistic symbiosis involving *Rhizobium*; (ii) associative symbiosis involving diazotrophs, most of them produce phytohormones, e.g., bacteria of genera *Pseudomonas*, *Bacillus*, etc. (Döbereiner and Pedroza 1987); and (iii) symbiotic association of *Frankia* with actinorhizal plants (Bargali 2011). According to their existence and association with plants, PGPR can be grouped into two types: (i) extracellular (ePGPR) existing in the rhizosphere, on the rhizoplane, or living between cells of the root cortex and (ii) intracellular existing inside the root cells as in nodules (iPGPR) (Figueiredo et al. 2010).

5.3 Plant Microbe Interaction Under Saline Condition

Monocotyledon and dicotyledon species can respond in a different way to microbial inoculation due to differences in root architecture, composition of root exudates, and the exudation activity (Munns and Tester 2008). Flavonoids secreted by plants play a key role in plant microbe interactions and can affect the salinity tolerance (Dardanelli et al. 2009). Miransari and Smith (2009) reported positive role of *nod* gene inducer genistein, a flavonoid, in mitigating effect of salinity on soybean-*Bradyrhizobium japonicum* symbiosis. Since plant age and physiological conditions determine the amount and type of signal molecules (Eckardt 2006), the salt-induced alteration of the plant metabolism may affect the type and amount of signal molecules and thus the molecular crosstalk between the two symbionts.

5.4 Salt Tolerance Potential of Rhizobacteria

Rhizobacteria growing in saline soil survive and tolerate salinity more efficiently than higher plants. The threshold level of salt tolerance in most plants is 40 mM NaCl though survival of barley can occur at 170 mM NaCl (Leonova et al. 2005). The salt tolerance potential of microbial isolates is significantly higher from salt range as compared to that of non-saline soil. The presence of osmoprotectant further enhances their salt tolerance potential. Hua et al. (1982) isolated *Rhizobium* spp. strain WR 1001 from Sonoran desserts, which could tolerate up to 500 mM NaCl, a concentration approaching the concentration of seawater. Hartmann and Zimmer (1994) reported that *Azospirillum* sp. can survive in seawater and can be associated with mangrove roots. *A. brasilense* (N040) isolated from hypersaline soil was found to tolerate up to 1,800 mM NaCl when grown in the basal medium (Omar et al. 2008). *A. brasilense* NH from saline soil could tolerate 300 m.mol/L NaCl in the absence of osmoprotectant, and the tolerance was doubled with the osmoprotectant (Chowdhury et al. 2007). Gontia-Mishra and Sharma (2012) isolated eight rhizobacterial species from rhizosphere of a halophyte *Salicornia brachiata*. These microbes were subjected to salt stress in minimal medium M9 to determine their osmotolerance properties. These rhizobacteria were capable of tolerating NaCl up to 0.714 M NaCl. The *Zhihengliuella* sp. and *Brachybacterium* sp. were reported to produce highest amount of proline as osmoprotective substance under salinity stress.

Bacillus subtilis SU47 and *Arthrobacter* sp. SU18 tolerated up to 8 % NaCl (Upadhyay et al. 2011). The *Frankia* spp. tolerated 1.5 mg NaCl per g soil with 60 % decrease in N fixation when inoculated to *Casuarina obesa* Miq. in field (Reddell et al. 1986). Tani and Sasakawa (2000) observed that growth of the strain of *Frankia* isolated from the root nodules of *E. macrophylla* was inhibited by 100 mM NaCl. However, the Na^+ in the cells was maintained below 20 mM. On return to salt-free medium, their growth was recovered and they were capable of multiplication. Gomaa et al. (2008) isolated 16 *Frankia* isolates, and their NaCl tolerance level was checked over a range of NaCl concentrations (2.0, 4.0, 6.0, 6.5, 7.0, 7.5, 8.0, 8.5, 9.0, 10.0, and 10.5 %). They observed that two isolates (K 03 and K 05) were highly tolerant to salt stress and tolerated NaCl up to 10 %. The isolates K 01, K 02, I 04, and I 05 tolerated NaCl up to 9.5 %.The tolerance of different species of *Rhizobium* to NaCl can range from 100 to 650 mM (Bernard et al. 1986). Tree rhizobia are relatively more salt tolerant than the rhizobia associated with annual grain legumes. Both *Prosopis* and *Acacia* spp. tree legumes are highly tolerant to salinity while grain legumes are either moderately tolerant or sensitive to salinity (Bouhmouch et al. 2005), perhaps due to better competition with soil microorganisms and greater exudation of phenolics and alkaloids from tree roots. El-Sheikh and Wood (1990) demonstrated that fast-growing rhizobia were more salt tolerant than slow growers which may be due to the differences in the use of C/N source. Role of the host plant in determining tolerance to salt appears less important with *Rhizobium* legume symbiosis than in the association with free-living bacteria.

5.5 Application of Rhizobacteria in Salt Tolerance of Crops

Leptochloa fusca, commonly known as Kallar grass, is a C_4 perennial halophytic forage plant highly tolerant to salt, and thus it can grow in coastal salt marsh. The roots of this plant harbor *Azospirillum halopreferans* (Reinhold-Hurek et al. 1987), a salt-tolerant PGPR, as an endophyte. Due to the high salt tolerance ability of the associated microsymbiont, the Kallar grass has been used to reclamate a wide range of unproductive land. In addition, this plant can be associated with N-fixing bacteria Cuealine and/or P-solubilizing bacteria (Phosphorein), and these bacteria also stimulated growth of the Kallar grass at different levels of salinity (Tawfik et al. 2006).

Plant growth-promoting rhizobacteria can contribute to the reforestation process of mangrove, which are important in coastal ecosystem and provide ecological niche for various economically and ecologically important marine species. Bashan and Holguin (2002) reported that mangrove rhizosphere bacteria can be inoculated to maintain arid tropical mangrove ecosystems.

Inoculation of oil seed halophyte *Salicornia bigelovii* with eight species of halotolerant bacteria from seawater significantly stimulated plant growth (Bashan et al. 2000). Inoculation of barley cvv. Giza 123 and Giza 2000 with *A. brasilense* (N040) mitigated the adverse effect of salinity on photosynthetic pigment, photosynthetic activity, stomatal conductance, transpiration rate, and accumulation of osmoregulant and proline. Noteworthy, the oxidative stress was mitigated as evidenced by the decreased activities of the antioxidant enzymes, viz., SOD, POD, and catalase. Aly et al. (2003) observed that inoculation of maize plants with *Azotobacter chroococcum* and *Streptomyces niveus* increased the content of total free amino acids, proline, total soluble sugars, total soluble proteins, DNA, and RNA in shoots and roots of the maize plants. Naz et al. (2009) reported that *Azotobacter vinelandii* Khsri isolated from roots of *Chrysopogon aucheri* growing in Khewra salt range augmented proline content in the shoots and roots of maize under salt stress. The isolates were capable of producing IAA, GA, t-Zr, and ABA in the culture medium. Production of grain legumes is particularly vulnerable because of their low tolerance to salinity. According to Nabizadeh et al. (2011), salinity affects the morphogenesis of nodule as well as the relative nitrogen fixation, which is affected more than plant growth and root nodulation. At later stages of legume-*Rhizobium* symbiosis, when the nodule is formed, the allocation of C to root nodules is not only reduced but also altered. Salt stress decreases the permeability of the nodule which results in the contraction of nodule inner-cortex cells and increases the levels of abscisic acid in the nodule (Irekti and Drevon 2003). Rao et al. (2002) reported that salinity and sodicity severely reduce the formation and functioning of root nodules in *Rhizobium*-legume symbiosis, resulting in significant reduction in yield. Ogutcu et al. (2010) reported that chickpea rhizobia exhibited diversity in their salt tolerance potential. It is further evident from their results that the rhizobial isolates from wild chickpea were more salt tolerant and the ability of chickpea to grow and survive in saline conditions improved following inoculation.

The microbial inoculant can also increase tolerance of seedlings at early phases of growth under salinity stress. Inoculation with *Azospirillum* imparts higher percent germination and better establishment of seedling (Barassi et al. 2006).

Creus et al. (1997) demonstrated that *Azospirillum brasilense* sp. 245 improved water status of wheat seedling under salt stress, stimulated coleoptile growth, produced large leaves, increased stomatal conductance of leaves, and increased hydraulic conductivity of roots (Okon 1985; Sarig et al. 1992). The PGPR can stimulate root proliferation with increase in root biomass, which may provide tolerance to plant (Mayak et al. 2004). Different response of plant organ to salt has been reported in nodulated beans (Verdoy et al. 2004; Zhang et al. 2008).

Mixtures of bacterial species were more effective than single bacterial species in stimulating the growth of halotolerant plants. Bashan et al. (2000) demonstrated that inoculation with a mixture, composed of *A. halopraeferens*, two *A. brasilense* strains, *Vibrio aestuarianus* strains, and *Vibrio proteolyticus* or a mixture of *Bacillus licheniformis* and *Phyllobacterium* spp. significantly improved plant growth and nutrient content of leaves. Inoculation of *Vicia faba* plants and subsequently exposed to five NaCl levels (0–6 dsm^{-1}) with Rabie and Almadivi (2005) reported better salt tolerance potential of *Azospirillum brasilense* when combined with arbuscular mycorrhizal fungus *Glomus clarum*.

5.6 Mechanism of Osmotic Tolerance in Rhizobacteria

The cellular adaptation to osmotic stress is important in order to determine the growth and survival efficiency of rhizobacteria in the saline ecosystems. The osmotic stress tolerance of rhizobacteria can be explained at molecular, biochemical, physiological, and morphological levels.

5.6.1 Molecular Adaptation (Proteomics and Transcriptomics)

The bacteria respond to salt stress by the induction of chaperons, chaperon-like proteins, and peptidases to rectify the errors produced by high salinity during the synthesis of protein and DNA (Hecker and Völker 2001). In these perspectives, many membrane proteins, like those involved in the transport of compatible solutes and ions, play a fundamental role in the adaptation of bacteria to salt stress (Höper et al. 2006). Modification of membrane proteins under salt shock has been reported in bacteria. The ompF and ompC are the membrane proteins that determine the permeability of the outer membranes. The total cellular levels of OmpF and OmpC usually remain constant under normal conditions, but the relative proportions of these two protein types vary. The conditions favoring the synthesis of OmpF repress OmpC and vice versa. Media of low osmolarity enhance the levels of OmpF with a subsequent reduction in the levels of OmpC. In contrast, media of high osmolarity reduce the levels of OmpF and increase the levels of OmpC (Csonka 1989). Talibart et al. (1994) demonstrated that salinity induced periplasmic glycine betaine-binding

Table 5.1 Profile of salt stress proteins in the cell membrane of *Pseudomonas pseudoalcaligenes* MSP-538

Character of salt stress protein	Molecular weight (kDa)
Newly induced	98
Repressed	41
Overexpressed	42, 43, 44, 92, 94, 96

Adapted from Paul et al. (2005)

protein of 32 kDa size, which was involved in the uptake of glycine betaine by *A. brasilense* SP7. Salt-induced changes in the protein profiles occurred in the halophytic strains of rhizobia (Zahran et al. 1994). Paul et al. (2005) found one newly induced protein (98 kDa) and one repressed protein (41 kDa) in the cell membrane of *Pseudomonas pseudoalcaligenes* MSP-538 under salt stress. Moreover, six different new proteins were overexpressed (Table 5.1).

Twenty-two different proteins were regulated (up or down) in *P. fluorescens* MSP-393 upon the imposition of salt stress. By using peptide mass fingerprinting and bioinformatics tools, function was assigned to 13 induced proteins and 2 repressed proteins. All of the proteins identified had molecular weights and isoelectric points between 12 and 77 kDa and between 4 and 7 pI, respectively. The induced proteins 26i and 42i were associated with membrane. The proteins 41i, 39i, and 71i were induced under salt shock. Moreover, a 60 kDa chaperon was identified that was upregulated in MSP-393 in response to salt shock (Paul et al. 2006).

Zhang et al. (2008) reported that *Bacillus subtilis* GB03 downregulates the expression of the sodium transporter gene HKTI in roots resulting in lower accumulation of Na^+ in the plant. Similarly, tagging of *Azospirillum lipoferum* with green fluorescence protein gene (*gfp*), i.e., *Azospirillum lipoferum* J A4::ngf15, minimized the adverse effects of high concentration of NaCl (Bacilio et al. 2004).

5.6.2 Morphological, Physiological, and Biochemical Adaptation

Microorganisms have developed various biochemical strategies to maintain structural and functional stability of the cells (Westover et al. 1997). About 86 % of the bacteria isolated from the rhizosphere of various plants produce secondary metabolites and enzymes. He et al. (2010) studied the *Desulfovibrio vulgaris* Hildenborough, adaptation to salt stress by taking into account the physiological, global transcriptional, and metabolite analyses. Decrease in carbon metabolism was observed under salt stress which was mainly due to downregulation of carbon utilization genes and putative carbon starvation genes. The expression of several genes such as *porA*, *porB*, *oorB*, and DVU3349-encoding pyruvate/ferredoxin oxidoreductase was significantly decreased under salt adaptation. Moreover, the genes *pta* and *ackA* were significantly down-expressed. The downregulation of genes involved in C-utilization and C-starvation occurred resulting in decreased C-metabolism under salt stress. Downregulation also occurred in genes encoding C-proteins in plant. Alloing et al. (2006) further reported that bacteria also have the capacity to use compatible solutes, e.g., proline and glycine betaine, as C and N sources in addition to osmoprotection. Proline also functions as molecular chaperons with ability to protect protein integrity and enhances the activities of different enzymes (Szabados and Savoure 2009).

Arcobacter sp. produces lipase, nitrate reductase, and collagenase under salt stress. The C/N source utilization pattern of this bacterium also differs (Egamberdiyeva and Islam 2008).

The cell membrane lipids are also altered with increasing levels of salinity. To cope with these higher levels of salinity, bacteria adjust the composition of their membrane lipids either by modifying the types of existing fatty acids or by changing the pre-existing phospholipids (Zhang and Rock 2009). When salt-tolerant or moderately halophilic Gram-positive bacteria are grown in the media having high salt concentrations, among other adaptive responses to salt stress, changes in composition of membrane lipids are also common. For example, in case of *Planococcus* representative, the number of short-chain fatty acids increased and alteration in the unsaturation levels of lipid chains were common which were closely related with salt tolerance of this microbe (Miller 1985).The induction of higher levels of cyclopropane fatty acid (ΔC19:0) and lower levels of oleic acid (C18:1) were observed in the lipid membrane of *L. lactis*, under salt stress (Guillot et al. 2000).

To maintain their cytoplasmic osmolarity, rhizobacteria produce various osmoprotectants such as trehalose, proline, glutamate, and glycine betaine (Bremer and Krämer 2000). The production of reducing sugars was common among PGPR isolates from the rhizosphere of wheat (Upadhyay et al. 2012). Evidences suggest that mechanism of osmoregulation of plants is similar in nitrogen-fixing bacteria and PGPR (Han and Lee 2005). The type and the amount of the osmolytes produced by rhizobacteria depend on the osmotic stress (Moat et al. 2002). Under hyperosmotic conditions when the osmolality of the surrounding environment increases, both turgor pressure of microbial cell and microbial growth decrease, the macromolecular biosynthesis is inhibited, and respiration rates get decreased. The *E. coli* produces higher K^+ ion influx with the concomitant decrease of the intracellular putrescine content (Moat et al. 2002). The divalent putrescine cation is replaced by the monovalent K^+ ion, and thus the cytoplasmic osmolality increases with minimal effect on the intracellular ionic strength. Osmoregulation is also achieved by glutamate, an osmolyte whose synthesis depends on the osmotic stress and the uptake of K^+.

Many Gram-negative bacteria maintain their osmotic balance by accumulating oligosaccharides, such as trehalose, amino acids, sugars, and sugar alcohol, which contain 6–12 glucose residues linked together via β(1–2) and β (1–6) linkages (El-Sheikh and Wood 1990). Table 5.2 shows a variety of B vitamins and amino acids produced by salt-tolerant bacterial strains under salt stress (Egamberdiyeva and Islam 2008).

The physiological mechanism underlying salt tolerance of *Azospirillum* is determined by the duration of salt stress; short-term salt stress can accumulate intracellular concentration of glutamate, whereas the 48-h stress causes an influx of K^+ (Rivarola et al. 1998). The accumulation of K^+ ions (Yelton et al. 1983) and organic compatible solutes such as amino acids, sugars, and betaines (Le Rudulier et al. 1984) have been reported.

Table 5.2 Production of B vitamins and amino acids by salt-tolerant bacterial strains

Bacterial isolates	B1	B3	B6	PP	Biotin	Norvaline	Histidine	Leicin	Valine	Glutamine
*P. alcaligenes*PsA15	+	–	–	+	–	+	+	–	+	+
P. denitrificans PsD6	+	+	–	–	+	–	–	–	+	–
P. mendocina Ps M13	+	–	+	–	–	–	–	–	+	–
B. laevolacticus Bc L28	+	+	–	–	–	–	+	–	–	–
B. megaterium BcM48	–	–	+	+	+	–	+	–	–	+
B. polymyxa BcP26	–	+	–	–	+	–	–	–	+	–
M. phlei MbP18	+	+	+	–	+	–	+	+	–	+
A. globiformis ArG1	+	–	–	–	+	+	–	–	+	–
A. simplex ArS50	+	+	–	+	+	+	+	–	–	–
A. tumescens ArT16	+	+	–	+	+	+	+	–	–	–

+ and – indicate the presence or absence of activity by a particular microorganism
Adapted from Egamberdieva and Islam (2008)

Putrescine, spermine, spermidine, and cadaverine were observed in the supernatants of *Azospirillum brasilense* strains upon exposure to salt stress (Perrig et al. 2007; Cassan et al. 2009).

Sharan et al. (2007) reported that improved tolerance of *Xanthomonas campestris* to salinity could be due to accumulation or secretion of xanthan. Llmas et al. (2006) reported synthesis of an exopolysaccharide, mauran, by the diazotrophic bacteria *Halomonas maura* under salt stress conditions.

Growth experiment indicated that betaine and its derivatives could be used as nitrogen and carbon sources for bacterial growth (Smith et al. 1988; Boncompagni et al. 1999). However, Bernard et al. (1986) reported that 15 *Rhizobium* species used glycine betaine for growth under low osmolarity but this compound did not restore growth rate of cells under medium of high osmolarity. This was further confirmed by Brhada et al. (2001) and Rudulier and Bernard (2006) regarding short-term accumulation of glycine betaine in *Rhizobium,* as an adaptive mechanism to salt stress.

Paul and Nair (2008) demonstrated that osmotolerance of *Pseudomonas fluorescence* MSP-393 was mediated by the synthesis of alanine, glycine, glutamic acid, serine, threonine, and aspartate in their cytosol and upregulation of salt stress proteins which imparted tolerance to salt stress.

Ecotine is produced by many halophilic (marine) bacteria and, unlike glycine betaine, does not accumulate intercellular and does not repress the synthesis of endogenous compatible solutes (Talibart et al. 1994).

Under hyposmotic shock or low osmolality conditions, the increase in turgor occurred due to influx of water into the cell, resulting in stretching of the cell envelop and possible cracks in the membrane; this can cause a transient increase in the permeability resulting in the leakage of solutes including amino acids and nucleotides (Moat et al. 2002).

Paul et al. (2005) reported that *Pseudomonas pseudoalcaligenes* MSP-538 synthesizes Ala, Gly, Glu, Ser, Thr, and Asp as osmolytes in response to salt stress. *Bacillus subtilis* produces glycine betaine (Lucht and Bremer 1994) which helps in lowering the water potential outside the cells.

The osmotic tolerance is also achieved in rhizobacteria with secretion of disaccharides such as sucrose, maltose, cellobiose, gentiobiose, turanose, and palatinose (Gouffi et al. 1999). *Frankia* Ema1 isolates tolerated salinity by preventing Na^+ influx into the cells which was mainly due to their short and thick hyphae (Tani and Sasakawa 2000).

5.6.3 Selectivity of Ion Uptake and Microbial Induced Salt Tolerance

One of the major effects of salinity is the ionic imbalance created due to Na^+ and Cl^- ions. The bacteria maintain the ion homeostasis but the efficiency and mechanism of ion sensitivity differ among the bacterial species and the sensitivity of the host plant. The inoculation with *Serratia proteamaculans* in soybean decreased the uptake of K^+ and Na^+ ions (Han and Lee 2005) while *Azospirillum* decreases the Na^+ uptake but increases K^+ and Ca^{2+} in salt-sensitive variety of maize cv. 323 (Hamdia et al. 2004).

The Na^+ ion inhibits the uptake and transport of Pi and Ca^{2+}; the Pi is required for ATP synthesis of the plant and Ca^{2+} acts as secondary messenger in signal transduction. Fatima (2007) reported significant increase in Na^+ accumulation in both the salt-sensitive and salt-tolerant cultivar of maize. The K^+/Na^+ ratio was found maximum in the salt-sensitive cultivar. The total ion contents (P+Ca+K) were greater in co-inoculation treatment under salt stress while the Na^+ uptake was decreased. Similar observations were made by Yue et al. (2007) in *Klebsiella oxytoca*. Similarly, inoculation with *Pseudomonas* sp. reduced the toxic ion uptake in wheat with a subsequent improvement in plant growth (Egamberdieva and Zulfiya 2009). High K^+/Na^+ ratio is considered an important indicator for salt tolerance of plants (Hamdia et al. 2004). In these perspectives, Shukla et al. (2012) found that rhizobacteria *Brachybacterium saurashtrense* (JG-06), *Brevibacterium casei* (JG-08), and *Haererohalobacter* (JG-11), significantly improved the K^+/Na^+ ratio and contents of Ca, P, and N in the leaves of *Arachis hypogaea* under salt stress. The decrease in Na^+ uptake resulted in better plant growth under salt stress. The selectivity of K over Na ions was recorded in several bacterial species, viz., *Pseudomonas putida*, *Enterobacter cloacae*, *Serratia ficaria*, and *Pseudomonas fluorescens*, when used as bioinoculant on wheat (Nadeem et al. 2013); which is mediated via the production of exo-polysaccharide (EPS) which binds Na^+, and thereby the availability of Na^+ to plant is decreased (Qurashi and Sabri 2012).

5.7 Antioxidant Production and Salt Tolerance

High concentration of salt normally impairs the cellular electron transport with the production of reactive oxygen species (Misra and Gupta 2006) which can trigger phytotoxic reactions and oxidative damage of macromolecules (Elkahoui et al. 2005). A correlation between these enzyme activities and salt tolerance has been reported (Hernandez et al. 2000; Mittova et al. 2003).

Certain strains of PGPR such as *S. proteamaculans* and *R. leguminosarum*, produce various antioxidant enzymes like SOD, POD, and CAT and non-enzymatic antioxidants such as ascorbic acid, glutathione, and tocopherol (Maheshwari 2012). *E. coli* produces cytoplasmic Mn-SOD (Sod A) and Fe-SOD (Sod B) that protect DNA and protein from oxidation. Mutants deficient in genes encoding these enzymes show reduced growth, enzyme inactivation, and DNA damage (Moat et al. 2002).

Han and Lee (2005) reported that inoculation of lettuce with *Serratia* spp. and *Rhizobium* spp. in saline soils ameliorated the inhibitory effects of salinity on antioxidant enzymes, such as ascorbate peroxidase (APX) and glutathione reductase (GR). Similarly, mechanism of salt tolerance was induced by *Piriformospora indica*, a root endophyte of barley. Scavenging of ROS prevents lipid peroxidation and fatty acid desaturation in leaves of sensitive barley cultivar (Baltruschat et al. 2008). Inoculation of *Lactuca sativa* L. cv. Tafalla with *Pseudomonas mendocina* Palleroni alone and in combination with an AM fungus, *Glomus intraradices* or *Glomus mosseae*, increased the activities of antioxidant enzymes such as peroxidase, catalase, and phosphatase and stimulated growth and mineral nutrient content of the plant (Kohler et al. 2009). Favorable effects of *Rhizobium* and *Pseudomonas* spp. co-inoculation on the activities of superoxide dismutase (SOD) and peroxidase (POD) of maize leaves have been reported by Bano and Fatima (2009). Gururani et al. (2012) revealed that the *Bacillus* isolates showed enhanced activity of ACC deaminase, phosphate solubilization, and siderophore production in potato and also enhanced mRNA expression levels of various ROS scavenging enzymes, higher proline production, and increased PSII photochemistry of potato plants.

5.8 Phosphate Solubilization and Salt Tolerance

Soils are usually P-deficient not only because free P concentration even in fertile soil is generally not higher than 10 μM but also because inorganic phosphate is precipitated in alkaline soil. Soil salinity significantly reduces plant uptake of mineral nutrients, especially phosphorus, because phosphate ions can be precipitated by Ca^{2+} ions (Grattan and Grieve 1999). Microorganisms can increase the P availability to plant by solubilizing precipitated phosphates (Goldstein 1986; Gyaneshwar et al. 2002). The phosphate compounds solubilized by the PGPR include tricalcium phosphate, dicalcium phosphate, hydroxyapatite, and rock phosphate (Rodriguez and Fraga 1999; Khan et al. 2009). The mechanism of P solubilization involves the release of chelates and organic acids (Whitelaw 2000; Richardson 2001; Vessey et al. 2004; Vikram et al. 2007); lowering of pH in the rhizosphere and helping in dissociation of bound forms of phosphate, release of phosphatase enzymes by plants and microbes, and proton extrusion during NH_4^+ assimilation and release of CO_2 during respiration which is converted to carbonic acid subsequently decreasing pH which help in solubilization. The phosphate solubilization ability has been documented for different PGPR genera such as *Piriformospora*, *Bacillus*, *Rhodococcus*, *Arthrobacter*, *Serratia*, *Chryseobacterium*, *Delftia*, *Gordonia*, *Phyllobacterium*, *Azotobacter*, *Xanthomonas*, *Pseudomonas*, and *Azospirillum* (Waller et al. 2005; Ivanova et al. 2006; Sharan et al. 2007; Rajabzadeh 2009).

Nautiyal et al. (2000) reported solubilization of phosphate in the presence of 10 % NaCl but solubilization activity decreases gradually with the increasing concentration of NaCl; possibly the higher concentration of Cl^- ions may chelate or neutralize proton ions or acid produced in the medium.

5.9 Interaction of PSB with Other Microorganisms

Beneficial association between plants and PSB is synergistic in nature because bacteria gives phosphate in soluble form, and plants supply carbon compounds which can be metabolized for bacterial growth (Khan et al. 2007). The PSB together with other beneficial rhizospheric microbes helps to improve production of crops. Co-inoculation of *Rhizobium* with PSB (Perveen et al. 2002) or arbuscular mycorrhizal fungi (Zaidi and Khan 2006) performed better than that of single application of any of the microbes.

5.10 Microbially Produced Phytohormones and Salt Tolerance

The PGPR induced salt tolerance involve: a) greater production of the hormone, abscisic acid (Kolb and Martin 1997; Cohen et al. 2007), b) increased level of jasmonic acid, a hormone involved in inducing systematic resistance in plants (Forchetti et al. 2007), c) enhancing the activity of ACC deaminase, which catalyzes the degradation of ethylene precursor, 1-aminocyclopropane-1-carboxylic acid (ACC), d) other bioactive metabolites produced by the PGPR, viz., polyamines, proline, glutamate, and the volatile organic compound (VOCs), can play significant role in the plant microbe interaction under stress (Ryu et al. 2004). Plant microbe interaction involves the exchange of signal molecules between plants and microorganisms. Ribaudo et al. (2006) reported that *Azospirillum* inoculation involves signaling with ethylene as a central positive regulator.

Cohen et al. (2007) reported the production of ABA by *Azospirillum* and *Pseudomonas* in culture media and also when these are used as bioinoculant on plants. Abscisic acid can be synthesized by several spp. of *Azospirillum* and *Rhizobium* (in chemically defined culture media) and, when these bacteria are used as inoculants, can increase the endogenous level of ABA in plant. Naz et al. (2009) reported that PGPR isolated from salt range produced higher quantities of ABA in culture media and also the isolates significantly improved the growth of soybean under salt stress. Root inoculation of wheat seedling cvs. Bhakkar 2000 and Inqalab 91 with *Rhizobium* sp. at 10^6 cells/ml/plant significantly increased ABA level of inoculated plants.

The production of other phytohormones like IAA, GAs, and cytokinins is widespread among free soil and plant-associated bacteria playing important roles in plant growth promotion, symbiotic associations, and pathogenesis (Baca and Elmerich 2003). *Rhizobium* inoculation can also increase IAA and GA concentrations of the inoculated plant under salt stress (Hadees 2009). Naz and Bano (2010) isolated rhizobacteria from the roots of halophytic weeds, viz., *Lactuca dissecta* D. Don, *Solanum surattense* Burm. f, and *Sonchus arvensis* L., collected from Khewra salt range. The rhizobacterial isolates *Pseudomonas stutzeri* Khsr3, compared to *Pseudomonas mendocina* Khsr2 and *Pseudomonas putida* Khsr4, proliferated the roots of inoculated wheat. These bacterial spp. also produced phytohormones IAA, GA, tzr, and ABA, and IAA has been reported (Vacheron et al. 2013) to augment root growth and facilitate nutrient uptake. Sachdev et al. (2009) and Egamberdieva (2009) reported IAA production under saline conditions by *Klebsiella* strains and *Pseudomonas putida* (W2) and *P. fluorescens* (W17) which was positively correlated with higher root growth. Nabti et al. (2010) further postulated that IAA produced by *A. brasilense* isolate NH under salt stress has significant contribution in the salt tolerance of inoculated wheat plants.

It is quite evident that under salt stress the plant growth-promoting phytohormones get conjugated (in storage form) and very little is available in free active form; on the contrary, the level of growth-inhibiting hormone particularly ABA increases many fold, redistribution release from conjugated form and de novo synthesis of ABA occurs under salt stress. The altered ratio of promoter to inhibitor in plants under salinity further modulation by PGPR helps to cope with salinity in much better way (Fig. 5.3).

The salicylic acid (SA) is a phytohormone in regulating plant growth and development and plays an active role in inducing systemic resistance in plants. Recently, SA has been reported to ameliorate adverse effect of abiotic stresses in plants. Mabood and Smith (2007) demonstrated involvement of SA during early stages of *Rhizobium*-legume symbiosis with inhibition of bacterial growth and the production of nod factor by rhizobia. Naz (2009) reported that the combined application of each *Azospirillum* and *Pseudomonas* with SA was more effective to combat salt stress in sunflower.

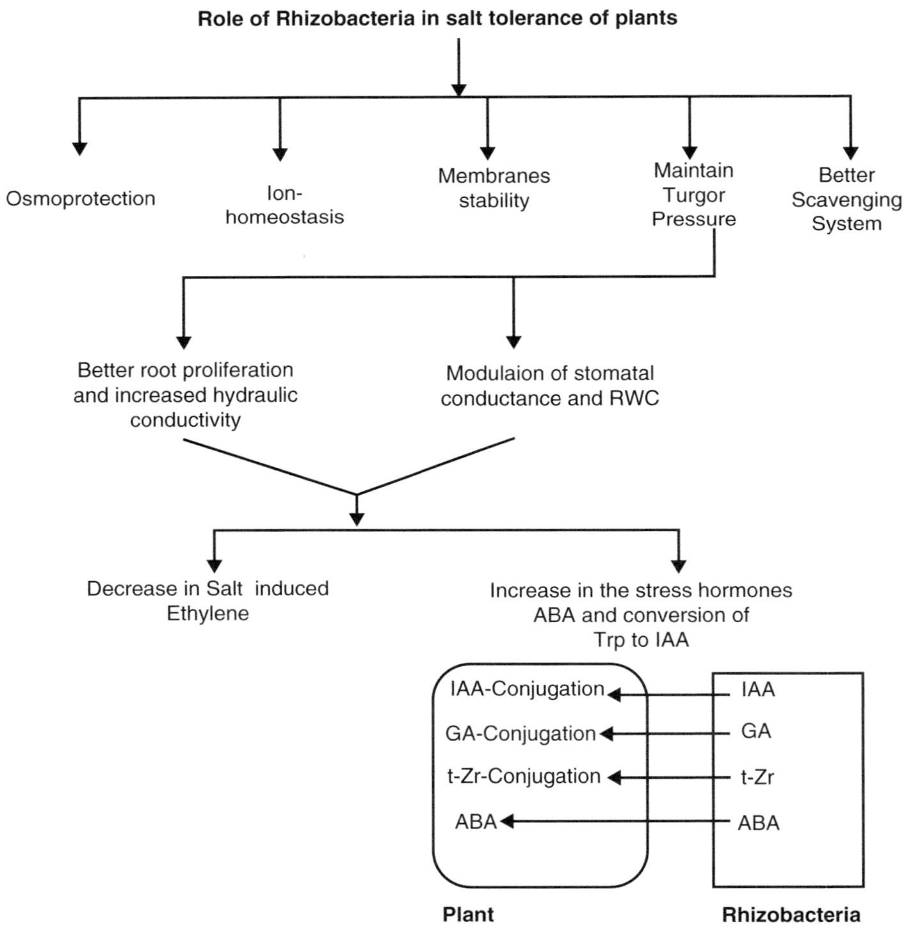

Fig. 5.3 Schematic representation of mechanism of action of rhizobacteria under salinity stress

Literature is documented on the bacterial production of enzyme 1-aminocyclopr opane-1-carboxylate (ACC) deaminase, which can cleave ACC (the immediate precursor of ethylene in plants) to ammonia and α-ketobutyrate (Fig. 5.4) and thus ameliorating the adverse effects of ethylene on plants (Glick 2005; Dimkpa et al. 2009). The ACC deaminase structural gene (*acdS*) has been reported in various rhizobacterial species like *Azospirillum* and various strains of *Burkholderia* (Blaha et al. 2006).

Enterobacter cloacae UW4 and CAL2 and *Pseudomonas putida* GR12-2 synthesize the enzyme ACC deaminase for the degradation of ACC, as their nitrogen source (Grichko and Glick 2001). Nadeem et al. (2007) was able to partially alleviate salt-induced inhibition in root and shoot length and yield of maize by using ACC deaminase containing rhizobacteria *Pseudomonas syringae*, S14 *Enterobacter aerogenes*, and S20 *Pseudomonas fluorescens*. Bal et al. (2013) have reported that inoculation with ACC deaminase producing rhizobacteria significantly decreased the ethylene production and improved the growth of rice under saline conditions.

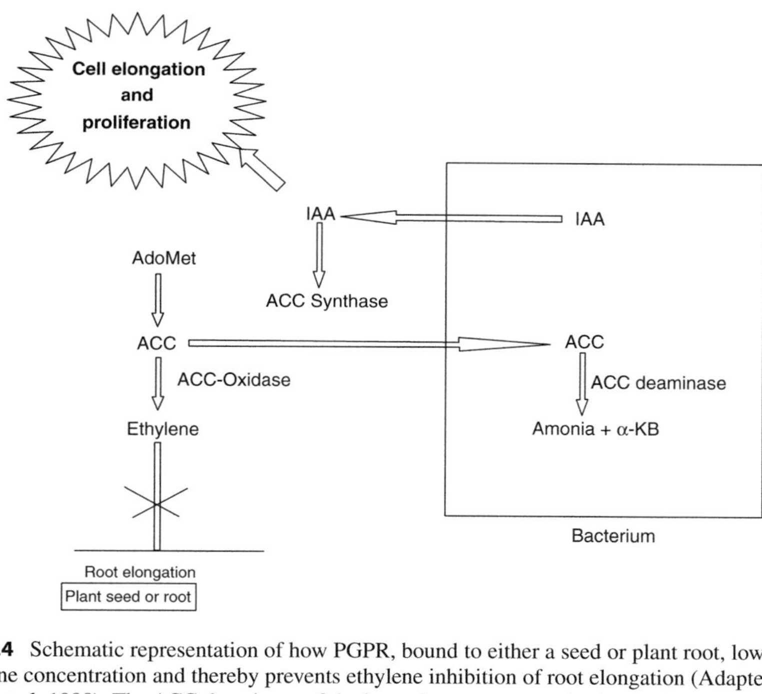

Fig. 5.4 Schematic representation of how PGPR, bound to either a seed or plant root, lowers the ethylene concentration and thereby prevents ethylene inhibition of root elongation (Adapted from Glick et al. 1998). The ACC deaminase of the bacterium causes a reduction in the ethylene levels in plants by degrading ACC to ammonia and α-ketobutyrate

5.11 Role of AMF in Plants Under Salinity Stress

Arbuscular mycorrhizal fungi have some special properties which make them useful to the host plant under various types of stresses. AM fungi colonize the plant roots resulting in the formation of structures known as arbuscules and vesicles which can significantly improve the ability of the root to absorb nutrients and water (Rillig and Mummey 2006). AM fungi can increase plant growth under salt stress, especially in soils having low P (Mohammad et al. 2003), and are able to increase tolerance in plants under salt stress by changing physiology of plant and increasing uptake of water and nutrients (Giri and Mukerji 2004; Miransari et al. 2008; Daei et al. 2009; Miransari 2011). Mycorrhizal inoculation reduces absorption of Na^+ and Cl^- and inhibit their transport toward the shoots (Scheloske et al. 2004) resulting in an increase in dry weight of cotton by 68 % under the salinity of 3 g/kg (Tian et al. 2004).

Arbuscular mycorrhiza symbiosis which is a natural association between the roots of higher plants and arbuscular mycorrhizal fungi (AMF) are known to improve water relations, host plant growth, and acquisition of nutrients especially P from soil (Maronek et al. 1981). Arbuscular mycorrhizal (AM) fungi can contribute to growth of plant under stress (Deressa and Schenk 2008). Fungi form symbiotic association with many species of plant (Smith and Read 2008) which ends up in the translocation of carbon and nutrients from the plant to the fungi and vice versa (Deressa and Schenk 2008). Mycorrhizal association mainly shows its effects on stomatal regulation rather than on resistance of root (Levy and Krikun 1980). In higher plants, metabolism of ROS such as hydrogen peroxide, superoxide, and hydroxyl radicals is in dynamic balance under normal conditions. Introduction of antioxidant enzymes in lettuce increases the efficiency of a PGPR more than sole use of arbuscular mycorrhizal (AM) fungi with respect to improve tolerance to high salt stress (Kohler et al. 2009). Evelin et al. (2009) studied the role of arbuscular mycorrhizal fungi in salt tolerance of plants. They reviewed biochemical changes such as accumulation of carbohydrates, amino acid proline, antioxidants, polyamines, and nutrient acquisition (Ca, P, Mg, N) and maintenance of the K^+: Na^+ ratio. They also investigated physiological changes (photosynthetic efficiency, membrane permeability, accumulation of ABA, and process of nitrogen fixation), molecular changes (the expression of genes: *PIP*, Na^+/H^+ antiporters, *Lsnced*, *Lslea*, and *LsP5CS*) and ultrastructural changes.

5.12 Future Perspective

The strategy adopted for improving plant productivity through the use of naturally occurring beneficial microbes should include selection of good-quality inoculants, awareness among the end user – the farmers – about inoculation technology, the persistence of inocula in the soil, the effective inoculant delivery systems, and formulation of the policy to exploit biofertilizers successfully.

References

Ahemad M, Khan MS (2011) Plant growth promoting fungicide-tolerant *Rhizobium* improves growth and symbiotic characteristics of lentil (*Lens esculentus*) in fungicide-applied soil. J Plant Growth Regul http://dx.doi.org/10.1007/s00344-011-9195-y

Akbari GA, Arab SM, Alikhani HA et al (2007) Isolation and selection of indigenous *Azospirillum* spp. and the IAA of superior strains effects on wheat roots. W J Agri Sci 3:523–529

Alloing G, Travers I, Sagot B et al (2006) Proline betaine uptake in *Sinorhizobium meliloti*: characterization of Prb, an Opp-like ABC transporter regulated by both proline betaine and salinity stress. J Bacteriol 188:6308–6317

Aly MM, El-Sabbagh S, El-Shouny W et al (2003) Physiological response of *Zea mays* to NaCl stress with respect to *Azotobacter chroococcum* and *Streptomyces niveus*. Pak J Biol Sci 6:2073–2080

Baca BE, Elmerich C (2003) Microbial production of plant hormones. In: Elmerich C, Newton WE (eds) Associative and endophytic nitrogen-fixing bacteria and cyanobacterial associations. Springer, Dordrecht, pp 113–143

Bacilio M, Rodriguez H, Moreno M et al (2004) Mitigation of salt stress in wheat seedlings by a *gfp*-tagged *Azospirillum lipoferum*. Biol Fertil Soils 40:188–193

Bal HB, Nayak L, Das S et al (2013) Isolation of ACC deaminase producing PGPR from rice rhizosphere and evaluating their plant growth promoting activity under salt stress. Plant Soil 366:93–105

Baltruschat H, Fodor J, Harrach BD et al (2008) Salt tolerance of barley induced by the root endophyte *Piriformospora indica* is associated with a strong increase in antioxidants. New Phytol 180:501–510

Bano A, Fatima M (2009) Salt tolerance in *Zea mays* L. following inoculation with *Rhizobium* and *Pseudomonas*. Biol Fertil Soils 45:405–413

Barassi CA, Ayrault G, Creus CM, Sueldo RJ, Sobrero MT (2006) Seed inoculation with *Azospirillum* mitigates NaCl effects on lettuce. Sci Hortic 109:8–14

Bargali K (2011) Actinorhizal plants of Kumaun Himalaya and their ecological significance. Afr J Plant Sci 5:401–406

Bashan Y, Holguin G (2002) Plant growth-promoting bacteria: a new tool in mangrove reforestation. Trees-Struc Funct 16:159–166

Bashan Y, Moreno M, Troyo E (2000) Growth promotion of the seawater-irrigated oilseed halophyte *Salicornia bigelovii* inoculated with mangrove rhizosphere bacteria and halotolerant *Azospirillum* spp. Biol Fertil Soils 32:365–272

Bernard T, Pocard J, Perroud B et al (1986) Variation in the response of salt stressed *Rhizobium* strains to betaine. Arch Microbiol 143:359–356

Blaha D, Prigent-Combaret C, Mirza MS et al (2006) Phylogeny of the 1-aminocyclopropane-1-carboxylic acid deaminase-encoding gene *acdS* in phytobeneficial and pathogenic Proteobacteria and relation with strain biogeography. FEMS Microbiol Ecol 56:455–470

Boncompagni E, Steras M, Poggi MC et al (1999) Occurrence of choline and glycine betaine uptake and metabolism in the family Rhizobiaceae and their roles in osmoprotection. Appl Environ Microbiol 65:2072–2077

Bouhmouch I, Souad-Mouhsine B, Brhada F et al (2005) Influence of host cultivars and *Rhizobium* species on the growth and symbiotic performance of *Phaseolus vulgaris* under salt stress. J Plant Physiol 162:1103–1113

Bremer E, Krämer R (2000) Coping with osmotic challenges: osmoregulation through accumulation and release of compatible solutes in bacteria. In: Storz G, Hengge-Aronis R (eds) Bacterial stress responses. ASM Press, Washington, DC, pp 79–97

Brhada F, Poggi MC, Vandesype G et al (2001) Osmoprotection mechanisms in rhizobia isolated from *Vicia faba* v. major and *Cicer arietinum*. Agronomie 21:583–590

Cassan F, Maiale S, Masciarelli O et al (2009) Cadaverine production by *Azospirillum brasilense* and its possible role in plant growth promotion and osmotic stress mitigation. Eur J Soil Biol 45:12–19

Chakraborty U, Roy S, Chakraborty AP et al (2011) Plant growth promotion and amelioration of salinity stress in crop plants by a salt-tolerant bacterium. Recent Res Sci Technol 3:61–70

Chowdhury SP, Nagarajan T, Tripathi R et al (2007) Strain-specific salt tolerance and osmoregulatory mechanisms in *Azospirillum brasilense*. FEMS Microbiol Lett 267:72–79

Cohen AC, Bottini R, Piccoli PN (2007) *Azospirillum brasilense* SP 245 produces ABA in chemically-defined culture medium and increase ABA content in *Arabidopsis* plants. Pl Growth Regul 54:97–103

Creus CM, Sueldo RJ, Barassi CA (1997) Shoot growth and water status in *Azospirillum*-inoculated wheat seedlings grown under osmotic and salt stresses. Pl Physiol Biochem 35:939–944

Csonka LN (1989) Physiological and genetic responses of bacteria to osmotic stress. Microbiol Rev 53:121–147

Daei G, Ardakani M, Rejali F et al (2009) Alleviation of salinity stress on wheat yield, yield components, and nutrient uptake using arbuscular mycorrhizal fungi under field conditions. J Plant Physiol 166:617–625

Dardanelli M, González P, Medeot D et al (2009) Effects of peanut rhizobia on the growth and symbiotic performance of *Arachis hypogaea* under abiotic stress. Symbiosis 47:175–180

Deressa TG, Schenk MK (2008) Contribution of roots and hyphae to phosphorus uptake of mycorrhizal onion (*Allium cepa* L.) – a mechanistic modeling approach. J Plant Nutr Soil Sci 171:810–820

Dimkpa C, Weinand T, Asch F (2009) Plant–rhizobacteria interactions alleviate abiotic stress conditions. Plant Cell Environ 32:1682–1694

Döbereiner J, Pedroza F (1987) Nitrogen-fixing bacteria in non-leguminous crop plants. Science Tech, Madison, pp 1–155

Eckardt NA (2006) The role of flavonoids in root nodule development and Auxin transport in Medicago truncatula. Plant Cell Online 18(7):1539–1540

Egamberdieva D (2009) Alleviation of salt stress by plant growth regulators and IAA producing bacteria in wheat. Acta Physiol Plant 1:861–864

Egamberdieva DE, Zulfiya K (2009) Selection for root colonizing bacteria stimulating wheat growth in saline soils. Biol Fertil Soils 45:563–571

Egamberdiyeva D, Islam KR (2008) Salt tolerant rhizobacteria: plant growth promoting traits and physiological characterization within ecologically stressed environment. In: Ahmad I, Pichtel J, Hayat S (eds) Plant-bacteria interactions: strategies and techniques to promote plant growth. Wiley-VCH, Weinheim, pp 257–281

Elkahoui S, Hernandez JA, Abdelly C et al (2005) Effect of salt on lipid peroxidation and antioxidant enzyme activities of *Catharanthus roseus* suspension cells. Plant Sci 168:607–613

El-Sheikh EAE, Wood M (1990) Rhizobia and bradyrhizobia under salt stress: possible role of trehalose in osmoregulation. Lett Appl Microbiol 10:127–129

Evelin H, Kapoor R, Giri B (2009) Arbuscular mycorrhizal fungi in alleviation of salt stress: a review. Ann Bot 104:1263–1280

FAO (2008) FAO Land and plant nutrition management service. Available online: http://www.fao.org/ag/agl/agll/spush. Accessed 25 Apr 2008

Fatima M (2007) Antioxidative response and alleviation of salt stress by inoculation of *Rhizobium* and phosphate solubilizing bacteria in maize (*Zea mays* L.). Dissertation, Quaid-i-azam University

Figueiredo MVB, Seldin L, Araujo FF et al (2010) Plant growth promoting rhizobacteria: fundamentals and applications. In: Maheshwari DK (ed) Plant growth and health promoting bacteria, Microbiology monographs, vol 18. Springer, Berlin, pp 21–43

Forchetti G, Masciareelli O, Alemano S et al (2007) Endophytic bacteria in sunflower (*Helianthus annuus* L.): isolation, characterization, and production of jasmonates and abcissic acid in culture medium. Appl Microbiol Biotechnol 76:1145–1152

Giri B, Mukerji KG (2004) Mycorrhizal inoculant alleviates salt stress in *Sesbania aegyptiaca* and *Sesbania grandiflora* under field conditions: evidence for reduced sodium and improved magnesium uptake. Mycorrhiza 14:307–312

Glick BR (2005) Modulation of plant ethylene levels by the bacterial enzyme ACC deaminase. FEMS Microbiol Lett 251:1–7

Glick BR, Penrose DM, Li J (1998) A model for the lowering of plant ethylene concentrations by plant growth-promoting bacteria. J Theor Biology 190:63–68

Goldstein AH (1986) Bacterial phosphate solubilization: historical perspective and future prospective. Am J Altern Agric 1:57–65

Gomaa AM, Abo-Aba SEM, Awad NS (2008) Isolation, characterization and genetic differentiation of *Frankia* sp. isolated from ecologically different Egyptian locations. Res J Cell Mol Biol 2:6–17

Gontia-Mishra I, Sharma A (2012) Exogenously supplied osmoprotectants confer enhanced salinity tolerance in rhizobacteria. J Ecobiotech 4:11–13

Gouffi K, Pica N, Pichereau V et al (1999) Disaccharides as a new class of non-accumulated osmoprotectants for *Sinorhizobium meliloti*. Appl Environ Microbiol 65:1491–1500

Grattan SR, Grieve CM (1999) Salinity mineral nutrient relations in horticultural crops. Sci Hortic 78:127–157

Grichko VP, Glick BR (2001) Amelioration of flooding stress by ACC deaminase-containing plant growth-promoting bacteria. Plant Physiol Biochem 39:11–17

Guillot A, Obis D, Mistou MY (2000) Fatty acid membrane composition and activation of glycine-betaine transport in *Lactococcus lactis* subjected to osmotic stress. Int J Food Microbiol 55:47–51

Gurmani AR, Bano A, Din J, Khan SU et al (2009) Effect of phytohormone on growth and ion accumulation of wheat under salinity stress. Afr J Biotechnol 8:1887–1894

Gururani MA, Upadhyaya CP, Baskar V et al (2012) Plant growth promoting rhizobacteria enhance abiotic stress tolerance in *Solanum tuberosum* through inducing changes in the expression of ROS-scavenging enzymes and improved photosynthetic performance. J Plant Growth Regul http://dx.doi.org/10.1007/s00344-012-9292-6

Gyaneshwar P, Naresh KG, Parekh LJ (2002) Effect of buffering on the phosphate solubilizing ability of microorganisms. World J Microbiol Biotechnol 14:669–673

Hadees M (2009) Effect of salt stress, abscisic acid and *Rhizobium* on physiology of wheat. Dissertation, Quaid-i-Azam University

Hamdia ME, Shaddad MAK, Doaa MM (2004) Mechanisms of salt tolerance and interactive effect of *Azospirillum brasilense* inoculation on maize cultivars grown under salt stress conditions. Plant Growth Regul 44:165–174

Han HS, Lee KD (2005) Plant growth promoting rhizobacteria effect on antioxidant status, photosynthesis, mineral uptake and growth of lettuce under soil salinity. Res J Agric Biol Sci 1:210–215

Hartmann A, Zimmer W (1994) Physiology of *Azospirillum*. In: Okon Y (ed) *Azospirillum*/plant association. CRC Press, Boca Raton, pp 15–39

Hayat R, Safdar Ali S, Amara U et al (2010) Soil beneficial bacteria and their role in plant growth promotion: a review. Ann Microbiol 60:579–559

He Z, Zhou A, Baidoo E et al (2010) Global transcriptional, physiological, and metabolite analyses of the responses of *Desulfovibrio vulgaris* Hildenborough to salt adaptation. Appl Environ Microbiol 76(5):1574–1586

Hecker M, Völker U (2001) General stress response of *Bacillus subtilis* and other bacteria. Adv Microb Physiol 44:35–91

Hernandez JA, Jiménez A, Mullineaux P et al (2000) Tolerance of pea (*Pisum sativum* L.) to long term salt stress is associated with induction of antioxidant defenses. Plant Cell Environ 23:853–862

Höper D, Bernhardt J, Hecker M (2006) Salt stress adaptation of *Bacillus subtilis*: a physiological proteomics approach. Proteomics 6:1550–1562

Hua ST, Tsai VY, Lichens SGM et al (1982) Accumulation of amino acids in *Rhizobium* sp. strain WR1001 in response to sodium chloride salinity. Appl Environ Microbiol 44:135–140
Irekti H, Drevon JJ (2003) Acide abcissique et conductance a` la diffusion de l'oxyge'ne dans les nodosite's de haricot soumises a` un choc salin. In: J.J. Drevon, B. Sifi (eds) Fixation Symbiotique de l'Azote et De'veloppement Durable dans le Bassin Me'diterrane'en, INRA Editions, les colloques, vol. 100
Ivanova R, Bojinova D, Nedialkova K (2006) Rock phosphate solubilization by soil bacteria. J Uni Chem Technol Metallurg 41:297–302
Jaleel CA, Gopi R, Sankar B et al (2007) Studies on germination, seedling vigor, lipid peroxidation and proline metabolism in *Catharanthus roseus* seedlings under salt stress. S Afr J Bot 73:190–195
Khan SM, Zaidi A, Wani PA (2007) Role of phosphate-solubilizing microorganisms in sustainable agriculture – a review. Agron Sustain Dev 27:29–43
Khan MS, Zaidi A, Wani PA et al (2009) Functional diversity among plant growth-promoting rhizobacteria. In: Khan MS, Zaidi A, Musarrat J (eds) Microbial strategies for crop improvement. Springer, Berlin, pp 105–132
Kohler J, Hernandez JA, Caravaca F et al (2009) Induction of antioxidant enzymes is involved in the greater effectiveness of PGPR versus AM fungi with respect to increasing the tolerance of lettuce to severe salt stress. Environ Exp Bot 65:245–252
Kolb W, Martin P (1997) Responses of plant roots to inoculation with *Azospirillum brasilense* and to application of indole acetic acid. In: Klinguller W (ed) *Azospirillum* III: genetics, physiology, ecology. Springer, Berlin, pp 215–221
Le Rudulier D, Strom AR, Danekar AM et al (1984) Molecular biology of osmoregulation. Science 244:1064–1068
Leonova TG, Goncharova EA, Khodorenko AV et al (2005) Characteristics of salt-tolerant and salt-susceptible cultivars of barley. Russ J Plant Physiol 52:774–778
Levy Y, Krikun J (1980) Effect of vesicular-arbuscular mycorrhiza on *Citrus jambhiri* water relations. New Phytol 85:25–31
Llamas I, del Moral, Martínez-Checa F et al (2006) *Halomonas maura* is a physiologically versatile bacterium of both ecological and biotechnological interest. Antonie Van Leeuwenhoek 89:395–403
Lucht JM, Bremer E (1994) Adaptation of *Escherichia coli* to high osmolarity environments: osmoregulation of the high-affinity glycine betaine transport system ProU. FEMS Microbiol Rev 14:3–20
Mabood F, Smith D (2007) The role of salicylates in *Rhizobium*-legume symbiosis and abiotic stresses in higher plants. Biomed Life Science, Springer, Heidelberg, pp 151–162. doi: 10.1007/1-4020-5184-0
Maheshwari DK (2012) Bacteria in agrobiology: plant probiotics. Springer, Heidelberg
Maronek DM, Hendrix JW, Kiernan J (1981) Mycorrhizal fungi and their importance in horticultural crop production. Hortic Rev 3:172–213
Mayak S, Tirosh T, Glick B (2004) Plant growth promoting bacteria that confer resistance to water stress in tomatoes and peppers. Plant Sci 166:525–530
Miller KJ (1985) Effects of temperature and sodium chloride concentration on the phospholipid and fatty acid compositions of a halotolerant *Planococcus* sp. J Bacteriol 162:263–270
Miransari M (2011) Arbuscular mycorrhizal fungi and nitrogen uptake. Review article. Arch Microbiol 193:77–81
Miransari M, Smith D (2009) Alleviating salt stress on soybean (*Glycine max* (L.) Merr.) – *Bradyrhizobium japonicum* symbiosis, using signal molecule genistein. Eur J Soil Biol 45:146–152
Miransari M, Bahrami HA, Rejali F et al (2008) Using arbuscular mycorrhiza to reduce the stressful effects of soil compaction on wheat (*Triticum aestivum* L.) growth. Soil Biol Biochem 40:1197–1206

Misra N, Gupta AK (2006) Effect of salinity and different nitrogen sources on the activity of antioxidant enzymes and indole alkaloid content in *Catharanthus roseus* seedlings. J Plant Physiol 163:11–18

Mittova V, Tal M, Volokita M, Guy M (2003) Up-regulation of the leaf mitochondrial and peroxisomal antioxidative systems in response to salt-induced oxidative stress in the wild salt-tolerant tomato species *Lycopersicon pennellii*. Plant Cell Environ 26:845–856

Moat GA, Foster JW, Spector MP (2002) Microbial stress response. In: Moat GA, Foster WJ, Spector PM (eds) Microbial physiology. Wiley, New York, pp 582–611

Mohammad MJ, Malkawi HI, Shibli R (2003) Effects of arbuscular mycorrhizal fungi and phosphorous fertilization on growth and nutrient uptake of barley grown on soils with different levels of salts. J Plant Nutr 26:125–137

Muneer S, Ahmad J, Bashir H et al (2012) Proteomics of nitrogen fixing nodules under various environmental stresses. Plant Omics J 5:167–176

Munns R, Tester M (2008) Mechanism of salinity tolerance. Annu Rev Pl Biol 59(1):651–681

Nabizadeh M, Tayeb SN, Mani M (2011) Effect of irrigation on the yield of mungbean cultivars. J Am Sci 7:86–90

Nabti E, Sahnoune M, Adjrad S (2010) Restoration of growth of durum wheat (*Triticumdurum var. waha*) under saline conditions due to inoculation with the rhizosphere bacterium (*Azospirillum brasilense* NH) and extracts of the marine alga (*Ulva lactuca)*. J Plant Growth Regul 29:6–22

Nadeem SM, Zahir ZA, Naveed M, Arshad M (2007) Preliminary investigations on inducing salt tolerance in maize through inoculation with rhizobacteria containing ACC deaminase activity. Can J Microbiol 53:1141–1149

Nadeem SM, Zahir ZA, Naveed M et al (2013) Mitigation of salinity-induced negative impact on the growth and yield of wheat by plant growth-promoting rhizobacteria in naturally saline conditions. Ann Microbiol 63:225–232

Nautiyal SC, Bhadauria S, Kumar P et al (2000) Stress induced phosphate solubilization in bacteria isolated from alkaline soils. FEMS Microbiol Lett 182:291–296

Naz R (2009) Effect of plant growth promoting rhizobacteria and salicylic acid on sunflower (*Helianthus annus* L.) under salt stress. Dissertation, Quaid-i-Azam University

Naz I, Bano A (2010) Biochemical, molecular characterization and growth promoting effects of phosphate solubilizing *Pseudomonas* sp. isolated from weeds grown in salt range of Pakistan. Plant Soil 334(1–2):199–207

Naz I, Bano A, Hassan T-UL (2009) Isolation of phytohormone producing plant growth promoting rhizobacteria from weeds growing in Khewra salt range, Pakistan and their implication in providing salt tolerance to *Glycine max*. L. Afr J Biotechnol 8:5762–5766

Öğütçü H, Kasımoğlu C, Elkoca E (2010) Effects of *Rhizobium* strains isolated from wild chickpeas on the growth and symbiotic performance of chickpeas (*Cicer arietinum L.*) under salt stress. Turk J Agric For 34:361–371

Okon Y (1985) *Azospirillum* is potential inoculant for agriculture. Trends Biotechnol 3:223–228

Omar MNA, Osman MEH, Kasim WA et al (2008) Improvement of salt tolerance mechanisms of barley cultivated under salt stress using *Azospirillum brasilense*. Biomed Life Sci, Part 44:133–147

Paul D (2012) Osmotic stress adaptations in rhizobacteria. J Basic Microbiol 52:1–10

Paul D, Nair S (2008) Stress adaptation in a plant growth promoting rhizobacterium (PGPR) with increasing salinity in the coastal agriculture soils. J Basic Microbiol 48(5):378–384

Paul D, Bharathkumar S, Nair S (2005) Osmotolerance in biocontrol strain of *Pseudomonas pseudoalcaligenes* MSP-538: a study using osmolyte, protein and gene expression profiling. Ann Microbiol 55:243–247

Paul D, Dineshkumar N, Nair S (2006) Proteomics of a plant growth-promoting rhizobacterium, *Pseudomonas fluorescens* MSP-393, subjected to salt shock. World J Microbiol Biotechnol. doi:10.1007/s11274-005-9043-y

Perrig D, Boiero ML, Masciarelli OA et al (2007) Plant growth promoting compounds produced by two strains of *Azospirillum brasilense* and implications for inoculant formation. J Appl Microbiol Biotechnol 75:1143–1150

Perveen S, Khan MS, Zaidi A (2002) Effect of rhizospheric microorganisms on growth and yield of green gram (*Phaseolus radiatus*). Ind J Agric Sci 72:421–423

Pospisilova J, Vagner M, Malbeck J et al (2005) Interactions between abscisic acid and cytokinins during water stress and subsequent rehydration. Biol Plant 49:533–540

Qurashi AW, Sabri AS (2012) Bacterial exopolysaccharide and biofilm formation stimulate chickpea growth and soil aggregation under salt stress. Braz J Microbiol 43:1183–1191

Rabie GH, Almadivi AM (2005) Role of bio-inoculants in development of salt-tolerance of *Vicia faba* plants under salinity stress. Afr J Biotechnol 4:210–222

Rajabzadeh F (2009). Isolation, identification and application of *Azospirillum* spp. (PGPR) on rice (*Oryza sativa*). Dissertation, Gorgan University of Agricultural Sciences and Natural Resources

Rao DLN, Giller KE, Yeo ER et al (2002) The effect of salinity and sodicity upon nodulation and nitrogen fixation in chickpea (*Cicer arietinum*). Ann Bot 89:563–570

Reddell P, Eoster RC, Bowen GD (1986) The effects of sodium chloride on growth and nitrogen fixation in *Casuarina obesa* MIQ. New Phytol 102:397–408

Reinhold-Hurek B, Hurek T, Frendrik I et al (1987) *Azospirillum halopraferens* sp. Nov., a nitrogen fixing organism associated with the roots of Kallar grass (*Leptochloa fusca* L. Kunth). Int J Syst Bacteriol 37:43–51

Ribaudo CM, Krumpholz EM, Cassan FD et al (2006) *Azospirillum* sp. promotes root hair development in tomato plants through a mechanism that involves ethylene. J Plant Growth Regul 24:175–185

Richardson AE (2001) Prospects for using soil microorganisms to improve the acquisition of phosphorus by plants. Aust J Plant Physiol 28:897–906

Rillig MC, Mummey DL (2006) Mycorrhizas and soil structure. New Phytol 171:41–53

Rivarola V, Castro S, Mori G et al (1998) Response of *Azospirillum brasilense* Cd to sodium chloride stress. Antonie Van Leeuwenhoek 73:255–261

Rodriguez H, Fraga R (1999) Phosphate solubilizing bacteria and their role in plant growth promotion. Biotechnol Adv 17:319–339

Rudulier DL, Bernard T (2006) Salt tolerance in *Rhizobium:* a possible role for betaines. FEMS Microbiol Lett 39:67–72

Ryu CM, Farag MA, Hu CH et al (2004) Bacterial volatiles induce systematic resistance in *Arabidopsis*. Plant Physiol 134:1017–1026

Sachdev D, Chaudhari H, Kasture V et al (2009) Isolation and characterization of indole acetic acid (IAA) producing *Klebsiella pneumoniae* strains from rhizosphere of wheat (*Triticum aestivum*) and their effect on plant growth. Indian J Exp Biol 47:993–1000

Sarig S, Okon Y, Blum A (1992) Effect of *Azospirillum brasilense* inoculation in growth dynamics and hydraulic conductivity of *Sorghum bicolour* roots. J Plant Nutr 15:805–819

Scheloske S, Maetz M, Schneider T et al (2004) Element distribution in mycorrhizal and nonmycorrhizal roots of the halophyte *Aster tripolium* determined by proton induced X-ray emission. Protoplasma 223:183–189

Sharan R, Ulitsky I, Shamir R (2007) Network-based prediction of protein function. Mol Syst Biol 3:88. doi:10.1038/msb4100129

Shukla D, Kesari R, Mishra S et al (2012) Expression of phytochelatin synthase from aquatic macrophyte *Ceratophyllum demersum* L. enhances cadmium and arsenic accumulation in tobacco. Plant Cell Rep 31:1687–1699

Smith LT, Pocard JA, Bernard T et al (1988) Osmotic control of glycine betaine biosynthesis and degradation in *Rhizobium meliloti*. J Bacteriol 170:3142–3149

Smith SE, Read DJ (2008) Mineral nutrition, toxic element accumulation and water relations of arbuscular mycorrhizal plants. In: Smith SE, Read DJ (eds) Mycorrhizal symbiosis, 3rd edn. Academic Press, London, pp 145–180

Szabados L, Savoure A (2009) Proline: a multifunctional amino acid. Trends Plant Sci 15:89–97

Talibart R, Jebbar M, Gouesbat G et al (1994) Osmoadaptation in rhizobia: ecotine-induced salt tolerance. J Bacteriol 176:5210–5217

Tani C, Sasakawa H (2000) Salt tolerance of *Elaeagnus macrophylla* and *Frankia* Ema1strain isolated from the root nodules of *E. macrophylla*. Soil Sci Plant Nutr 46:927–937

Tawfik MM, Amany A, Bahr AKM et al (2006) Response of Kallar grass (*Leptochloa fusca* L.) to biofertilizer inoculation under different levels of seawater irrigation. J Appl Sci Res 2:1203–1211
Thrall PH, Broadhurst LM, Hoque S et al (2009) Diversity and salt tolerance of native *Acacia* rhizobia isolated from saline and non- saline soils. Austral Ecol 34:950–963
Tian CY, Feng G, Li XL et al (2004) Different effects of arbuscular mycorrhizal fungal isolates from saline or non-saline soil on salinity tolerance of plants. Appl Soil Ecol 26:143–148
Upadhyay SK, Singh JS, Singh DP (2011) Exopolysaccharide-producing plant growth-promoting rhizobacteria under salinity condition. Pedosphere 21(2):214–222
Upadhyay SK, Singh JS, Saxena AK et al (2012) Impact of PGPR inoculation on growth and antioxidants status of wheat plant under saline condition. Plant Biol 14:605–611
Vacheron J, Marie-Lara Bouffaud G, Touraine B et al (2013) Plant growth-promoting rhizobacteria and root system functioning. Front Plant Sci 4:356
Verdoy D, Lucas MM, Manrique E et al (2004) Differential organ-specific response to salt stress and water deficit in nodulated bean (*Phaseolus vulgaris*). Plant Cell Environ 27:767–767
Vessey JK, Pawlowski K, Bergman B (2004) Root based N_2-fixing symbioses: legumes, actinorhizal plants, *Parasponia* sp. and cycads. Plant Soil 266:205–230
Vikram A, Alagawadi AR, Amzehzarghani H et al (2007) Factors related to the occurrence of phosphate solubilizing bacteria and their isolation in vertisols. Int J Agric Res 2:571–580
Waller F, Achatz B, Baltruschat H et al (2005) The endophytic fungus *Piriformospora indica* reprograms barley to salt-stress tolerance, disease resistance, and higher yield. Proc Nat Acad Sci USA 102: 13386–13391
Walsh UF, Morrissey JP, O'Gara F (2001) *Pseudomonas* for biocontrol of phytopathogens: from functional genomics to commercial exploitation. Curr Opin Biotechnol 12:289–295
Wei L, Kloepper JW, Tuzun S (1996) Induced systemic resistance to cucumber diseases and increased plant growth by plant growth promoting rhizobacteria under field conditions. Phytopathology 86:221–224
Westover KM, Kennedy AC, Kelley SE (1997) Patterns of rhizosphere microbial community structure associated with co-occurring plant species. J Ecol 85:863–873
Whitelaw MA (2000) Growth promotion of plants inoculated with phosphate-solubilizing fungi. Adv Agron 69:99–151
Yelton MM, Yang SS, Edie SA et al (1983) Characterisation of an effective salt tolerant, fast growing strain of *Rhizobium japonicum*. J Gen Microbiol 129:1547–1573
Yue GJ, Wu GQ, Hao XM (2007) The status and prospects of fuel ethanol process technology in China. Prog Chem 19:1084–1090
Zahran HH, Rasanen LA, Karsisto M et al (1994) Alteration of polysaccharide and protein profiles in SDS-PAGE of rhizobia by osmotic and heat stress. World J Microbial Biotechnol 10:100–105
Zaidi A, Khan MS (2006) Co-inoculation effects of phosphate solubilizing microorganisms and *Glomus fasciculatum* on green Gram-*Bradyrhizobium* symbiosis. Turk J Agric 30:223–230
Zhang H, Kim MS, Sun Y et al (2008) Soil bacteria confer plant salt tolerance by tissue-specific regulation of sodium transporter HKI1. Mol Plant Microbe Interact 21:737–744
Zhang YM, Rock CO (2009) Transcriptional regulation in bacterial membrane lipid synthesis. J Lipid Res 50:115–119

Challenges Faced in Field Application of Phosphate-Solubilizing Bacteria

6

Abdul Aziz Eida, Heribert Hirt, and Maged M. Saad

Abstract

The general inaccessibility of soil phosphorous (P) to plants in combination with the depletion of global P reserves provides an incentive for researchers to find sustainable solutions to sustain food security for the ever-increasing world population. Bio-fertilizers based on bacteria and fungi able to solubilize endogenous P in soils have a high potential for increasing nutrient availability in agriculture. However, the inconsistency of bio-fertilizer performance in the field poses a major challenge for farmers. This discrepancy is thought to stem from the complexity of the interactions between crop plants, microbes, and their soil environments, as well as our lack of understanding of the processes involved. For farmers, a clear beneficial effect across different soil types, crop species, environmental conditions, and microbial communities will be required to make it worth to adopt bio-fertilizer technology based on phosphate-solubilizing microbes (PSMs). Here, we attempt to review the current knowledge of the complexity of the P-solubilization mechanisms used by PSMs and how they may be affected by interactions in the field. We also identify possible explanations for the inconsistent performance of P-solubilizing bacteria in the field and ways to solve these obstacles.

A.A. Eida • H. Hirt (✉) • M.M. Saad
Center for Desert Agriculture, 4700 King Abdullah University of Science and Technology (KAUST), Thuwal 23955-6900, Kingdom of Saudi Arabia
e-mail: heribert.hirt@kaust.edu.sa

S. Mehnaz (ed.), *Rhizotrophs: Plant Growth Promotion to Bioremediation*, Microorganisms for Sustainability 2, DOI 10.1007/978-981-10-4862-3_6

6.1 Introduction

Phosphorus (P) is an essential macronutrient that plays a pivotal role in all living organisms. It is a fundamental component of nucleic acids (RNA and DNA), proteins, phospholipids, metabolites, and many of cofactors such as ATP and NADP. Hence, it is central to major energy-dependent metabolic and regulating processes (e.g., photosynthesis, respiration, and protein phosphorylation), and its deficiency significantly affects plant growth, development, reproduction, and the formation of seeds and fruits (Duca 2015). P is the 11th-most abundant element in the Earth's lithosphere. Yet, it is the second most-important element when it comes to limiting crop plant productivity owing to its general inaccessibility to plants in the forms it is present in soil (Vitousek and Howarth 1991). The low biological availability of P due to biological and chemical immobilization, runoff, and leaching poses a limitation for crop plant growth and development.

P fertilizers such as phosphate rock (PR) or its mined derivatives are used to mitigate P unavailability. Large-scale production and application of P fertilizers may seem to be a solution to maintaining global food supplies in the light of a growing human population. However, these practices have a high potential to cause detrimental agronomic, environmental, geopolitical, and socioeconomic impacts in the future. Depletion of P reserves is emerging as a serious challenge for global food security (Cordell and White 2015). Assessments of global P reserves, production rates, and global supply and demand, indicate that the reserves may be depleted in 50–100 years and that peak P production, considering current P extraction technologies, may occur well within the coming 100 years (Van Vuuren et al. 2010). Therefore, global P depletion may lead to increased extraction and production costs of P raw materials, rises in fertilizer and food prices, and ultimately economic instability and famine.

In contrast to P fertilizers applied exogenously, endogenous P resources present in soils may very well last up to several centuries, if they could be rendered accessible by sustainable approaches. Development of soil management practices, plant genetic engineering strategies, and bio-fertilizer systems to enhance P solubilization, mobilization, assimilation, and utilization are being promoted for sustainable agricultural practices (Tian et al. 2012; Nesme et al. 2014). P-solubilizing microbes (PSMs) are emerging as key factors for P dynamics in soils and are widely considered as a potentially cost-effective, sustainable, and environmentally benign approach to overcome P limitations in soil (Sharma et al. 2013). However, plant growth responses and P-solubilizing effects observed in field trials involving PSMs often do not reflect those observed under controlled laboratory or greenhouse conditions (Weller and Thomashow 1994; Yarzábal 2010; Jones and Oburger 2011). This applies in particular to P-solubilizing bacteria. These inconsistent plant responses are a result of the complex interactions with the soil and their complex microbial structures. Understanding the different forms of P in soil and how microbes solubilize inaccessible P, is essential for improved selection of bio-fertilizer, P-solubilizing bacteria (PSB), that function successfully under field conditions. Furthermore, symbiotic associations with other beneficial microbes such as fungi may improve the performance of the PSB.

6.2 Forms of P in Soils and P-Solubilization Mechanisms Used by PSMs

P in soil can exist as inorganic phosphate (P_i) and phosphate bound to organic matter (P_o). Primary mineral forms of P_i such as apatite, berlinite, strengite, and variscite that are very stable, are converted to soil solution P via weathering (Fig. 6.1) (Osman 2012). Plant accessible soluble orthophosphates $H_2PO_4^-$ and HPO_4^{2-}, however, are found only at very low concentration (<10 μM) in most soil solutions (Bieleski 1973). They are fixed into insoluble P forms through various mechanisms that are predominantly influenced by soil pH: (1) reacting rapidly with cations, such as calcium (Ca^{2+}) in alkaline, calcareous soils, and aluminum (Al^{3+}) or iron (Fe^{3+}) in acidic soils, to form precipitates; (2) adsorption to surfaces of clay or aluminum and iron (hydr) oxides particles; and (3) complexing or immobilization with organic molecules to form P_o, which can account for 30–65% of total soil P in mineral soils and up to 90% in organic soils (Fig. 6.1) (Schachtman et al. 1998). The majority of P_o is found as inositol phosphates, but phosphate esters (e.g., phospholipids, humic acid, or nucleic acids) are present as well (Turner et al. 2002). PSB and phosphate-solubilizing fungi (PSF) play a fundamental role in the cycling of P in soil directly by (1) altering soil sorption and precipitation equilibria for the release of free P into soil solution (solubilization) and (2) facilitating the mobilization of poorly available organic P (mineralization) and indirectly by (3) inducing biochemical, molecular, and physiological changes in plants, which in turn lead to alteration of the P availability in soil (Khan et al. 2014) (Fig. 6.1).

Solubilization of P_i (e.g., Ca, Al, and Fe phosphates) by PSMs is commonly associated with the production of low molecular mass organic acids (OAs), such as gluconic acid, and proton (H^+) release resulting in acidification of the microbial cells and their surroundings. OAs chelating metal cations can efficiently compete with P_i for sorption sites, thus releasing soluble P_i from minerals (Goldstein 1994; Filius et al. 1997). PSMs have been shown to release a variety of OAs, including acetic, oxaloacetic, citric, oxalic, lactic, succinic, gluconic, glutamic, 2-ketogluconic, 2-ketoglutamic, formic, tartaric, malic, malonic, and glycolic acid (Rashid et al. 2004; Vyas and Gulati 2009; Mardad et al. 2013). Various bacterial (e.g., *Pseudomonas*, *Bacillus*, *Burkholderia*, *Rhizobium*) and fungal (e.g., *Aspergillus*, *Penicillium*) genera, have been shown to possess P-solubilizing abilities (Khan et al. 2010, 2013).

Solubilization of P_o, also called mineralization, by bacteria and fungal genera is mediated by intracellular, membrane-bound, or extracellular-free phosphatases which catalyze the hydrolysis of esters and anhydrides of phosphoric acid (Eivazi and Tabatabai 1977; Nannipieri et al. 2011). PSMs can secrete both, acid and alkaline phosphatases, depending on pH conditions in the surrounding soil (Eivazi and Tabatabai 1977; Kim et al. 1998). Phytases release P from phytate, a compound for P storage in plants, which can also be found in considerable amounts in soil (Lim et al. 2007; Maougal et al. 2014). Other P-solubilization strategies include the release of microbial compounds such as exopolysaccharides or siderophores able to chelate metal ions in soil and, thus, can influence P soil availability (Crowley and Kraemer 2007; Yi et al. 2008). Bacteria, however, are also able to immobilize P_i by

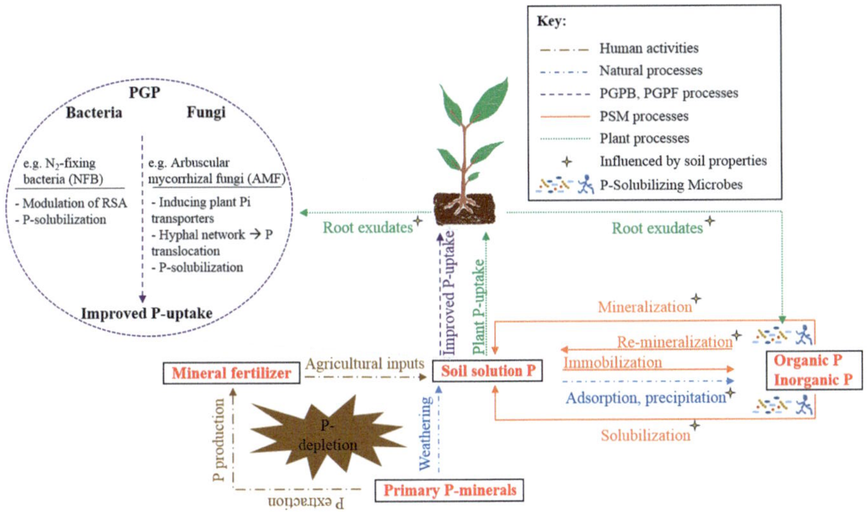

Fig. 6.1 Schematic representation of factors mediating the distribution and cycling of P between soil, microbes, and plants. Extraction of P from primary minerals for fertilizer inputs lead to P depletion. Soil solution P from P fertilizers or from the weathering of primary P minerals is either taken up by the plants or converted to P_i and P_o by adsorption, precipitation, or incorporation. Exudates released from plant roots recruit microbes to establish symbiotic associations. PSMs are able to convert P_i and P_o back to soil solution P. PGPB and PGPF are also able to assist the plant in solubilizing and increasing P uptake efficiency via direct and indirect pathways. Various processes are color-coded and line-patterned as indicated in the key

converting it into organic forms and incorporating it into microbial biomass. This may occur through high-affinity P transporters as shown for *Escherichia coli*, which are activated during internal P depletion (Medveczky and Rosenberg 1971). The incorporated P_i may then be released into the soil through re-mineralization processes, or microbial turnover or decomposition, in response to environmental change, starvation, or predation (Grierson et al. 1998; Oberson et al. 2005).

P-solubilization and P-mineralization efficiencies vary among bacterial or fungal species and are affected by soil physicochemical properties and particular compounds such as energy sources. For example, pH, carbon (C), and nitrogen (N) sources in the medium modulated the efficiency of P solubilization by *Penicillium purpurogenum* via the alteration of OA production, with the highest P-solubilization activity corresponding to highest OA production under high C and low N source availability (Scervino et al. 2011). Complex carbohydrate sources in otherwise nutrient-deficient soils were shown to positively influence the exudation of OAs from two fungal strains, the zygomycete fungus *Mucor hiemalis* and the ascomycete fungus *Penicillium chrysogenum* (Brunner et al. 2014). P solubilization by *Pseudomonas fluorescence* RAF15 was also significantly promoted when glucose was available as a C source in comparison to fructose as C source (Park et al. 2009). The available organic energy sources, e.g., glucose or starch vs. phytic acid or glycerol-2-phosphate, were also shown to affect the activity of acid and alkaline phosphatases in the PSB *Enterobacter agglomerans* (Kim et al. 1998). Soil pH can affect the surface charge of phosphatases and thus their adsorption to soil particles. For example, at high pH, the adsorption of two extracellular acidic phosphatases of the ectomycorrhizal fungus *Hebeloma cylindrosporum* to montmorillonite clay was lower than at low pH (Leprince and Quiquampoix 1996). Temperature and pH also play crucial roles in influencing the activity of phytases as shown for three commercial microbial phytases from *Aspergillus oryzae*, *A. niger*, and *Saccharomyces cerevisiae* (Naves et al. 2012). The quality and concentration of OAs and the activity of phosphatases and phytases are, thus, determinants of the P-solubilization and P-mineralization efficiency, respectively, and are greatly influenced by soil pH, C and N sources, and temperature.

The presence of different P forms in soil, diverse P-solubilization mechanisms of bacteria, fungi, and the influence of the physicochemical properties of the soil and nutritional composition are critical factors to assess when screening and selecting PSB for field applications. A high proportion of the publications reporting the isolation and identification of PSB, however, is solely based on in vitro studies applying unsuitable screening procedure and selection criteria. The most common method to screen for potential PSB is based on defined selective media, most often Pikovskaya (PVK) or a version modified according to India's National Botanical Research Institute (NBRIP) (Pikovskaya 1948; Nautiyal 1999). These media assess the ability of PSB to solubilize solely tricalcium phosphate (TCP; $Ca_3(PO_4)_2$) as an insoluble P compound using a standard concentration of C and N sources and a pH around 7. Other metal-P compounds commonly found in nature are rarely employed for this type of assessment. The bias introduced by relying solely on TCP is discussed thoroughly by Bashan et al. (2013), with the authors proposing the use of a combination of different metal-P substrates for the selection procedure depending on the pH and composition of the soil in which the bacteria are going to be applied.

The inconsistent behavior of PSB in field conditions compared with in vitro tests might in large parts be due to such biased testing, as soil properties in fields encompass a broad range of P sources and soil properties that are not examined in vitro. In order to obtain more consistency in the performance of PSB and increase the success of field trials, we propose that the screening and selection of PSB prior to inoculation requirement (1) implement a thorough assessment of soil properties of the field, (2) employ several metal-P compounds as a source of inaccessible P_i, and (3) in addition assess the ability of PSB to mineralize P_o by analysis of the activity of the relevant enzymes prior to inoculation. Zarei et al. (2006) performed a co-inoculation study of arbuscular mycorrhizal fungi (AMF) of the *Glomus* genus, *G. mosseae* and *G. intraradices*, with *Rhizobium leguminosarum bv. viciae* and *Mesorhizobium cicero* which are strains of rhizobial bacteria possessing P-solubilizing activity. The study was conducted on lentil (*Lens culinaris*) using two different P fertilizer sources, superphosphate and PR, in a greenhouse. Both P fertilizers enhanced the plant growth and P and N uptake compared to the unfertilized control, while they decreased AMF colonization of lentil roots. Nevertheless, the superphosphate performed better than the PR in all parameters except the root colonization of the AMF, leading to lower infection compared with the PR. These results clearly indicate that the different P sources used will have a significant effect on the overall enhancement of plant growth. It also demonstrates that co-inoculation with other microbial partners (e.g., AMF) may influence the P-solubilization mechanisms. The effects of other microbes on the enzyme activity and concentration of OAs are demonstrated in a study by Kim et al. (1997) that showed alkaline phosphatase activity, and concentration of oxalic, gluconic, and citric acids is highest when the PSB *E. agglomerans* was co-inoculated with an AMF *G. etunicatum*. In another study, dual inoculation of the PSB *Pseudomonas alcaligenes* and the AMF *Rhizophagus irregularis* showed the highest acid phosphatase activity, enhanced mineralization of phytate, and led to the lowest residual phytate-P concentration in soil, compared to single inoculation (Zhang et al. 2014a). Therefore, contributions from crop plants and associated fungi may also play critical roles in P dynamics and thus must not be ignored when assessing PSB performance in the field.

6.3 Rhizosphere Interactions Involving Plants, Bacteria, and Fungi

Plants have evolved efficient biochemical, molecular, and physiological mechanisms to respond and adapt to P-deficiency stress that has been the subject of recent reviews (Liang et al. 2014; Zhang et al. 2014b). One strategy, which resembles the mechanism used by PSM, is the secretion of organic acids from the roots to the soil to solubilize P as seen in alfalfa (*Medicago sativa*), maize (*Zea mays*), and white lupin (*Lupinus albus*) (Lipton et al. 1987; Johnson et al. 1994; Carvalhais et al. 2011). The plant root system is the main site for the secretion of organic compounds, signaling molecules, and, thus, is a "hotspot" for microbial activity, colonization, and symbiotic associations. The continuous production and secretion into the rhizosphere of a wide range of compounds and nutrients, known in summary as root exudates, play a significant role in shaping of soil structure and microbial communities (Fig. 6.1). Root exudates are the primary source of organic C and energy for microorganisms and can add up to a substantial C cost to the plant (Nguyen 2003; Uren 2007).

The provision of carbohydrates via root exudates enables plants to recruit beneficial microbial partners and to establish symbiotic interactions with them. Specific compounds in the root exudates have been identified and shown to be involved in recruitment and establishment of associations with microbes, including flavonoids, isoflavonoids, malic acid, and strigolactones (Steinkellner et al. 2007; Yoneyama et al. 2007; Rudrappa et al. 2008). Most legumes have the ability to form root nodules that accommodate symbiotic nitrogen-fixing bacteria, collectively named rhizobia. The rhizobia fix atmospheric N_2 and converted into ammonia (NH_3) that the plant can use to fulfill its nitrogen requirement; in return plants provide the rhizobia with necessary carbon source. Root-exuded flavonoids act as signal components, chemoattractants, and inducers of essential genes for establishing a successful plant–rhizobia dialogue (Peters et al. 1986; Downie 2014). Depending on the physicochemical properties of the soil, crop species, and genotype, the plant developmental stage and the microbial community in the rhizosphere, the composition, and the quantity of root exudates can vary greatly (Neumann and Römheld 2007; Uren 2007; Badri and Vivanco 2009). In turn, the quantity and composition of the root exudates strongly influence the microbial community in a species-dependent manner (Kamilova et al. 2006). However, the plant genotype has proven to be the most critical factor in shaping the rhizosphere microbiome (Marques et al. 2014).

Beneficial rhizosphere microbes are essential partners of plants for survival and development of stress tolerance under harsh conditions (de Zelicourt et al. 2013). Symbiotic rhizosphere microbes have become of great interest for their use in agriculture as bio-fertilizers, such as mycorrhizal fungi and rhizobia for nutrient acquisition and N_2-fixation (Hodge et al. 2010; Geurts et al. 2012). Plant growth-promoting bacteria (PGPB) or plant growth-promoting fungi (PGPF) can stimulate plant growth and/or enhance the resistance and confer tolerance to biotic and abiotic stresses, including increasing nutrient availability (Fig. 6.1) (Van Wees et al. 2008; Lugtenberg and Kamilova 2009; Yang et al. 2009). Plant-microbe interactions and the mechanisms of plant growth-promoting activities and stress tolerance abilities are well documented (Adesemoye and Kloepper 2009; Bonfante and Genre 2010; Glick 2015).

It is conceivable that PGPB can act by assisting in plant nutrient acquisition and increasing the availability of P to plants through mechanisms other than solubilization of P, for example, via an adjustment of root system architecture (RSA). 3D adjustment and modulation of the RSA in order to forage and mine for limiting P_i are well documented (Lambers et al. 2006; Péret et al. 2014). The various strategies of modifying root traits, such as (1) primary/principal root length, (2) lateral root angle, density and length, and (3) root hair density and length, enable the plant to increase the surface area for the absorption of nutrients. Typically, PGPB enhances the shoot growth by increasing lateral root formation, root hair density, and length, and inhibiting primary root growth (Verbon and Liberman 2016). These effects often occur via modulation of the hormonal balance of the host plant (e.g., auxin and cytokinin) by either producing the hormones themselves (e.g., IAA; indole-3-acetic acid) or indirectly affecting the hormonal pathway (e.g., ACC deaminase which degrades the ethylene precursor ACC, thereby lowering ethylene levels) (Sukumar et al. 2013; Vacheron et al. 2013; Glick 2014). The modulation of RSA can also occur via production of volatile organic compounds (VOCs) which do not require physical contact between the bacteria and plant (Gutiérrez-Luna et al. 2010). The phytohormonal biosynthesis, metabolism, regulation, physiological role, and agronomical impact of phytohormones produced by the PGPB model system of the genus *Azospirillum* are reviewed in detail by Gutiérrez-Luna et al. (2010). Several PGP abilities/activities may be possessed by one bacterial species. For example, *Bacillus thuringiensis* and *Pseudomonas aeruginosa* were shown to possess P-solubilizing activity in addition to the ability to produce IAA in the presence of L-tryptophan (Shahab et al. 2009).

Most of the PGPF constitute arbuscular mycorrhizal fungi (AMF) which are found to interact with approximately 80% of terrestrial plant species in nature and are also able to assist plants in acquiring nutrients, such as P, from the soil. The acquisition of P occurs via two pathways: (1) direct uptake of P_i from the rhizosphere by plant P_i transporters in the epidermis and root hairs and (2) indirect uptake via the AM pathway by which P is taken up by the fungal P_i transporters in the hyphae, translocated to fungal structures, and transferred to the plant at intracellular interfaces (Smith et al. 2011). AMF can influence the direct uptake of P_i in plants via alterations in the expression of P_i transporters, and these effects are crop species dependent. For example, the expression levels of alfalfa (*M. truncatula*) P_i transporter MtPT1 increased in the roots under P starvation but decreased after colonization of AMF *G. versiforme* and *G. intraradices* (Chiou et al. 2001). However, the expression levels of rice (*Oryza sativa*) P_i transporter OsPT11 was upregulated when colonized by *G. intraradices* resulting in the highest P concentration in plant tissue (Glassop et al. 2007; Chen et al. 2013). Plants can also indirectly take up P_i via associations with AMF. After root colonization, AMF develop an external mycelium made up of hyphae that extend the rooting zone of plants, exploiting large soil volumes, and allowing them to access more P_i. The fungi form symbiotic interfaces with the plant via arbuscules which are hypothesized to be the main site of transfer of C and lipids from the host root to the fungi and P in the opposite direction (Bago et al. 2003; Smith et al. 2010). Variations in the ability of different AMF to interact and colonize, as well as their development in root systems of the same species exist (Gao et al. 2001).

PGPB or AMF indigenous to the soil, in which the PSB inoculant is to be used, may possess PGP activities that give them a competitive advantage over the inoculant. The inoculation of the endophytic fungus *Piriformospora indica* resulted in an increased population of the PSB *Pseudomonas striata* at 20 days after sowing to flowering, followed by a decline in the bacterial population at crop maturity of chickpea (Meena et al. 2010). The authors suggested that this might be due to reprogramming of root exudates of the host plant with the advancement of growth stages under the colonization of the endophytic fungus. Moreover, the microbes may be responding differently to the root exudate composition. Thus, indigenous beneficial microbes may also respond better to signals of plant root exudates and establish a niche quicker than the inoculant, leading to the decreased fitness of the later. Therefore, we suggest screening for the various PGP activities (e.g., P solubilization, siderophore production, N_2 fixation, hormone production) and selecting the fittest and most beneficial bacteria to ensure survival and persistence of the inoculant in the rhizosphere.

PSB and AMF may very well compete for the resources in the rhizosphere, but they may also interact with each other forming symbiotic associations where their survival depends on their synergistic effect. Interactions between AM fungi and bacteria may be mutually beneficial for their development and survival (de Boer et al. 2005). AMF can form various interactions with bacteria including N_2-fixing bacteria, phytostimulators (e.g., *Azospirillum*), and PSB (Barea et al. 2005). It has been already noted that some rhizobacteria, called mycorrhiza helper bacteria (MHB), can help in the establishment and functioning of mycorrhiza (e.g., root colonization) (Duponnois and Garbaye 1991). Some of the MHB have been shown to stimulate fungal growth, development, and gene expression (Deveau et al. 2007). Different types of interactions between plants, mycorrhizal fungi, and bacteria were reviewed in detail by Bonfante and Anca (2009). As the interactions between bacteria and fungi may be symbiotic and beneficial, co-inoculation of these microbes for crops under P limitations may improve their efficiency in field trials.

6.4 Co-inoculation of Bacteria and Fungi Under P Limitations

Co-inoculation studies of PSB and AMF have received great interest since the pioneering work of Azcón et al. (1976), due to their synergistic interactions that result in the enhancement of plant P nutrition. Examples of synergistic interactions between bacteria and AMF have been discussed in several reviews (Artursson et al. 2006; Sharma et al. 2013). Co-inoculation of bacteria and fungi have shown a significant selectivity in the enhancement of plant growth, nutrient uptake, mycorrhizal root infection, and microbial biomass, where particular combinations of fungi and bacteria produce a positive effect, while other combinations do not, for example, single or dual inoculation of bacteria with AMF *G. fasciculatum* increased N, P, and K content and enhanced the growth of tomato plants (Azcón 1989). In this study, there was a selective interaction between the AMF and bacteria. One bacterial species was ineffective in increasing growth of tomato when co-inoculated with *G. mosseae*. In another study, the single inoculation of a PGP rhizobacteria or dual inoculation with *Rhizobium* species increased the percentage of mycorrhizal root infection of a mycotrophic legume by *G. intraradices*, while dual inoculation of two bacterial species negatively affected the colonization by *G. coronatum* (Requena et al. 1997). Furthermore, inoculation of *G. intraradices* and *A. niger*, a PSF, resulted in the highest increase of microbial biomass which was positively correlated with shoot biomass of lettuce (Kohler et al. 2007). However, the highest increase in shoot dry weight was observed when *G. intraradices* was combined with the PGPB, *Bacillus subtilis*.

Selectivity can be observed in several studies involving PSB and AMF under P limitations. Responses of lentil to the dual inoculation of AMF and rhizobial strains depended on the compatibility of the two partners, where the strain with P-solubilizing ability enhanced plant growth more than the rhizobial strain without this activity (Zarei et al. 2006). The PSB strain may have a negative or no effect unless co-inoculated with another partner. The single inoculation of PSB *P. striata* exhibited a negative effect on plant growth and yield of chickpea, but dual inoculation with AMF *P. indica* showed an increase of plant dry biomass and grain yield (Meena et al. 2010). Different AMF partners may function differently either as single inoculants or with PSB partners. For example, the single inoculation of *G. intraradices* or *G. mosseae* performed better than dual inoculation of a local *G. mosseae* with *P. striata* (Suri et al. 2011) on maize. *G. intraradices* showed better crop yield and grain quality than two other AMF cultures, but the highest grain and stover yield and root weight was obtained when *G. mosseae* was co-inoculated with *P. striata*.

Further selectivity has been demonstrated when four different microbes were tested: AMF *G. fasciculatum*, PSF *A. awamori*, N_2-fixing bacteria (NFB) *Bradyrhizobium* sp., and PSB *B. subtilis* on mung bean (*Vigna radiata*) under P limitations (Zaidi and Khan 2006). This study revealed compatibility of some bacterial strains with each other or with fungi in improving plant growth and other growth parameters. For example, dual inoculation of PSF with the other inoculants negatively impacted plant growth, while all other dual combinations resulted in enhancement. However, triple inoculation of the PSF, PSB, and AMF significantly enhanced the growth of the mung bean. The triple inoculation of the AMF, NFB, and PSB showed the highest increase in dry matter yield, chlorophyll content, and N and P uptake. Typically, co-inoculation of PSB and AMF under P limitation has shown positive effects on the plant growth of several crop species (Table 6.1).

Competitive and cooperative behavior discussed earlier exists between PSB and AMF under P limitations. A study demonstrated that the dual inoculation of the PSB *Enterobacter* sp. with the AMF *G. mosseae* led to the highest increase of alfalfa (*M. sativa*) shoot dry weight and shoot P content. It was suggested that the *Enterobacter* release P ions from PR into the soil from which the *G. mosseae* translocate it to the plant using isotopically labeled P (Barea et al. 2007). Another study investigated the effects of an AMF *R. irregularis* with a PSB *P. alcaligenes* on phytate mineralization and subsequent transfer to alfalfa plants (Zhang et al. 2014a). It was observed that P mobilized from phytate was not absorbed by the AMF or the plant but instead was converted into microbial biomass by the PSB, indicating that competition exists between the two microbes in soil with low available P. A recent study by the same author suggested that an AMF and a free-living PSB cooperated with each other in providing a key source to each other, where P

Table 6.1 Examples of the synergistic effects of PSB and AMF co-inoculation on different crop species

Fungal species	Bacterial species	Plant species	P source	Experimental scale	Effect	References
G. *etunicatum*	*E. agglomerans* (PSB)[a]	Tomato (*Lycopersicum esculentum*)	Hydroxyapatite	N.S.[b] (pot)	↑ swt.[c]	Kim et al. (1997)
					↑ N, P uptake	
G. *mosseae*; G. *intraradices*	*Azotobacter chroococcum* (NFB)[a]; *Bacillus megaterium* (PSB); *Bacillus mucilaginous* (KSB)[a]	Maize (*Z. mays*)	PR[d]	Greenhouse (pot)	↑ total wt.[c]	Wu et al. (2005)
					↑ N, P uptake	
G. *fasciculatum*; *A. awamori*	*Bradyrhizobium* sp. (NFB); *B. subtilis* (PSB)	Mung bean (*V. radiata*)	PR	N.S. (pot)	↑ total wt.	Zaidi and Khan (2006)
					↑ seed mass	
					↑ chl. [c]	
					↑ N, P uptake	
G. *mosseae*; G. *intraradices*	*R. leguminosarum* (NFB); *M. cicero* (NFB + PSB)	Lentil (*L. culinaris*)	SSP[d]; PR	Greenhouse (pot)	↑ swt.	Zarei et al. (2006)
					↑ N, P uptake	
G. *mosseae*	*Enterobacter* sp. (PSB)	Alfalfa (*M. sativa*)	PR	Greenhouse (microcosms)	↑ swt.	Barea et al. (2007)
					↑ P uptake	
P. indica	*P. striata* (PSB)	Chickpea (*Cicer arietinum*)	N.S.	Growth chamber (pot)	↑ total wt.	Meena et al. (2010)
					↑ gy. [c]	
G. *mosseae*; G. *intraradices*	*P. striata* (PSB)	Maize (*Z. mays*)	SSP	Greenhouse (pot)	↑ rwt. [c]	Suri et al. (2011)
					↑gy. sy.[c]	

(continued)

Table 6.1 (continued)

Fungal species	Bacterial species	Plant species	P source	Experimental scale	Effect	References
G. *etunicatum*	*Burkholderia cepacia* BAM-6 (PSB)	Wheat (*Triticum aestivum*)	SSP	N.S. (wire housed pots)	↑ swt. Rwt.	Saxena and Jha (2014)
					↑ gy. gwt.	
					↑ spl., spwt.[c]	
					↑ N, P uptake	
R. irregularis	*P. alcaligenes* (PSB)	Alfalfa (*M. sativa*)	KH_2PO_4; Sodium-phytate	Greenhouse (Microsoms)	↑ swt.	Zhang et al. (2014a)
					↑ P uptake	

[a]*PSB* P-solubilizing bacteria, *NFB* N_2-fixing bacteria, *KSB* K-solubilizing bacteria
[b]*N.S.* not stated
[c]*wt.* weight, *swt.* shoot weight, *chl.* chlorophyll content, *rw* root weight, *gy.* grain yield, *sy.* stover yield, *gw.* grain weight, *spl.* spike length, *spwt.* spike weight
[d]*PR* phopshate rock, *SSP* single superphosphate

status plays an important role in this interaction (Zhang et al. 2016). It demonstrated that the AMF *Rhizophagus intraradices* release C compounds into the hydrosphere for the PSB *Rahnella aquatilis* to utilize and increase their growth and activity. In return, the PSB phosphatases result in organic-P hydrolysis (mineralization), increasing inorganic-P availability for the AMF to acquire and enhance their hyphal growth. However, when soil available P was low, the PSB competed with the AMF for P, and its activity was not increased, demonstrating again a competition between PSB and AMF.

The effects of PSB and AMF on each other are further demonstrated regarding mycorrhizal root colonization, where fertilization also plays a role. Generally, PSB inoculation tends to increase the percentage of root mycorrhizal colonization of tomato, maize, and wheat (Zarei et al. 2006; Suri et al. 2011; Saxena and Jha 2014). In some cases, the inoculation had no effect on the root infection of tomato (Kim et al. 1997). In other cases, fertilizer application decreased the root colonization of maize and lentil (Wu et al. 2005; Zarei et al. 2006). On the other hand, the AMF had an effect on the PSB population, where the PSB population significantly increased with AMF inoculation of tomato and wheat crops (Suri et al. 2011; Saxena and Jha 2014).

Regarding the nutrient uptake, co-inoculation methods exhibited variability for the different species. Inoculation of AMF *G. etunicatum* or PSB *E. agglomerans* showed increased uptake of N and P but did not show any difference between single or dual inoculation (Kim et al. 1997). Using a combination of PSB, KSB, and NFB co-inoculated with the AMF *G. intraradices* showed higher P and K assimilation by maize than with the AMF *G. mosseae* (Wu et al. 2005). In another study, dual inoculation of a PSF *Aspergillus awamori* and PSB *B. subtilis* exhibited higher increases in P contents at 45 and 60 days after sowing of mung bean compared to other combinations including NFB and AMF. However, triple inoculation of the PSF, PSB, and the AMF *G. fasciculatum* showed the highest increase of P in mung bean (Zaidi and Khan 2006). *G. intraradices* showed superiority over *G. mosseae* in nutrient uptake when co-inoculated with PSB, but co-inoculation with *G. mosseae* showed better P solubilization (Zarei et al. 2006; Suri et al. 2011).

6.5 Conclusions and Recommendations

P is a vital nutrient in agriculture for sustaining high productivity of crops to feed the ever-growing human population. PSB have a great potential in alleviating the damaging effects of P starvation in plants as a result of P unavailability. However, the development of robust PSB bio-fertilizer technology for broad use in agriculture is hindered by the complexity of interactions with the soil and biota and the inappropriate screening, selection, and inoculation processes. We recommend considering several points when developing PSB for application in field conditions. Firstly, the screening of PSB must assess their ability to solubilize and mineralize P sources that resemble P sources in the soil of interest. Accordingly, selection should rely not solely on TCP but needs to include other metal-P and also organic-P compounds.

Secondly, these assays must be performed under varying conditions of pH, temperature, and C and N sources in order to resemble diverse field conditions. Depending on the soil context of the field, PSB may function differently. Therefore, assessing soil properties and available P sources in the relevant area will greatly facilitate the selection of a compatible PSB. Due to a competition of biota in the rhizosphere, selection of the plant-beneficial PSB fittest under the given conditions will greatly increase its chances of persistence in soil and its ability to develop and remain in the niche. Soil microbes can exert beneficial effects on plants independent from P solubilization. Thus, we recommend selection for PSB with further plant growth-promoting activities. As revealed by co-inoculation studies, cooperative behavior also exists between bacteria and fungi. Typically, the co-inoculation of AMF has shown to enhance the P-solubilization efficiencies of PSB and increase plant growth of several crops (Table 6.1). We suggest using dual or triple inoculation of PSB with other beneficial microbes to enhance the likelihood of success in the field. However, the compatibility between the PSB and the selectivity of the crop plant must be assessed prior to inoculation. Ideally, each inoculant should possess a plant growth-promoting ability with the potential to act synergistically.

Acknowledgments The authors would like to thank KAUST for their financial support, Florian Mette for his highly useful comments, the members of the Center for Desert Agriculture, and the Hirt's lab group members for their support and fruitful discussion.

References

Adesemoye AO, Kloepper JW (2009) Plant–microbes interactions in enhanced fertilizer-use efficiency. Appl Microbiol Biotechnol 85(1):1–12. doi:10.1007/s00253-009-2196-0

Artursson V, Finlay RD, Jansson JK (2006) Interactions between arbuscular mycorrhizal fungi and bacteria and their potential for stimulating plant growth. Environ Microbiol 8(1):1–10

Azcón R (1989) Selective interaction between free-living rhizosphere bacteria and vesicular arbuscular mycorrhizal fungi. Soil Biol Biochem 21(5):639–644

Azcon R, Barea J, Hayman D (1976) Utilization of rock phosphate in alkaline soils by plants inoculated with mycorrhizal fungi and phosphate-solubilizing bacteria. Soil Biol Biochem 8(2):135–138

Badri DV, Vivanco JM (2009) Regulation and function of root exudates. Plant Cell Environ 32(6):666–681

Bago B, Pfeffer PE, Abubaker J et al (2003) Carbon export from arbuscular mycorrhizal roots involves the translocation of carbohydrate as well as lipid. Plant Physiol 131(3):1496–1507

Barea JM, Azcón R, Azcón-Aguilar C (2005) Interactions between mycorrhizal fungi and bacteria to improve plant nutrient cycling and soil structure. In: Buscot F, Varma A (eds) Microorganisms in soils: roles in genesis and functions, vol 3. Springer, Berlin, pp 195–212

Barea J, Toro M, Azcón R (2007) The use of ^{32}P isotopic dilution techniques to evaluate the interactive effects of phosphate-solubilizing bacteria and mycorrhizal fungi at increasing plant P availability. In: Velazquez E, Rodriguez-Barrueco C (eds) First international meeting on microbial phosphate solubilization, vol 102. Springer, Dordrecht, pp 223–227

Bashan Y, Kamnev AA, de-Bashan LE (2013) Tricalcium phosphate is inappropriate as a universal selection factor for isolating and testing phosphate-solubilizing bacteria that enhance plant growth: a proposal for an alternative procedure. Biol Fertil Soils 49(4):465–479. doi:10.1007/s00374-012-0737-7

Bieleski R (1973) Phosphate pools, phosphate transport, and phosphate availability. Annu Rev Plant Physiol 24(1):225–252

de Boer W, Folman LB, Summerbell RC, Boddy L (2005) Living in a fungal world: impact of fungi on soil bacterial niche development. FEMS Microbiol Rev 29(4):795–811

de Zelicourt A, Al-Yousif M, Hirt H (2013) Rhizosphere microbes as essential partners for plant stress tolerance. Mol Plant 6(2):242–245

Bonfante P, Anca I-A (2009) Plants, mycorrhizal fungi, and bacteria: a network of interactions. Annu Rev Microbiol 63:363–383

Bonfante P, Genre A (2010) Mechanisms underlying beneficial plant–fungus interactions in mycorrhizal symbiosis. Nat Commun 1:48

Brunner I, Goren A, Schlumpf A (2014) Patterns of organic acids exuded by pioneering fungi from a glacier forefield are affected by carbohydrate sources. Environ Res Lett 9(2):025002

Carvalhais LC, Dennis PG, Fedoseyenko D et al (2011) Root exudation of sugars, amino acids, and organic acids by maize as affected by nitrogen, phosphorus, potassium, and iron deficiency. J Plant Nut Soil Sci/Z Pflanzenernähr Bodenkd 174(1):3

Chen XW, Wu FY, Li H et al (2013) Phosphate transporters expression in rice (*Oryza sativa* L.) associated with arbuscular mycorrhizal fungi (AMF) colonization under different levels of arsenate stress. Environ Exp Bot 87:92–99. doi:http://dx.doi.org/10.1016/j.envexpbot.2012.08.002

Chiou T-J, Liu H, Harrison MJ (2001) The spatial expression patterns of a phosphate transporter (MtPT1) from *Medicago truncatula* indicate a role in phosphate transport at the root/soil interface. Plant J 25(3):281–293. doi:10.1046/j.1365-313x.2001.00963.x

Cordell D, White S (2015) Tracking phosphorus security: indicators of phosphorus vulnerability in the global food system. Food Sec 7(2):337–350

Crowley DE, Kraemer SM (2007) Function of siderophores in the plant rhizosphere. In: Pinton R, Varanini Z, Nannipieri P (eds) The rhizosphere: biochemistry and organic substances at the soil–plant interface, 2nd edn. CRC, Boca Raton, pp 173–200

Deveau A, Palin B, Delaruelle C et al (2007) The mycorrhiza helper *Pseudomonas fluorescens* BBc6R8 has a specific priming effect on the growth, morphology and gene expression of the ectomycorrhizal fungus *Laccaria bicolor* S238N. New Phytol 175(4):743–755

Downie JA (2014) Legume nodulation. Curr Biol 24(5):R184–R190

Duca M (2015) Mineral nutrition of plants. In: Mohr H, Schopfer P (eds) Plant physiology. Springer, Berlin, pp 149–185

Duponnois R, Garbaye J (1991) Mycorrhization helper bacteria associated with the Douglas fir-*Laccaria laccata* symbiosis: effects in aseptic and in glasshouse conditions. Ann Sci For 3:239–251

Eivazi F, Tabatabai M (1977) Phosphatases in soils. Soil Biol Biochem 9(3):167–172

Filius JD, Hiemstra T, Van Riemsdijk WH (1997) Adsorption of small weak organic acids on goethite: modeling of mechanisms. J Colloid Interface Sci 195(2):368–380

Gao LL, Delp G, Smith SE (2001) Colonization patterns in a mycorrhiza-defective mutant tomato vary with different arbuscular-mycorrhizal fungi. New Phytol 151(2):477–491. doi:10.1046/j.0028-646x.2001.00193.x

Geurts R, Lillo A, Bisseling T (2012) Exploiting an ancient signalling machinery to enjoy a nitrogen fixing symbiosis. Curr Opin Plant Biol 15(4):438–443

Glassop D, Godwin RM, Smith SE, Smith FW (2007) Rice phosphate transporters associated with phosphate uptake in rice roots colonised with arbuscular mycorrhizal fungi. Can J Bot 85(7):644–651. doi:10.1139/B07-070

Glick BR (2014) Bacteria with ACC deaminase can promote plant growth and help to feed the world. Microbiol Res 169(1):30–39

Glick BR (2015) Beneficial plant-bacterial interactions. Springer, Cham

Goldstein AH (1994) Involvement of the quinoprotein glucose dehydrogenase in the solubilization of exogenous phosphates by gram-negative bacteria. In: Torriani-Gorini A, Yagil E, Silver S (eds) Phosphate in microorganisms: cellular and molecular biology. ASM Press, Washington, DC, pp 197–203

Grierson P, Comerford N, Jokela E (1998) Phosphorus mineralization kinetics and response of microbial phosphorus to drying and rewetting in a Florida Spodosol. Soil Biol Biochem 30(10):1323–1331

Gutiérrez-Luna FM, López-Bucio J, Altamirano-Hernández J et al (2010) Plant growth-promoting rhizobacteria modulate root-system architecture in *Arabidopsis thaliana* through volatile organic compound emission. Symbiosis 51(1):75–83

Hodge A, Helgason T, Fitter A (2010) Nutritional ecology of arbuscular mycorrhizal fungi. Fungal Ecol 3(4):267–273

Johnson JF, Allan DL, Vance CP (1994) Phosphorus stress-induced proteoid roots show altered metabolism in *Lupinus albus*. Plant Physiol 104(2):657–665

Jones DL, Oburger E (2011) Solubilization of phosphorus by soil microorganisms. In: Bunemann E, Oberson A, Frossard E (eds) Phosphorus in action, vol 26. Springer, Berlin, pp 169–198

Kamilova F, Kravchenko LV, Shaposhnikov AI, Makarova N, Lugtenberg B (2006) Effects of the tomato pathogen *Fusarium oxysporum* f. sp. radicis-lycopersici and of the biocontrol bacterium *Pseudomonas fluorescens* WCS365 on the composition of organic acids and sugars in tomato root exudate. Mol Plant Microbe Interact 19(10):1121–1126. doi:10.1094/MPMI-19-1121

Khan MS, Zaidi A, Ahemad M, Oves M, Wani PA (2010) Plant growth promotion by phosphate solubilizing fungi–current perspective. Arch Agron Soil Sci 56(1):73–98

Khan M, Ahmad E, Zaidi A, Oves M (2013) Functional aspect of phosphate-solubilizing bacteria: importance in crop production. In: Maheshwari DK, Saraf M, Aeron A (eds) Bacteria in agrobiology: crop productivity. Springer, Berlin/Heidelberg. doi:10.1007/978-3-642-37241-4_10

Khan MS, Zaidi A, Ahmad E (2014) Mechanism of phosphate solubilization and physiological functions of phosphate-solubilizing microorganisms. In: Khan MS, Zaidi A, Musarrat J (eds) Phosphate solubilizing microorganisms. Springer, Cham, pp 31–62. doi:10.1007/978-3-319-08216-5_2

Kim K, Jordan D, McDonald G (1997) Effect of phosphate-solubilizing bacteria and vesicular-arbuscular mycorrhizae on tomato growth and soil microbial activity. Biol Fertil Soils 26(2):79–87

Kim KY, Jordan D, McDonald G (1998) *Enterobacter agglomerans*, phosphate solubilizing bacteria, and microbial activity in soil: effect of carbon sources. Soil Biol Biochem 30(8):995–1003

Kohler J, Caravaca F, Carrasco L, Roldan A (2007) Interactions between a plant growth-promoting rhizobacterium, an AM fungus and a phosphate-solubilising fungus in the rhizosphere of *Lactuca sativa*. Appl Soil Ecol 35(3):480–487

Lambers H, Shane MW, Cramer MD, Pearse SJ, Veneklaas EJ (2006) Root structure and functioning for efficient acquisition of phosphorus: matching morphological and physiological traits. Ann Bot 98(4):693–713

Leprince F, Quiquampoix H (1996) Extracellular enzyme activity in soil: effect of pH and ionic strength on the interaction with montmorillonite of two acid phosphatases secreted by the ectomycorrhizal fungus *Hebeloma cylindrosporum*. Eur J Soil Sci 47(4):511–522. doi:10.1111/j.1365-2389.1996.tb01851.x

Liang C, Wang J, Zhao J, Tian J, Liao H (2014) Control of phosphate homeostasis through gene regulation in crops. Curr Opin Plant Biol 21:59–66

Lim BL, Yeung P, Cheng C, Hill JE (2007) Distribution and diversity of phytate-mineralizing bacteria. ISME J 1(4):321–330

Lipton DS, Blanchar RW, Blevins DG (1987) Citrate, malate, and succinate concentration in exudates from P-sufficient and P-stressed *Medicago sativa* L. seedlings. Plant Physiol 85(2):315–317

Lugtenberg B, Kamilova F (2009) Plant-growth-promoting rhizobacteria. Annu Rev Microbiol 63:541–556

Maougal R, Brauman A, Plassard C, Abadie J, Djekoun A, Drevon J-J (2014) Bacterial capacities to mineralize phytate increase in the rhizosphere of nodulated common bean (*Phaseolus vulgaris*) under P deficiency. Eur J Soil Biol 62:8–14

Mardad I, Serrano A, Soukri A (2013) Solubilization of inorganic phosphate and production of organic acids by bacteria isolated from a Moroccan mineral phosphate deposit. Afr J Microbiol Res 7:626–635

Marques JM, da Silva TF, Vollu RE et al (2014) Plant age and genotype affect the bacterial community composition in the tuber rhizosphere of field-grown sweet potato plants. FEMS Microbiol Ecol 88(2):424–435. doi:10.1111/1574-6941.12313

Medveczky N, Rosenberg H (1971) Phosphate transport in *Escherichia coli*. Biochim Biophys Acta Biomembr 241(2):494–506

Meena KK, Mesapogu S, Kumar M et al (2010) Co-inoculation of the endophytic fungus *Piriformospora indica* with the phosphate-solubilising bacterium *Pseudomonas striata* affects population dynamics and plant growth in chickpea. Biol Fertil Soils 46(2):169–174

Nannipieri P, Giagnoni L, Landi L, Renella G (2011) Role of phosphatase enzymes in soil. In: Bunemann E, Oberson A, Frossard E (eds) Phosphorus in action, vol 26. Springer, Heidelberg, pp 215–243

Nautiyal CS (1999) An efficient microbiological growth medium for screening phosphate solubilizing microorganisms. FEMS Microbiol Lett 170(1):265–270

Naves LP, Corrêa A, Bertechini A, Gomide E, Cd S (2012) Effect of pH and temperature on the activity of phytase products used in broiler nutrition. Rev Bras Ciências Avícola 14:181–185

Nesme T, Colomb B, Hinsinger P, Watson CA (2014) Soil phosphorus management in organic cropping systems: from current practices to avenues for a more efficient use of P resources. In: Bellon S, Penvern S (eds) Organic farming, prototype for sustainable agriculture. Springer, Dordrecht, pp 23–45

Neumann G, Römheld V (2007) The release of root exudates as affected by the plant physiological status. In: Pinton R, Varanini Z, Nannipieri P (eds) The rhizosphere: biochemistry and organic substances at the soil-plant interface, 2nd edn. CRC, Boca Ratoon, pp 23–72

Nguyen C (2003) Rhizodeposition of organic C by plants: mechanisms and controls. Agron Sci Prod Veg Environ 23(5–6):375–396

Oberson A, Joner EJ, Turner B, Frossard E, Baldwin D (2005) Microbial turnover of phosphorus in soil. In: Turner BL, Frossard E, Baldwin DS (eds) Organic phosphorus in the environment. CABI publishers, Wallingford, pp 133–164

Osman KT (2012) Soils: principles, properties and management. Springer Science & Business Media, Chittagong

Park KH, Lee CY, Son HJ (2009) Mechanism of insoluble phosphate solubilization by *Pseudomonas fluorescens* RAF15 isolated from ginseng rhizosphere and its plant growth-promoting activities. Lett Appl Microbiol 49(2):222–228

Péret B, Desnos T, Jost R, Kanno S, Berkowitz O, Nussaume L (2014) Root architecture responses: in search of phosphate. Plant Physiol 166(4):1713–1723

Peters NK, Frost JW, Long SR (1986) A plant flavone, luteolin, induces expression of *Rhizobium meliloti* nodulation genes. Science 233(4767):977–980

Pikovskaya R (1948) Mobilization of phosphorus in soil in connection with vital activity of some microbial species. Mikrobiologiya 17(362):e370

Rashid M, Khalil S, Ayub N, Alam S, Latif F (2004) Organic acids production and phosphate solubilization by phosphate solubilizing microorganisms (PSM) under *in vitro conditions*. Pak J Biol Sci 7(2):187–196

Requena N, Jimenez I, Toro M, Barea J (1997) Interactions between plant-growth-promoting rhizobacteria (PGPR), arbuscular mycorrhizal fungi and *Rhizobium* spp. in the rhizosphere of *Anthyllis cytisoides*, a model legume for revegetation in mediterranean semi-arid ecosystems. New Phytol 136(4):667–677

Rudrappa T, Czymmek KJ, Paré PW, Bais HP (2008) Root-secreted malic acid recruits beneficial soil bacteria. Plant Physiol 148(3):1547–1556

Saxena J, Jha A (2014) Impact of a phosphate solubilizing bacterium and an arbuscular mycorrhizal fungus (*Glomus etunicatum*) on growth, yield and P concentration in wheat plants. Clean Soil Air Water 42(9):1248–1252

Scervino JM, Papinutti VL, Godoy MS et al (2011) Medium pH, carbon and nitrogen concentrations modulate the phosphate solubilization efficiency of *Penicillium purpurogenum* through organic acid production. J Appl Microbiol 110(5):1215–1223. doi:10.1111/j.1365-2672.2011.04972.x
Schachtman DP, Reid RJ, Ayling SM (1998) Phosphorus uptake by plants: from soil to cell. Plant Physiol 116(2):447–453
Shahab S, Ahmed N, Khan NS (2009) Indole acetic acid production and enhanced plant growth promotion by indigenous PSBs. Afr J Agric Res 4(11):1312–1316
Sharma SB, Sayyed RZ, Trivedi MH, Gobi TA (2013) Phosphate solubilizing microbes: sustainable approach for managing phosphorus deficiency in agricultural soils. Springer Plus 2:587
Smith SE, Facelli E, Pope S, Andrew Smith F (2010) Plant performance in stressful environments: interpreting new and established knowledge of the roles of arbuscular mycorrhizas. Plant Soil 326(1):3–20. doi:10.1007/s11104-009-9981-5
Smith SE, Jakobsen I, Grønlund M, Smith FA (2011) Roles of arbuscular mycorrhizas in plant phosphorus nutrition: interactions between pathways of phosphorus uptake in arbuscular mycorrhizal roots have important implications for understanding and manipulating plant phosphorus acquisition. Plant Physiol 156(3):1050–1057
Steinkellner S, Lendzemo V, Langer I et al (2007) Flavonoids and strigolactones in root exudates as signals in symbiotic and pathogenic plant-fungus interactions. Molecules 12(7):1290–1306
Sukumar P, Legue V, Vayssieres A, Martin F, Tuskan GA, Kalluri UC (2013) Involvement of auxin pathways in modulating root architecture during beneficial plant–microorganism interactions. Plant Cell Environ 36(5):909–919
Suri V, Choudhary AK, Chander G, Verma T, Gupta M, Dutt N (2011) Improving phosphorus use through co-inoculation of vesicular arbuscular mycorrhizal fungi and phosphate-solubilizing bacteria in maize in an acidic Alfisol. Commun Soil Sci Plant Anal 42(18):2265–2273
Tian J, Wang X, Tong Y, Chen X, Liao H (2012) Bioengineering and management for efficient phosphorus utilization in crops and pastures. Curr Opin Biotechnol 23 (6):866–871. doi:http://dx.doi.org/10.1016/j.copbio.2012.03.002
Turner BL, McKelvie ID, Haygarth PM (2002) Characterisation of water-extractable soil organic phosphorus by phosphatase hydrolysis. Soil Biol Biochem 34(1):27–35
Uren NC (2007) Types, amounts, and possible functions of compounds released into the rhizosphere by soil-grown plants. In: Pinton R, Varanini Z, Nannipieri P (eds) The rhizosphere biochemistry and organic substances at the soil-plant interface, 2nd edn. CRC Press, Boca Raton, pp 1–21
Vacheron J, Desbrosses G, Bouffaud M-L et al (2013) Plant growth-promoting rhizobacteria and root system functioning. Front Plant Sci 4:356. doi:10.3389/fpls.2013.00356
Van Vuuren DP, Bouwman A, Beusen A (2010) Phosphorus demand for the 1970–2100 period: a scenario analysis of resource depletion. Glob Environ Chang 20(3):428–439
Van Wees SC, Van der Ent S, Pieterse CM (2008) Plant immune responses triggered by beneficial microbes. Curr Opin Plant Biol 11(4):443–448
Verbon EH, Liberman LM (2016) Beneficial microbes affect endogenous mechanisms controlling root development. Trends Plant Sci 21(3):218–229
Vitousek PM, Howarth RW (1991) Nitrogen limitation on land and in the sea: how can it occur? Biogeochemistry 13(2):87–115
Vyas P, Gulati A (2009) Organic acid production in vitro and plant growth promotion in maize under controlled environment by phosphate-solubilizing fluorescent *Pseudomonas*. BMC Microbiol 9:174. doi:10.1186/1471-2180-9-174
Weller DM, Thomashow LS (1994) Current challenges in introducing beneficial microorganisms into the rhizosphere. In: O'Gara F, Dowling DN, Boesten B (eds) Molecular ecology of rhizosphere microorganisms: biotechnology and the release of GMOs. VCH, Weinheim, pp p1–18
Wu S, Cao Z, Li Z, Cheung K, Wong M (2005) Effects of biofertilizer containing N-fixer, P and K solubilizers and AM fungi on maize growth: a greenhouse trial. Geoderma 125(1):155–166
Yang J, Kloepper JW, Ryu C-M (2009) Rhizosphere bacteria help plants tolerate abiotic stress. Trends Plant Sci 14(1):1–4

Yarzábal L (2010) Agricultural development in tropical acidic soils: potential and limits of phosphate-solubilizing bacteria. In: Dion P (ed) Soil biology and agriculture in the tropics, Soil biology, vol 21. Springer, Berlin/Heidelberg, pp 209–233. doi:10.1007/978-3-642-05076-3_10

Yi Y, Huang W, Ge Y (2008) Exopolysaccharide: a novel important factor in the microbial dissolution of tricalcium phosphate. World J Microbiol Biotechnol 24(7):1059–1065. doi:10.1007/s11274-007-9575-4

Yoneyama K, Yoneyama K, Takeuchi Y, Sekimoto H (2007) Phosphorus deficiency in red clover promotes exudation of orobanchol, the signal for mycorrhizal symbionts and germination stimulant for root parasites. Planta 225(4):1031–1038

Zaidi A, Khan MS (2006) Co-inoculation effects of phosphate solubilizing microorganisms and *Glomus fasciculatum* on green gram-*Bradyrhizobium symbiosis*. Turk J Agri Forest 30(3):223–230

Zarei M, Saleh-Rastin N, Alikhani HA, Aliasgharzadeh N (2006) Responses of lentil to co-inoculation with phosphate-solubilizing rhizobial strains and arbuscular mycorrhizal fungi. J Plant Nutr 29(8):1509–1522

Zhang L, Fan J, Ding X, He X, Zhang F, Feng G (2014a) Hyphosphere interactions between an arbuscular mycorrhizal fungus and a phosphate solubilizing bacterium promote phytate mineralization in soil. Soil Biol Biochem 74:177–183

Zhang Z, Liao H, Lucas WJ (2014b) Molecular mechanisms underlying phosphate sensing, signaling, and adaptation in plants. J Integr Plant Biol 56(3):192–220

Zhang L, Xu M, Liu Y, Zhang F, Hodge A, Feng G (2016) Carbon and phosphorus exchange may enable cooperation between an arbuscular mycorrhizal fungus and a phosphate-solubilizing bacterium. New Phytol 210(3):1022–1032. doi:10.1111/nph.13838

Corn and Its Interactions with Bacterial Communities

7

Shimaila Ali, Saveetha Kandasamy, Soledad Saldias, and George Lazarovits

Abstract

Corn (maize) hosts a huge array of microbial communities in its root zone (rhizosphere) as well as inside the plant, as endophytes. It is assumed that the most intimate interactions impacting both the host and the microbes occur inside the plant tissues. The microorganism communities associated with corn are now well characterized, and many of their biological functions that impact the plant have been identified, as have factors that modulate the induction and extent of the plants' responses. This review focuses on the impacts that endophytic bacteria have on corn plants and how they may be utilized to maximize crop health and yield. The source of corn endophytes, how they develop and establish inside the plant tissues and seeds, and the biological functions they possess that impact plant growth are discussed. We focus on key functions such as nitrogen fixation by diazotrophic microbes, plant growth promotion via synthesis of hormones, and production of antibiotics that protect plants against both pests and diseases. The influences of transgenes on the corn microbial communities are identified as well as how soil characteristics, agronomic practices, and environmental factors impact the relationship of the corn-microbe interactions. Recent advances in the use of remote sensing technology in corn microbial research are introduced and discussed as to how it can be used to better identify the role of microorganism in crop health and productivity.

S. Ali • S. Kandasamy • S. Saldias
A&L Biologicals, Agroecological Research Services Centre, N5V 3P5 London, ON, Canada

G. Lazarovits (✉)
A&L Biologicals, Agroecological Research Services Centre, N5V 3P5 London, ON, Canada

Department of Biology, University of Western Ontario, London, ON, Canada
e-mail: Lazarovitsg@alcanada.com

S. Mehnaz (ed.), *Rhizotrophs: Plant Growth Promotion to Bioremediation*, Microorganisms for Sustainability 2, DOI 10.1007/978-981-10-4862-3_7

7.1 Introduction

Corn (*Zea mays*) is one of the major cereal crops, worldwide. It is an essential crop to improve if we have to overcome the increasing food demands globally for livestock and biofuel production (Montañez et al. 2012). Many studies have documented the ability of the corn plant to be able to establish a beneficial relationship with a variety of microbial flora. It is important to identify these microorganisms so that we can explore how they can be used to get higher corn productivity. Corn plants can actively interact with the microbiota that is present inside as endophytes or on/in the root as rhizosphere flora. This chapter deals mainly with the endophytes; however, some insights about the rhizospheric microbes are also provided as these all could be potentially utilized as biofertilizers or biocontrol agents.

Endophytes are microbes that can colonize the internal plant tissues without causing any obvious damage or disease symptoms. These microorganisms have many direct impacts on plant growth and health. Due to their location, it is surmised that they interact with their host more effectively than their rhizospheric counterparts (Reiter and Sessitsch 2006; Johnston-Monje and Raizada 2011b; Rashid et al. 2012). Endophytes residing within seed are readily transferred throughout the plant via vascular connections resulting in subsequent colonization of embryos and endosperm (Malfanova et al. 2013). While microbiological studies for most part have focused on the role, microorganisms play as plant or animal pathogens, or for their use in industrial production, currently, the focus is on their role in ecosystem health. The effort on the role bacteria play in the human ecosystem called "the human microbiome" has provided an impetus to study similar roles of bacteria in plants. With the major reductions in the cost and speed of sequencing, we can for the first time examine whole communities and their functions.

7.2 Fungal Endophytes

A group of 63 fungal endophytes were isolated from corn var. *Pulut* (waxy corn) roots; they belong to six major genera, namely, *Trichoderma*, *Fusarium*, *Aspergillus*, *Penicillium*, *Acremonium*, and *Botryodiplodia* (Amin 2013). The population of fungal endophytes in corn is mediate in a major way by the production of defense compounds within the host plants. Fungi, however, have a tendency to evolve fairly rapid tolerance to host defense compounds. Saunders and Kohn (2009) experimentally proved that the corn varieties which produce benzoxazinoids (BXs), a group of host defense compounds, had 35 times higher isolation frequency of *Fusarium* species than nonproducers. Some BX-detoxifying fungal species, such as *Fusarium verticillioides*, *F. subglutinans*, *F. proliferatum*, and *F. graminearum*, cause diseases in corn but are also common endophytes. Their overall study indicated that the host plant production of defense compounds can potentially alter the associating endophytic fungal community structure and composition. While such compounds restrict fungal agents, they appear to be attractants to plant-beneficial rhizobacteria such as *Pseudomonas putida* during the young and vulnerable growth stages of maize (Neal et al. 2012).

7.3 Actinomycetes Endophytes

Actinomycetes represent a dominant part of soil and rhizosphere microbial populations (Araújo et al. 2000). However, both the presence and isolation of this important genus in corn have been neglected, although the presence of endophytic actinomycetes has been reported in many other plants (Sardi et al. 1992; Matsukuma et al. 1994; Matsumoto et al. 1998). Only one report was found that describes the isolation of 53 actinomycetes in *Zea mays* variety Piranão. These actinomycetes were isolated from surface-sterilized corn leaves and root tissues and belong to genera *Microbispora*, *Streptomyces*, and *Streptosporangium*, being *Microbispora* the prevailing genus (Araújo et al. 2000).

7.4 Bacterial Endophytes

Bacterial endophytes have been isolated and identified from seed, root, stem, leaf, and kernel of corn plants. The various plant tissues show differential levels and types of bacterial colonization with the preferences being related to the plant age, environmental conditions, sampling time, corn ontogeny, and corn cultivar (Roesch et al. 2006; Montañez et al. 2009). A healthy corn plant's tissues carry from 10^4 to 10^{10} cfu/g of fresh weight (fw) from emergence to postharvest stages, respectively, with an average of 10^3–10^7 cfu/g fw throughout the growing season (Fisher et al. 1992; McInroy and Kloepper 1995a; Rai et al. 2007). However, the endophytic population also depends on the variety of corn used and the region of cultivation, i.e., temperate vs. tropical (Rai et al. 2007). A tropical corn variety, Pusa Early Hybrid Makka-1 (PEHM-1), when pretreated at the seedling stage with a bacterial suspension and then grown under greenhouse conditions, had bacterial densities in plants at 28 days after emergence (DAE) that were slightly less when compared to that of field-grown plants of the same age and variety (Rai et al. 2007). Likewise, Palus et al. (1996) reported the presence of cultivated endophytes in the apoplastic fluid of maize plants. Colonization of host tissues by endophytes under gnotobiotic conditions is determined by the inoculum density, temperature, and host genotype (Pillay and Nowak 1997). A wide variety of bacterial endophytes were isolated from various local varieties of Brazilian corn roots (Ikeda et al. 2013) that revealed that the plants discriminate as to which genera they tolerate or harbor. Hybrid varieties generally showed the highest bacterial diversity. The bacterial endophytes belonging to the genera *Klebsiella*, *Burkholderia*, *Bacillus*, and *Pantoea* were most common, with *Pantoea* sp. being present in all corn genotypes studied. Szilagyi-Zecchin et al. (2014) identified seven individual endophytes from corn, of which six belonged to the genus *Bacillus* and one to the genus *Enterobacter.* All bacteria were antagonists of corn pathogens. The *Enterobacter* isolates also facilitated seed germination and early seedling development and were considered as growth promoters under laboratory conditions. Figueiredo et al. (2009) isolated 42 bacterial endophytes that all turned out to be *Bacillus* species. The group claimed that the most prevalent bacterial endophytic species in Brazilian sweet corn is *Bacillus*, which was supported by the work published in China (Gao et al. 2004) on 14 different Chinese corn cultivars.

Montanez et al. (2012) isolated bacterial endophytes from a variety of commercial corn cultivars. The isolates belonged to the genera *Pantoea*, *Pseudomonas*, *Rhanella*, *Herbaspirillum*, *Azospirillum*, *Rhizobium* (*Agrobacterium*), *Enterobacter*, and *Brevundimonas*. An ensuing characterization of the functions of these bacteria revealed that many were able to fix nitrogen (diazotrophs; Montañez et al. 2012); produced the hormone indole acetic acid (IAA), as well as siderophores; and could solubilize phosphorous. However, only three isolates *Enterobacter* spp., *P. fluorescens*, and *Herbaspirillum frisingens* significantly promoted corn growth in vitro (Montañez et al. 2012). Pereira et al. (2011) suggested that culture-dependent and culture-independent methods for the analysis of bacterial community diversity were complementary to each other (Hardoim et al. 2008). The endophytic bacterial flora of maize cv. Monsanto DK684RR2 root was isolated through culture-independent method and found to be composed of γ-Proteobacteria (79.5%), with the prominent representation by *Enterobacter*, *Klebsiella*, *Erwinia*, and *Pseudomonas*, followed by Firmicutes (20.5%) with *Bacillus* as the prevailing genera. However, culturable bacterial endophytes isolated from the same sample showed mostly *Bacillus* (Firmicutes 81.5%), followed by *Achromobacter* (β-Proteobacteria 11.1%), and the remaining was identified as *Pseudomonas* (γ-Proteobacteria 7.4%) (Pereira et al. 2011). *Rhizobium etli*, considered to be a corn endophyte (Rosenblueth and Martínez-Romero 2004), has been found to colonize both the root and stem (Gutiérrez-Zamora and Martínez-Romero 2001).

7.5 Sources of Corn Endophytes

Microbial endophytes enhance the plants' ability to acquire nutrients, grow better via production of growth-promoting hormones, and enhance resistance by antagonizing pathogens or activating resistance genes (Rosenblueth and Martínez-Romero 2006; Johnston-Monje and Raizada 2011a). Modern maize, *Zea* was domesticated 9,000 years ago in Mexico from a wild grass called teosintes (Matsuoka et al. 2002). The edible crop gave rise to diverse traditional landraces by continuous breeding, and it was spread by native growers throughout the American continent (Matsuoka et al. 2002). Today we have created an enormous diversity of highly productive commercial inbreds and hybrids (Duvick et al. 2010). Modern maize cultivars are considered to be less resistant to pests and diseases than their wild relatives as a result of changes in their genetic makeup brought about by continuous selection for various traits (Wang et al. 2005; Lange et al. 2014).

There are conflicting reports on the primary source of endophytes; are they inherited (vertically transmitted) and/or is the relative proportion derived from the surrounding environment through the process of crop production (Hallmann et al. 1997). Through 16S-rDNA gene profiling, culturing, and microbial trait analysis, the effect of soil swapping on founder bacterial endophyte communities has been tested using phenotyping. The results indicated that soil swaps did not affect the bacterial endophyte composition, richness, and diversity. At the same time, the endophyte population of maize gets changed slowly over a long period of time so it is inferred that the wild maize (teosinte) are less susceptible to pests and diseases than their modern maize (corn) relatives, maybe because of the presence of pathogen combating microbial endophyte population. Mousa et al. (2015) tested a variety of endophytes isolated from diverse maize genotypes including wild teosintes, traditional landraces, and modern varieties and their ability to antagonize *F. graminearum*, which causes *Gibberella* ear rot (GER) in maize. The authors concluded that the wild relatives of the crop plants used today may serve as a reservoir for endophytes for use in the ongoing fight against disease threats to modern agriculture. Johnston-Monje et al. (2016) proved that the most common bacterial communities in early stage maize rhizosphere are seed transmitted by planting two genotypes of Brazilian maize into sterile sand, a deep underground subsoil, and a topsoil from the Amazon jungle.

7.6 Seed-Borne Corn Endophytes

Seeds are the source of life continuity, for both plants and microbial population that they carry to next generation. Only a few studies have been reported on seed-borne corn endophytes (Rijavec et al. 2007; Johnston-Monje and Raizada 2011a; Singh et al. 2016). In a study conducted on four genetically related parental lines and their hybrid varieties of Chinese corn (Liu et al. 2012), a noncultural approach was used to investigate bacterial communities in parental lines and their hybrid offspring. The bacterial endophytic population of genetically related corn hybrid seeds shared common microflora to their parental lines, especially the dominant genera composition was similar to those found in either or both of parents (Liu et al. 2012). This suggests that there is a direct relationship between endophytic bacterial content of offspring seed and their parental bacterial strain. The prevailing bacterial genera found common in all genetically related Chinese corn varieties included *Pseudomonas*, *Acinetobacter*, *Bacillus*, *Sphingomonas*, *Enterobacter*, *Burkholderia*, *Klebsiella*, *Sphingomonas*, *Leclercia*, *Pantoea*, and *Roseateles*, mainly belong to Proteobacteria.

A number of bacterial endophytes were isolated from kernels of four different corn varieties (Rijavec et al. 2007). Molecular identification by 16S rRNA gene amplification revealed that the isolates belonged to *Pantoea*, *Microbacterium*, *Frigoribacterium*, *Bacillus*, and *Sphingomonas* species, whereas the genus *Pantoea* was found to be corn cultivar specific. The corn cultivar (W22) was susceptible to fungus, i.e., *Lecanicillium aphanoladii*. Moreover, corn kernels that harbor *Pantoea* species did not show any fungal deterioration symptoms, and the isolated *Pantoea* strain exhibited antifungal activity against *Lecanicillium aphanoladii* when tested in vitro (Rijavec et al. 2007).

A study by Johnston-Monje and Raizada (2011a) discusses the diversity and conservation of corn seed-borne endophytes, in ten different varieties of corn. The culturable method endorsed the presence of *Enterobacter*, *Methylobacteria*, *Pantoea*, and *Pseudomonas* species with γ-Proteobacteria being the dominant class. However, the culture-independent method, terminal restriction fragment length polymorphism

(TRFLP) analysis of 16S rRNA gene, demonstrated the prediction of *Clostridium* and *Paenibacillus* species presence in all corn genotypes (Johnston-Monje and Raizada 2011a). It was also argued on the basis of TRFLP data-based analysis that there are corn species-specific seed-borne endophytes present in different subgroup of same corn species (i.e., *Zea mays*) and include *Chloroflexi*, *Bradyrhizobium* or uncultured bacterium TX1A8, and *Paenibacillus caespitis*. These microbes (or related peaks) were not found in the other species of corn seeds used in that study (Johnston-Monje and Raizada 2011a). Corn species-specific seed-borne endophytes are also able to colonize plant's roots and rhizosphere (Johnston-Monje and Raizada 2011a).

7.7 Nitrogen-Fixing Corn Endophytes

The presence of diazotrophic bacteria associated with corn is considered to improve the capacity of corn to garner nitrogen required for root growth, plant health, and productivity. The extent of nitrogen fixation in maize, however, has been shown to be highly variable although sometimes substantial. It appears that the choice of cultivars is very critical to get a response to this trait for the inoculum used to treat plants. Modern maize and inbreds varied greatly in their response to diazotrophs. Araujo et al. (2014) tested the effects of inoculating three hybrids with the diazotrophic bacteria *H. seropedicae* under greenhouse conditions with different rates of nitrogen fertilization. Inoculation coupled with fertilization increased shoot N contents 32% and 62% for the hybrids P3646H and BRS3035, respectively. However, with no added fertilizer, nitrogen use efficiency was increased by 34% and 64%, for the two respective hybrids with increased plant biometrics. There was no effect of inoculation on the morphology, productivity, or chlorophyll content of DeKalb (Monsanto) hybrid "DKB 390" grown for 30 days after emergence in a similar greenhouse inoculation trial with *Azospirillum brasilense* (AbV5) and *H. seropedicae* (SmR1). Araujo et al. (2014) screened 35 different Brazilian cultivars for their response to *H. seropedicae* inoculation. Only nine of the commercial hybrids responded. Alves et al. (2015) screened different strains under controlled conditions of the *H. seropedicae* for their effects on maize. Only one strain (ZAE94) responded well, with the field inoculation of that strain increasing yield up to 34%, depending on the maize genotype, and fixed 37% of the N in one of the hybrids (SHS5050). Although nitrogen fixation has not been clearly documented in corn, only a few reports described the possibility of the presence of endophytic nitrogen fixers in corn. A study by Montanez et al. (2012) described the isolation and identification of diazotrophic bacterial endophytes in a variety of commercial corn cultivars by ^{15}N isotope dilution method. The nitrogen fixation capabilities of the isolated bacterial endophytes were checked by acetylene reduction assay and then verified by amplifying the *nif*H genes. Only 6.2% of isolates (11 of 178) had this activity and hence are considered as corn endophytic diazotrophs. The isolated endophytic diazotrophs belonged to the genera *Pantoea*, *Pseudomonas*, *Rhanella*, *Herbaspirillum*, *Azospirillum*, *Rhizobium* (*Agrobacterium*), *Enterobacter*, and *Brevundimonas*. *Burkholderia tropica* sp. nov. has been reported to be a corn nitrogen-fixing endophyte under in vitro conditions (Reis et al. 2004). *Rhizobium etli*, a corn endophyte, was reported to fix nitrogen in common bean (Martínez-Romero 2003) but not in corn, although it is widely present in all tissues of corn (Rosenblueth and Martínez-Romero 2004). While bacteria in corn are not considered as having a major role in nitrogen fixation (Chelius and Triplett 2000), the *Burkholderia* sp. found in corn cultivated in Mexico (Estrada et al. 2002) was confirmed to possess this activity by the acetylene reduction assay, PCR-based amplification of *nif*H genes, and corn seed bacterial inoculation assays, under laboratory conditions. *Klebsiella pneumoniae* isolated from surface-sterilized tissues of corn (Palus et al. 1996) was also found to be able to fix nitrogen and to effectively colonize roots (zone of root hair formation) and stems (intracellular space of stem cortex) (Chelius and Triplett 2000). *H. seropedicae* was also documented as corn root diazotrophic endophyte (Baldani et al. 1992; Dobereiner et al. 1993) but a pathogen to other plants such as *Sorghum* and *Pennisetum* (Dobereiner et al. 1993). Additionally, it cannot utilize sucrose as carbon source so its use as biofertilizer cannot be greatly exploited since corn is a sucrose-rich environment (Bertolini et al. 1993; Palus et al. 1996). Many other studies have described biological nitrogen fixation associated with corn in vitro by means of acetylene reduction assay (Von Bülow and Döbereiner 1975), by direct nitrogen isotope incorporation assays (Rennie 1980), and by inoculation (Salamone et al. 1996), but none clearly stated the involvement of endophytic bacteria in the nitrogen fixation activity.

7.8 Microbial Treatments to Corn

Bacillus subtilis seed treatment increased the nutrient availability to corn plants and its growth potential (Canbolat et al. 2006). Corn yield was increased by 24–30% by the inoculation of *A. brasilense* (Hungria et al. 2010). A fungal endophyte, *Acremonium zeae*, originally isolated from corn kernels harvested in North Carolina in 1977 (Wicklow et al. 2005), exhibited strong antifungal antibiotics against *Aspergillus flavus* and *Fusarium verticillioides*, in vitro. When the culture was used as biocontrol, it produced pyrrocidine antibiotic in preharvest corn kernels, which was active against *F. verticillioides* and *A. flavus* (Wicklow et al. 2005).

The inoculation of the plant growth-promoting rhizobacteria (PGPR), *A. brasilense* strain Az, into corn (hybrid 310) significantly increased growth of 2-week-old seedlings under greenhouse conditions. The treatments were applied to seed as free living bacteria or as immobilized bacterium in calcium alginate pellets (El-Katatny and Idres 2014). The effect of co-inoculation of *A. brasilense* with the biocontrol fungus *Trichoderma harzianum* (T24) increased shoot and root growth and the mineral contents significantly compared to un-inoculated plants. The largest differences were observed when the seeds were pre-inoculated with immobilized cells of *A. brasilense* alone (El-Katatny and Idres 2014).

The mechanisms by which bacteria colonize plants are starting to be studied (Rosenblueth and Martínez-Romero 2006). The colonization abilities of bacteria are dependent on various genetic factors, and loss of certain secreted proteins in bacterial mutants limits colonization ability. Molecular studies on the host defense responses provided information that the host plants can limit the microbial populations inside the host plants.

7.9 Biocontrol

7.9.1 Biocontrol of Fungal Pathogens

Bacterial endophytes (*Paenibacillus* sp. and *Citrobacter* sp.) isolated from wild corn exhibited growth suppression of a number of crop fungal pathogens in vitro including *Alternaria alternata*, *A. arborescens*, *Aspergillus flavus*, *A. niger*, *Bionectria ochroleuca*, *Davidiella* (*Cladosporium*) *tassiana*, *Diplodia pinea*, *Epicoccum nigrum*, *Fusarium lateritium*, *F. sporotrichioides*, *F. graminearum*, *F. avenaceum*, *Gibberella avenacea*, *Nigrospora oryzae*, *N. sphaerica*, *Paraconiothyrium brasiliense*, *Penicillium expansum*, *P. afellutanum*, and *Rosellinia corticium* (Mousa et al. 2015). When these wild-maize endophytes were applied as a seed treatment to a modern maize hybrid variety P35F40, which is susceptible to *F. graminearum* and causes *Gibberella* ear rot and produces mycotoxin, deoxynivalenol (DON), the bacterial treatment significantly suppressed fungal diseases as well as significantly reduced the concentration of mycotoxins during storage (Mousa et al. 2015). This suggests that the ancestors of modern maize (wild maize, including teosinte) acquired microbes that enhanced resistance to a diverse group of pathogens that were likely lost as a consequence of domestication, migration, and breeding (Wang et al. 2005; Johnston-Monje and Raizada 2011b; Johnston-Monje et al. 2014; Lange et al. 2014; Mousa et al. 2015). There are other studies done in greenhouse experiments or under gnotobiotic conditions that provide data indicating that *Fusarium*-mycotoxin-based infections and accumulations of the toxins can be controlled by treating corn with a combination of *Bacillus mojavensis* (Bacon et al. 2008) and *Trichoderma* sp. as seed treatments (Bacon et al. 2001). However, when this study was expanded to the field, the results showed it was the biocontrol agent (i.e., *Bacillus mojavensis*) itself that was suppressed by the *Fusarium* sp. (Bacon et al. 2008) as a result of potent phytotoxic (Arias 1985) and antibacterial compound such as fusaric acid, which inhibited *B. mojavensis* and other biocontrol agents (Bacon and Hinton 1996; Landa et al. 2002; Bacon et al. 2008). Mercado-Flores et al. (2014), demonstrated biological control of corn head smut, caused by *Sporisorium reilianum*, by *B. subtilis* strain 160 under field conditions. The bacterium was isolated from soil where corn had a high incidence of this disease (Petatán-Sagahón et al. 2011). A suspension of this bacterium was applied to seeds of hybrid AS150, highly susceptible to head smut, and was subsequently sown into a field with a history of smut disease in Mexico. *B. subtilis* significantly decreased the incidence of smut disease by 47.6% in the first year and 31.9% for the second year compared to untreated corn. Inoculated plants were healthier and yielded significantly higher amounts of grain (Mercado-Flores et al. 2014).

Brevibacillus reuszeri has potent antagonistic activity against a number of pathogenic fungi, mainly mycotoxigenic *Fusarium* sp. (Joo et al. 2015). When fully grown corn ears were co-inoculated with *B. reuszeri* culture filtrate and fungal conidia of 46 different mycotoxigenic *Fusarium* species known to produce trichothecenes, zearalenone, fumonisin, beauvericin, gibberellins, and moniliformin (Proctor et al. 2006), the filtrate suppressed corn ear rot produced by 39 of 46 mycotoxigenic *Fusarium* species (Joo et al. 2015). Furthermore, this isolate has since been shown to suppress other fungi including members of *Alternaria* sp., *Aspergillus* sp., *Penicillium* sp., and *Epicoccum* sp. in dual-culture plate assay (Joo et al. 2015).

Soil microorganisms produce volatile organic compounds (VOC) that significantly influence plant growth and disease control (Bailly and Weisskopf 2012). One of the very early-discovered VOC was 2,3-butanediol (2,3-BD) that is believed to confer plant resistance against a number of biological stresses (Ryu et al. 2004; Han et al. 2006; Cortes-Barco et al. 2010; Rudrappa et al. 2010). When surface-sterilized maize seeds (*Z. mays* var. Delprim) were treated with a 2,3-BD producing *E. aerogenes*, the emerging seedlings exhibited increased resistance to the Northern corn leaf blight fungus *Setosphaeria turcica*, and a significant decrease was found to the attractiveness of the plant toward the wasp parasitoid *Cotesia marginiventris* (D'Alessandro et al. 2014).

Many plant-associated microorganisms produce antibiotics that can be utilized by their host plants to overcome attack by various phytopathogens. *Acremonium zeae*, a fungal endophyte of corn, produces antibiotic pyrrocidine A, which exhibits potent antifungal activity against corn kernel-rotting and mycotoxin-producing fungi *A. flavus* and *F. verticillioides* in vitro (Wicklow and Poling 2009). Cells of *A. zeae* are also able to prevent postharvest corn seeds from aflatoxin contamination by *A. flavus* (Wicklow et al. 2005). Moreover, pyrrocidine A is highly effective in reducing seed-borne pathogens *Stenocarpella maydis*, *F. graminearum*, and *Clavibacter michiganense* subsp. *nebraskensis*, causal agents of severe seedling blights and vascular wilts of corn (Wicklow and Poling 2009).

7.9.2 Corn Insect (Pest) Biocontrol

Like other grasses, corn is susceptible to a number of insect pests. The South American corn rootworm, *Diabrotica speciosa* (Germer), is a polyphagous herbivore that causes severe damage to corn and other plants (Ventura et al. 2001). The effects of the bacterium *A. brasilense* on the impact of *D. speciosa* larvae infestation on the corn cultivar *Z. mays* variety Delprim were done in Brazil. Ten-day-old seedlings derived from treated seed were exposed to larval infestation. The results showed that seedlings pretreated with *A. brasilense* were protected against attack by *D. speciosa* possibly due to elevated levels of a volatile compound, (E)-ß-caryophyllene, found in roots of the inoculated corn (Santos et al. 2014). Volatiles produced by bacteria are known to trigger host resistance mechanisms (Ryu et al. 2004). The indirect biocontrol of corn rootworm by *A. brasilense* can be considered as potential tool for insect pest management and control.

7.10 Microbial Communities Associated with Transgenic vs. Non-transgenic Corn

The rhizospheric microbial population is influenced by many factors including plant genetic makeup, environmental conditions, and soil type. The community structure of transgenic corn mainly depends on the type of the gene modified, the level of transgene expression, and the insertion site of transgene within the genome or plasmid (Brusetti et al. 2005). The transgenic Bt corn cultivar (Bt 176) was found to exhibit a different rhizospheric microbial community than its non-transgenic counterpart (Brusetti et al. 2005). Bt 176 expresses Cry protein, a crystalline toxin of *B. thuringiensis*, that is secreted as root exudate (Saxena et al. 1999). The root exudation pattern of the transgenic plant with Cry protein was significantly different from the non-transgenic corn (Brusetti et al. 2005) and likely explains the differences in the rhizospheric microbial structures of the corn plants (Saxena and Stotzky 2000). However, the bacterial community profile was not significantly affected in genetically modified corn with pat gene confers glufosinate herbicidal resistance (Schmalenberger and Tebbe 2002). Plants such as potato and cotton that were also genetically modified by insertion of the Cry protein gene showed no significant differences in their rhizospheric microbial structures (Donegan et al. 1995, 1996).

7.11 Impact of Tillage Practices on Corn Microbial Ecology

The soil hosts and sustains a large array of soil micro- and macroorganisms with various ranges of abundance and diversity as a result of the variable soil chemical, physical, and textural properties. Intensive agricultural practices potentially effect soil properties thereby its functioning. Sustainable agricultural practices such as minimal tillage, cover cropping, crop residue retention, fence row farming, and regular crop rotations will be an alternative to intensive agricultural practices that will retain the soil health and productivity (Alvear et al. 2005; Ussiriet al. 2009).

Researchers compared the influence of various conventional and conservative farming practices on the microbial community structure and diversity through extensive phylogenetic analysis and found that the zero tillage had more influence on the microbial community structure and diversity than the traditional practices. The conventional soil tilling influenced the communities such as Actinobacteria, Betaproteobacteria, and Gammaproteobacteria. Additionally, they found a significant positive correlation of community structure with the soil organic matter and the proportion of its clay content. Bacteroidetes, Betaproteobacteria, Cyanobacteria, and Gemmatimonadetes communities were mostly affected by residue management with the conventional farming than no tillage farming. Crop rotation or the monoculturing did not appear to affect the bacterial communities (Navarro-Noya et al. 2013).

Mr. Dean Glenney (Haldimand, ON) developed a cropping system where corn and soybeans are planted in the exact same place on alternating years using a no-till production system and precision planters. While he did not find benefits of the rotation for the first 5 years, in the 6 year, corn production increased and has now reached 300 bu/A (20 tons/Ha) in a region where the average yield is 150 bu/A (10 tons/Ha). We have shown that the biological factors likely contribute to the yield increases in Mr. Glenney's soils. The cultural practices he developed likely enriched the beneficial microbial communities in the soil and quite possible that fluorescent pseudomonads are involved. Management of beneficial soil microbial communities by cultural practices has enormous promise as a means to improve plant health and reduce costs of production (Islam et al. 2015). In the last few years of study, we have compared the no-till, minimal-till, and conventional tillage practices and the effects of cover crops on the microbial communities of corn which was planted in rotation with corn; we found a significant difference in corn sap microbial communities between no-till and conventional tillage farms but no significant change between zero and minimal tillage sites. Among the four cover crops such as millet, garden pea, soybean, and mustard tested, only the mustard significantly changed the community structure and diversity of corn sap, root, or the rhizosphere soil, while the rest of the cover crops did not affect both the structure and diversity (Islam et al. and Saveetha et al. unpublished).

7.12 Factors Determining Microbial Population Inside the Plants

Plant genotype is a significant factor in determining which microflora is established in the close vicinity (rhizosphere) (Michiels et al. 1989; Gomes et al. 2001; Rosenblueth and Martínez-Romero 2004; Pereira et al. 2011) and inside the plant including reproductive parts (i.e., seed) (Johnston-Monje and Raizada 2011a). Even the wild-type plant may carry different set of microorganisms than the mutant of the same plant cultivar in its rhizosphere (Neal et al. 1973). The plants' physiological characteristics control the type of microbial populations that reside inside and on the surface of the plant (Hardoim et al. 2008; Van Overbeek and Van Elsas 2008). Additionally, seed shape, different part of tissue and development, and germination stage are also considered important factors in harboring microbial population in plants (Zou et al. 2011; Liu et al. 2012). The diversity of endophytes (bacterial and fungal) in corn also depends on the age of the plant (Baudoin et al. 2002; Pereira et al. 2011) and the method (culturable or nonculturable) chosen for such analysis (Pereira et al. 2011). The age of the plant is also a determinant of the number of microflora a plant may carry; the older the plant is, the lower the number of microbes it will possess. Upon maturity, corn plants carry less number of bacteria, and the prominent bacterial flora includes *Bacillus* and *Pseudomonas* strains (McInroy and Kloepper 1995a). Agricultural practices, such as the use and type of fertilizers, also have impact on the endophytic communities of corn (Seghers et al. 2004).

Soil environment is the major source that contributes a plant's microbiome, significantly (Buyer et al. 1999). The most obvious portal of entry for soilborne endophytes is root and its cracks. Plant roots are in direct contact with the microbes present in the soil and have greater chances to become the endophytic population of the respective plant; however, certain genes (of the plants or/and the bacteria) also control the microbial colonization of different plants (Liu et al. 2012).

Plants that are more genetically diverse than their parents have a better chance to carry higher and more diverse microbial communities compared to parental plants. Picard and Bosco (2006) found that the plants from the filial generation of hybrid maize attracted more *Pseudomonas* as a consequence of having more crude protein contents in their roots. It has been shown that hybrid plants of rice (Xiao et al. 1995) acquire genetic traits that make them superior microbial hosts compared to their parents as a result of dominance complementation. In our current studies on the corn microbiome, we found that plants with the same genetic background showed greater diversity and abundance when grown in low vs. high productive sites of the same land (Saveetha et al. unpublished).

7.13 Application of Remote Sensing Technology in Corn Microbial Research

Corn is one of the major grains grown in Canada, particularly Ontario. The proposed maximum theoretical yield of corn is 400–450 bushels per acre in Canada, but the growers in Ontario are producing on an average 150 bushels per acre. In the course of our studies, we discovered that almost all corn production sites have zones of high, average, and low yield, within the fields. Based on this aspect, we designed a study that aims to identify the relationships between the soil chemical properties and the plant endophytic microbial communities, in order to determine the factors influencing the yield. If we could identify the drivers of yield between high- and low-yielding zones, we could devise means to make the field more uniform toward higher productivity. To separate such zones, we have been using aerial imaging systems in this study.

A remote sensing unmanned aerial vehicle (UAV) was flown over each field during V10 to R1 stage (60–80 days after planting – DAP) of the crop to generate a visual representation of healthy and stressed spots. A Normalized Difference Vegetation Index (NDVI) is then created of the field (Fig. 7.1). The UAV bears two cameras mounted on each wing; one camera collects full-color imagery, while the other camera collects infrared imagery. NDVI is calculated based on contrasting intense chlorophyll pigment absorption in the red (R) visible light against the high reflectance of plant materials in the near infrared (NIR) NDVI = [NIR−R]/[NIR+R] (Moriondo et al. 2007). Theoretical and experimental studies have proven NDVI to be linked to several vegetation properties including percentage cover, green leaf area index, and active green biomass. NDVI is also an indirect indicator of primary productivity.

The ratio of different soil chemical parameters affects microbial community richness and diversity in many ways. Although some studies exist reporting gross measures of soil microbial parameters and processes, limited information is available on how the soil chemistry affects the richness, diversity, and composition of plant microbiome and vice versa. In our study, we thus far compared the structure

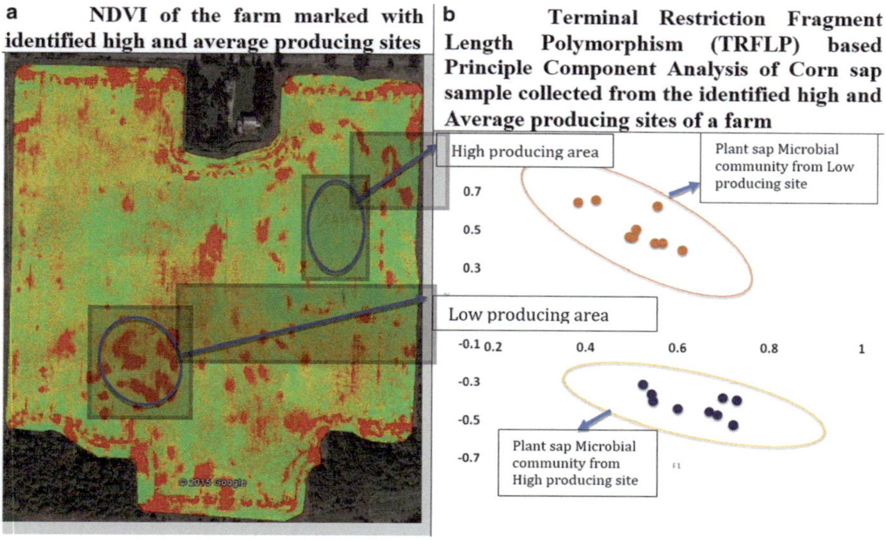

Fig. 7.1 (**a**) NDVI of the farm marked with identified high- and average-producing sites; (**b**) terminal restriction fragment length polymorphism (TRFLP)-based principal component analysis of corn sap sample collected from the identified high and average production sites of a farm

and diversity of bacterial communities with soil and plant chemical parameters in high- and low-producing sites (identified based on remote sensing technology) of 15 different farms across Ontario and determined what soil and plant chemical properties influence bacterial communities. In all fields examined thus far, we have found that the endophytic microbial communities were distinct between low- and high-producing sites. The communities in high-producing sites were rich in their populations of *Pseudomonas* species, while the community of low-producing sites was populated with a gram-positive bacterium. The *Rhizobium* population was greatly influenced by general fertility index. In general, the high-producing zones had significantly greater bacterial richness and less diversity than the zones with low yields. Redundancy analysis revealed that high soil bacterial diversity correlated with a high respiratory quotient, indicating stressed microbial communities. By contrast, soil bacterial communities in high-producing sites positively correlated with high soil P levels and optimal K:Mg range (Saveetha et al. unpublished). Taken together, the corn sap, root or rhizosphere soil, bacterial community composition, and richness which may indicate shifts in their functionality despite equal levels of total colony-forming units were greatly influenced by soil and physical and chemical properties apart from genetics and other cultivation practices; these changes could be easily identified with the recent advances in the technology. A significant outcome of this work is that there is a great variability in the diversity of the bacterial communities within corn plants growing in a field. If one is sampling across a transect for the microbial populations, an average of the populations will be obtained. However, if the plants under stress can be segregated from those that are not stressed, significant differences can be found between the groups. What factors are driving the differences in populations should emerge from such studies. It is hope that we can also identify specific biological activities of the populations at the low- and high-producing zones.

7.14 Conclusion

The endophytic microbiome of a plant is a function of the bacterial communities that (1) originated from the seed and (2) came from surrounding soil environments and have selective and differential advantages on others to become colonized inside plant. Moreover, a successful endophyte must be a competent rhizospheric microbe in order to get inside the plant. Similarly, rhizospheric microbial communities are better adopted for the specific niches and generally provide benefits to plants so that a mutual relationship may establish.

The higher diversity of corn endophytes has been documented by a number of authors (Chelius and Triplett 2001; Roesch et al. 2008; Montañez et al. 2009; Liu et al. 2012). The most prominent corn bacterial endophytes belong to Proteobacteria class with varied findings of bacteria belonging to α-, β-, and the γ-subdivisions of Proteobacteria. For example, Chelius and Triplett (2001) documented the predominance of α-Proteobacteria followed by the β-Proteobacteria in corn. Roesch et al. (2008) described the preference of α-Proteobacteria and β-Proteobacteria in corn. Other groups found as prominent corn bacterial endophytes are mainly γ-Proteobacteria, which are followed by β-Proteobacteria (McInroy and Kloepper 1995b; Seghers et al. 2004) that corn tissues were colonized mainly by γ-Proteobacteria and then α-Proteobacteria, and the β-Proteobacteria were found the least in number (Montañez et al. 2009). A number of field trials have been done on corn biocontrol with the aim to find an environment friendly bio-cultivar that can suppress various diseases of corn and can be utilized a tool for sustainable agriculture.

Acknowledgments We are greatly thankful to the Agriculture and Agri-Food Canada, through the Agricultural Innovation Program, Grain Farmers of Ontario and Natural Sciences and Engineering Research Council through Industrial Fellowships, IRAP, NSERC Engage, Ontario Genomics, and Ontario Centres of Excellence for their financial support. We are also thankful to the Ontario Corn Growers who were involved in the study program and allowed us to collect the samples from their farm. We greatly appreciate the A&L Canada Laboratories Inc., students, and the staff members who contributed their time, knowledge, and skills for this innovative work.

References

Alvear M, Rosas A, Rouanet JL, Borie F (2005) Effects of three soil tillage systems on some biological activities in an Ultisol from southern Chile. Soil Tillage Res 82(2):195–202

Alves GC, Videira SS, Urquiaga S, Reis VM (2015) Differential plant growth promotion and nitrogen fixation in two genotypes of maize by several *Herbaspirillum* inoculants. Plant Soil 387(1–2):307–321

Amin N (2013) Diversity of endophytic fungi from root of maize var. Pulut (waxy corn local variety of south sulawesi, Indonesia). Int J Curr Microbiol Appl Sci 2:148–154

Araújo JMD, Silva ACD, Azevedo JL (2000) Isolation of endophytic actinomycetes from roots and leaves of maize (*Zea mays* L.). Braz Arch Biol Technol 43(4):0–0

Araujo EO, Mercante FM, Vitorino ACT, Paim LR (2014) Inoculation of *Herbaspirillum seropedicae* in three corn genotypes under different nitrogen levels. Afr J Agri Res 9(21):1628–1634

Arias JA (1985) Secretory organelle and mitochondrial alterations induced by fusaric acid in root cells of *Zea mays*. Physiol Plant Pathol 27(2):149–158

Bacon CW, Hinton DM (1996) Symptomless endophytic colonization of maize by *Fusarium moniliforme*. Can J Bot 74(8):1195–1202

Bacon CW, Yates IE, Hinton DM, Meredith F (2001) Biological control of *Fusarium moniliforme* in maize. Environ Health Perspect 109(Suppl 2):325

Bacon CW, Glenn AE, Yates IE (2008) *Fusarium verticillioides*: managing the endophytic association with maize for reduced fumonisins accumulation. Toxin Rev 27(3–4):411–446

Bailly A, Weisskopf L (2012) The modulating effect of bacterial volatiles on plant growth: current knowledge and future challenges. Plant Signal Behav 7(1):79–85

Baldani VLD, Baldani JI, Olivares FL, Dobereiner J (1992) Identification and ecology of *Herbaspirillum seropedicae* and the closely related *Pseudomonas rubrisubalbicans*. Symbiosis 13:65–73

Baudoin E, Benizri E, Guckert A (2002) Impact of growth stage on the bacterial community structure along maize roots, as determined by metabolic and genetic fingerprinting. Appl Soil Ecol 19(2):135–145

Bertolini M, Lorenzoni C, Marocco A, Maggiore T (1993) Soluble solids content in the stalk of maize (*Zea mays* L.) lines and hybrids. Maydica 38(4):321–324

Brusetti L, Francia P, Bertolini C (2005) Bacterial communities associated with the rhizosphere of transgenic Bt 176 maize (*Zea mays)* and its non transgenic counterpart. Plant Soil 266(1–2):11–21

Buyer JS, Roberts DP, Russek-Cohen E (1999) Microbial community structure and function in the spermosphere as affected by soil and seed type. Can J Microbiol 45(2):138–144

Canbolat MY, Bilen S, Çakmakçı R, Şahin F, Aydın A (2006) Effect of plant growth-promoting bacteria and soil compaction on barley seedling growth, nutrient uptake, soil properties and rhizosphere microflora. Biol Fertil Soils 42(4):350–357

Chelius MK, Triplett EW (2000) Immunolocalization of dinitrogenase reductase produced by *Klebsiella pneumoniae* in association with *Zea mays* L. Appl Environ Microbiol 66(2):783–787

Chelius MK, Triplett EW (2001) The diversity of archaea and bacteria in association with the roots of *Zea mays* L. Microbial Ecol 41(3):252–263

Cortes-Barco AM, Goodwin PH, Hsiang T (2010) Comparison of induced resistance activated by benzothiadiazole, (2R,3R)-butanediol and an isoparaffin mixture against anthracnose of *Nicotiana benthamiana*. Plant Pathol 59(4):643–653

D'alessandro M, Erb M, Ton J (2014) Volatiles produced by soil-borne endophytic bacteria increase plant pathogen resistance and affect tritrophic interactions. Plant Cell Environ 37(4):813–826

Döbereiner J, Reis VM, Paula MA, Olivares FD (1993) Endophytic diazotrophs in sugarcane, cereals and tuber plants. In: Palacios R, Mora J, Newton WE (eds) New horizons in nitrogen fixation. Springer, Dordrecht, pp 671–676

Donegan KK, Palm CJ, Fieland VJ et al (1995) Changes in levels, species and DNA fingerprints of soil microorganisms associated with cotton expressing the *Bacillus thuringiensis* var. kurstaki endotoxin. Appl Soil Ecol 2(2):111–124

Donegan KK, Schaller DL, Stone JK (1996) Microbial populations, fungal species diversity and plant pathogen levels in field plots of potato plants expressing the *Bacillus thuringiensis* var. tenebrionis endotoxin. Trans Res 5(1):25–35

Duvick DN, Smith JSC, Cooper M (2010) Long term selection in a commercial hybrid maize breeding program. Janick I Plant Breed Rev 2(24):109–152

El-Katatny MH, Idres MM (2014) Effects of single and combined inoculations with *Azospirillum brasilense* and *Trichoderma harzianum* on seedling growth or yield parameters of

wheat (*Triticum vulgaris* L., Giza 168) and corn (*Zea mays* L., hybrid 310). J Plant Nutr 37(12):1913–1936
Estrada P, Mavingui P, Cournoyer B et al (2002) A N_2-fixing endophytic *Burkholderia* sp. associated with maize plants cultivated in Mexico. Can J Microbiol 48(4):285–294
Figueiredo JEF, Gomes EA, Guimarães CT (2009) Molecular analysis of endophytic bacteria from the genus *Bacillus* isolated from tropical maize (*Zea mays* L.) Braz J Microbiol 40(3):522–534
Fisher PJ, Petrini O, Scott HL (1992) The distribution of some fungal and bacterial endophytes in maize (*Zea mays* L.) New Phytol 122(2):299–305
Gao Z, Zhuang J, Chen J, Liu X, Tang S (2004) Population of entophytic bacteria in maize roots and its dynamic analysis. J Appl Ecol 15(8):1344–1348
Gomes NCM, Heuer H, Schönfeld J et al (2001) Bacterial diversity of the rhizosphere of maize (*Zea mays*) grown in tropical soil studied by temperature gradient gel electrophoresis. Plant Soil 232(1–2):167–180
Gutierrez-Zamora ML, Martınez-Romero E (2001) Natural endophytic association between *Rhizobium etli* and maize (*Zea mays* L.) J Biotechnol 91(2):117–126
Hallmann J, Quadt-Hallmann A, Mahaffee WF, Kloepper JW (1997) Bacterial endophytes in agricultural crops. Can J Microbiol 43(10):895–914
Han SH, Lee SJ, Moon JH et al (2006) *GacS*-dependent production of 2R, 3R-butanediol by *Pseudomonas chlororaphis* O6 is a major determinant for eliciting systemic resistance against *Erwinia carotovora* but not against *Pseudomonas syringae* pv. tabaci in tobacco. Mol Plant-Microbe Interact 19(8):924–930
Hardoim PR, van Overbeek LS, van Elsas JD (2008) Properties of bacterial endophytes and their proposed role in plant growth. Trends Microbiol 16(10):463–471
Hungria M, Campo RJ, Souza EM, Pedrosa FO (2010) Inoculation with selected strains of *Azospirillum brasilense* and *A. lipoferum* improves yields of maize and wheat in Brazil. Plant Soil 331(1–2):413–425
Ikeda AC, Bassani LL, Adamoski D et al (2013) Morphological and genetic characterization of endophytic bacteria isolated from roots of different maize genotypes. Microbial Ecol 65(1):154–160
Islam R, Glenney DC, Lazarovits G (2015) No-till strip row farming using yearly maize-soybean rotation increases yield of maize by 75%. Agron Sust Develop 35(2):837–846
Johnston-Monje D, Raizada MN (2011a) Conservation and diversity of seed associated endophytes in *Zea* across boundaries of evolution, ethnography and ecology. PLoS One 6(6):e20396
Johnston-Monje D, Raizada MN (2011b) Plant and endophyte relationships: nutrient management. Compre Biotech 2:713–727
Johnston-Monje D, Mousa WK, Lazarovits G, Raizada MN (2014) Impact of swapping soils on the endophytic bacterial communities of pre-domesticated, ancient and modern maize. BMC Plant Biol 14(1):1
Johnston-Monje D, Lundberg DS, Lazarovits G, Reis VM, Raizada MN (2016) Bacterial populations in juvenile maize rhizospheres originate from both seed and soil. Plant Soil 405(1):337–355
Joo HJ, Kim HY, Kim LH, Lee S, Ryu JG, Lee T (2015) A *Brevibacillus* sp. antagonistic to mycotoxigenic *Fusarium* spp. Biological Cont 87:64–70
Landa BB, Cachinero-Díaz JM, Lemanceau P, Jiménez-Díaz RM, Alabouvette C (2002) Effect of fusaric acid and phytoanticipins on growth of rhizobacteria and *Fusarium oxysporum*. Can J Microbiol 48(11):971–985
Lange ES, Balmer D, Mauch-Mani B, Turlings TC (2014) Insect and pathogen attack and resistance in maize and its wild ancestors, the teosintes. New Phytol 204(2):329–341
Liu Y, Zuo S, Xu L, Zou Y, Song W (2012) Study on diversity of endophytic bacterial communities in seeds of hybrid maize and their parental lines. Arch Microbiol 194(12):1001–1012
Malfanova N, Lugtenberg B, Berg G (2013) Bacterial endophytes: who and where, and what are they doing there. Endophytic bacteria with plant growth promoting and biocontrol abilities. In: Bruijn FJ (ed) Molecular microbial ecology of the rhizosphere. Wiley-Blackwell, Hoboken, pp 15–37

Martinez-Romero E (2003) Diversity of *Rhizobium-Phaseolus vulgaris* symbiosis: overview and perspectives. Plant Soil 252(1):11–23

Matsukuma S, Okuda T, Watanabe J (1994) Isolation of actinomycetes from pine litter layers. 日本放線菌学会誌 8(2):57–65

Matsumoto A, Takahashi Y, Mochizuki M, Seino A, Iwai Y, Omura S (1998) Characterization of actinomycetes isolated from fallen leaves. 日本放線菌学会誌 12(1):46–48

Matsuoka Y, Vigouroux Y, Goodman MM, Sanchez J, Buckler E, Doebley J (2002) A single domestication for maize shown by multilocus microsatellite genotyping. Proc Natl Acad Sci 99(9):6080–6084

McInroy JA, Kloepper JW (1995) Population dynamics of endophytic bacteria in field-grown sweet corn and cotton. Can J Microbiol 41(10):895–901

Mercado-Flores Y, Cárdenas-Álvarez IO, Rojas-Olvera AV et al (2014) Application of *Bacillus subtilis* in the biological control of the phytopathogenic fungus *Sporisorium reilianum*. Biol Cont 76:36–40

Michiels K, Vanderleyden J, Van Gool A (1989) *Azospirillum*—plant root associations: a review. Biol Fertil Soils 8(4):356–368

Montañez A, Abreu C, Gill PR, Hardarson G, Sicardi M (2009) Biological nitrogen fixation in maize (*Zea mays* L.) by ^{15}N isotope-dilution and identification of associated culturable diazotrophs. Biol Fertil Soils 45(3):253–263

Montañez A, Blanco AR, Barlocco C, Beracochea M, Sicardi M (2012) Characterization of cultivable putative endophytic plant growth promoting bacteria associated with maize cultivars (*Zea mays* L.) and their inoculation effects in vitro. Appl Soil Ecol 58:21–28

Moriondo M, Maselli F, Bindi M (2007) A simple model of regional wheat yield based on NDVI data. Eur J Agron 26(3):266–274

Mousa WK, Shearer CR, Limay-Rios V, Zhou T, Raizada MN (2015) Bacterial endophytes from wild maize suppress *Fusarium graminearum* in modern maize and inhibit mycotoxin accumulation. Front Plant Sci 6:805. doi:10.3389/fpls.2015.00805

Navarro-Noya YE, Gómez-Acata S, Montoya-Ciriaco N et al (2013) Relative impacts of tillage, residue management and crop-rotation on soil bacterial communities in a semi-arid agroecosystem. Soil Biol Biochem 65:86–95

Neal JL Jr, Larson RI, Atkinson TG (1973) Changes in rhizosphere populations of selected physiological groups of bacteria related to substitution of specific pairs of chromosomes in spring wheat. Plant Soil 39(1):209–212

Neal AL, Ahmad S, Gordon-Weeks R, Ton J (2012) Benzoxazinoids in root exudates of maize attract *Pseudomonas putida* to the rhizosphere. PLoS One 7(4):e35498

Palus JA, Borneman J, Ludden PW, Triplett EW (1996) A diazotrophic bacterial endophyte isolated from stems of *Zea mays* L. and *Zea luxurians* Iltis and Doebley. Plant Soil 186(1):135–142

Pereira P, Ibáñez F, Rosenblueth M, Etcheverry M, Martínez-Romero E (2011) Analysis of the bacterial diversity associated with the roots of maize (*Zea mays* L.) through culture-dependent and culture-independent methods. ISRN Ecol, 2011, 938546, 10 pages 10.5402/2011/938546

Petatán-Sagahón I, Anducho-Reyes MA, Silva-Rojas HV et al (2011) Isolation of bacteria with antifungal activity against the phytopathogenic fungi *Stenocarpella maydis* and *Stenocarpella macrospora*. Int J Mol Sci 12(9):5522–5537

Picard C, Bosco M (2006) Heterozygosis drives maize hybrids to select elite 2, 4-diacethylphlor oglucinol-producing *Pseudomonas* strains among resident soil populations. FEMS Microbiol Ecol 58(2):193–204

Pillay VK, Nowak J (1997) Inoculum density, temperature, and genotype effects on in vitro growth promotion and epiphytic and endophytic colonization of tomato (*Lycopersicon esculentum* L.) seedlings inoculated with a pseudomonad bacterium. Can J Microbiol 43(4):354–361

Proctor RH, Desjardins AE, Brown DW et al (2006) Biosynthesis of *Fusarium* mycotoxins and genomics of *Fusarium verticillioides*. Mycotoxin Res 22(2):75–78

Rai R, Dash PK, Prasanna BM, Singh A (2007) Endophytic bacterial flora in the stem tissue of a tropical maize (*Zea mays* L.) genotype: isolation, identification and enumeration. World J Microbiol Biotechnol 23(6):853–858

Rashid S, Charles TC, Glick BR (2012) Isolation and characterization of new plant growth-promoting bacterial endophytes. Appl Soil Ecol 61:217–224

Reis VM, Estrada-De Los Santos P, Tenorio-Salgado S (2004) *Burkholderia tropica* sp. nov., a novel nitrogen-fixing, plant-associated bacterium. Int J Syst Evol Microbiol 54(6):2155–2162

Reiter B, Sessitsch A (2006) Bacterial endophytes of the wildflower *Crocus albiflorus* analyzed by characterization of isolates and by a cultivation-independent approach. Can J Microbiol 52(2):140–149

Rennie RJ (1980) ^{15}N-isotope dilution as a measure of dinitrogen fixation by *Azospirillum brasilense* associated with maize. Can J Bot 58(1):21–24

Rijavec T, Lapanje A, Dermastia M, Rupnik M (2007) Isolation of bacterial endophytes from germinated maize kernels. Can J Microbiol 53(6):802–808

Roesch LFW, Olivares FL, Passaglia LMP et al (2006) Characterization of diazotrophic bacteria associated with maize: effect of plant genotype, ontogeny and nitrogen-supply. World J Microbiol Biotechnol 22(9):967–974

Roesch LFW, Camargo FA, Bento FM, Triplett EW (2008) Biodiversity of diazotrophic bacteria within the soil, root and stem of field-grown maize. Plant Soil 302(1–2):91–104

Rosenblueth M, Martínez-Romero E (2004) *Rhizobium etli* maize populations and their competitiveness for root colonization. Arch Microbiol 181(5):337–344

Rosenblueth M, Martínez-Romero E (2006) Bacterial endophytes and their interactions with hosts. Mol Plant-Microbe Interact 19(8):827–837

Rudrappa T, Biedrzycki ML, Kunjeti SG et al (2010) The rhizobacterial elicitor acetoin induces systemic resistance in *Arabidopsis thaliana*. Commun Integ Biol 3(2):130–138

Ryu CM, Farag MA, Hu CH (2004) Bacterial volatiles induce systemic resistance in *Arabidopsis*. Plant Physiol 134(3):1017–1026

Salamone IG, Döbereiner J, Urquiaga S, Boddey RM (1996) Biological nitrogen fixation in *Azospirillum* strain-maize genotype associations as evaluated by the ^{15}N isotope dilution technique. Biol Fertil Soils 23(3):249–256

Santos F, Peñaflor MFG, Paré PW et al (2014) A novel interaction between plant-beneficial rhizobacteria and roots: colonization induces corn resistance against the root herbivore *Diabrotica speciosa*. PLoS One 9(11):e113280

Sardi P, Saracchi M, Quaroni S, Petrolini B, Borgonovi GE, Merli S (1992) Isolation of endophytic *Streptomyces* strains from surface-sterilized roots. Appl Environ Microbiol 58(8):2691–2693

Saunders M, Kohn LM (2009) Evidence for alteration of fungal endophyte community assembly by host defense compounds. New Phytol 182(1):229–238

Saxena D, Stotzky G (2000) Insecticidal toxin from *Bacillus thuringiensis* is released from roots of transgenic Bt corn in vitro and in situ. FEMS Microbiol Ecol 33(1):35–39

Saxena D, Flores S, Stotzky G (1999) Insecticidal toxin in root exudates from Bt corn. Nature 402(6761):480

Schmalenberger A, Tebbe CC (2002) Bacterial community composition in the rhizosphere of a transgenic, herbicide-resistant maize (*Zea mays*) and comparison to its non-transgenic cultivar Bosphore. FEMS Microbiol Ecol 40(1):29–37

Seghers D, Wittebolle L, Top EM, Verstraete W, Siciliano SD (2004) Impact of agricultural practices on the *Zea mays* L. endophytic community. Appl Environ Microbiol 70(3):1475–1482

Singh DP, Singh HB, Prabha R (eds) (2016) Microbial inoculants in sustainable agricultural productivity. Springer, New Delhi

Szilagyi-Zecchin VJ, Ikeda AC, Hungria M (2014) Identification and characterization of endophytic bacteria from corn (*Zea mays* L.) roots with biotechnological potential in agriculture. AMB Express 4(1):1

Ussiri DA, Lal R, Jarecki MK (2009) Nitrous oxide and methane emissions from long-term tillage under a continuous corn cropping system in Ohio. Soil Tillage Res 104(2):247–255

Van Overbeek L, Van Elsas JD (2008) Effects of plant genotype and growth stage on the structure of bacterial communities associated with potato (*Solanum tuberosum* L.) FEMS Microbiol Ecol 64(2):283–296

Ventura MU, Mello EP, Oliveira AR et al (2001) Males are attracted by female traps: a new perspective for management of *Diabrotica speciosa* (Germar) (Coleoptera: Chrysomelidae) using sexual pheromone. Neotropical Entom 30(3):361–364

Von Bülow JF, Döbereiner J (1975) Potential for nitrogen fixation in maize genotypes in Brazil. Proc Natl Acad Sci 72(6):2389–2393

Wang H, Nussbaum-Wagler T, Li B et al (2005) The origin of the naked grains of maize. Nature 436(7051):714–719

Wicklow DT, Poling SM (2009) Antimicrobial activity of pyrrocidines from *Acremonium zeae* against endophytes and pathogens of maize. Phytopathology 99(1):109–115

Wicklow DT, Roth S, Deyrup ST, Gloer JB (2005) A protective endophyte of maize: *Acremonium zeae* antibiotics inhibitory to *Aspergillus flavus* and *Fusarium verticillioides*. 11 Dedicated to John Webster on the occasion of his 80th birthday. Mycolog Res 109(5):610–618

Xiao J, Li J, Yuan L, Tanksley SD (1995) Dominance is the major genetic basis of heterosis in rice as revealed by QTL analysis using molecular markers. Genetics 140(2):745–754

Zou Y, Liu Y, Wang J, Song W (2011) Advances in plant seed-associated microbial ecology. Acta Ecol Sin 31(10):2906–2914

Bacteriocin-Producing Rhizosphere Bacteria and Their Potential as a Biocontrol Agent

8

Naheed Mojgani

Abstract

An alarming rise in environmental pollution due to unsafe use of chemically toxic pesticides and fertilizers in agriculture industry has stimulated researchers to search for alternative control agents with wide spectrum of actions and lesser side effects. In this context, the antimicrobial polypeptide bacteriocins produced by rhizosphere bacteria have been evaluated as potential bio-stimulants of plant growth and as biocontrol agents. Bacteriocins are ribosomally synthesized peptides demonstrating bacteriostatic or bactericidal activity against other related and unrelated microorganisms. A wide range of rhizosphere- and plant-associated bacteria have been identified as potential bacteriocin producers demonstrating wide range of inhibitory spectrum toward economically important plant pathogens. To date, approximately 500 bacteriocins have been identified and characterized of which majority are produced by rhizosphere bacteria. These antimicrobials characterized as highly potent toxins with powerful killing action, high stability, and low toxicity to humans have been considered as a viable option and a suitable alternative to chemically toxic agents used in many industrial applications. The importance of bacteriocins is well recognized in agriculture industry for their role in reducing the use of fertilizers and chemical inputs such as fungicides and insecticides. This review presents an overview of bacteriocins, their nature, mode of action, resistance, and genetics with special emphasis on bacteriocin-producing rhizosphere bacteria and their possible potential as biocontrol agents for combating bacterial plant diseases.

N. Mojgani (✉)
Biotechnology Department, Razi Vaccine and Serum Research Institute-Agriculture Research Education and Extension Organization, Karaj, Islamic Republic of Iran

Research and Development Department, Nature Biotech Co., Karaj, Islamic Republic of Iran
e-mail: dnmoj@yahoo.com

S. Mehnaz (ed.), *Rhizotrophs: Plant Growth Promotion to Bioremediation*, Microorganisms for Sustainability 2, DOI 10.1007/978-981-10-4862-3_8

8.1 Introduction

Agriculture industry has been considered among one of the most dangerous industries, responsible for polluting the ground, river, and ocean water, leading to polluted food, soil, water, and air. The extended use of synthetic fertilizers and pesticides including insecticides, fungicides, herbicides, rodenticides, molluscicides, nematicides, plant growth regulators, and antibiotics not only pose dangers to the consuming public but is of high threat to the farmers and workers in the field handling these chemicals. Additionally, the widespread use of antibiotics has led to nature's creation of stronger, more antibiotic-resistant strains of bacteria which is significantly increasing the economic burden on the entire healthcare system. One strategy to combat or overcome the hazards is to consider an alternative to augment or replace these chemically toxic agents with more safe and efficacious antimicrobial agents. In this regard, antibiotics, bacteriocins, and bioactive peptides that originated from prokaryotic or eukaryotic cells with potential antimicrobial activity have been explored vastly as an alternative strategy to replace the chemically used control agents (Cintas et al. 2001; Zhang et al. 2009). The mentioned antimicrobial agents share some similarities but are distinctly different in their mode of action, host cell immunity, mechanism of target cell resistance, toxicity, and side effect mechanisms. In contrast to antibiotics which are the broad-spectrum metabolites produced by multi-enzyme complexes, bacteriocins are ribosomally synthesized narrow-spectrum proteinaceous substances produced by a wide range of bacterial species (Holtsmark et al. 2008; Sieiro et al. 2016); while, bioactive peptides show hormone-like activity in their effect, a mechanism which distinguishes them from all of the bacteriocins. In brief, the emergence of antibiotic resistance among microbial population due to extended use of antibiotics and the undesirable toxic properties of several bioactive peptides made bacteriocins highly superior as natural safe antimicrobial agents. Additionally, exploitation of narrow-spectrum antimicrobial compounds like bacteriocins might be an attractive strategy for the targeted control of bacterial plant diseases (Montesinos 2007).

In order to exploit bacteriocins in the industry, it is essential to understand their nature. This review describes the fundamental issues concerning bacteriocins including their definition, origin, mode of action, resistance mechanisms, and recent perspectives as future biocontrol agents.

8.2 History of Bacteriocins

Colicins produced by Gram-negative *E. coli* are the first ever bacteriocin described and the most widely studied bacteriocin to date. Following the discovery of colicins in 1925, nisin (N group inhibitory substance), a bacteriocin produced by Gram-positive bacteria, i.e., *Lactococcus lactis*, was reported for the first time in England by Rogers and Whittier (1928). The nontoxic property and wide inhibitory spectrum of nisin made them the first ever bacteriocin being approved for food applications by the Food and Agriculture Organization and World Health Organization

(FAO/WHO) in 1969. This bacteriocin was first marketed in England in 1953 and since then approved for use in more than 48 countries (Ruiz-Larrea et al. 2007). Furthermore, nisin was also accepted as bio-preservative ingredient in the European food additive list (No. E234).

Production of colicin-like substance by a number of other strains of family *Enterobacteriaceae* was also reported by Fredericq (1946). Based on his findings, colicins were proteinaceous compounds with inhibitory action dependent on the specific cell surface receptors present on the sensitive cells. Further distinguishing features of the colicins included their relatively high molecular weight and plasmid-associated production. Owing to their highly selective killing spectrum and potent cytotoxicity, colicins have been considered to have potential as targeted next-generation antibiotics for medical and agriculture use (Pugsley 1984; Riley and Wertz 2002).

Additionally, *E. coli* strains produce a second type of bacteriocin known as microcins. Microcins are smaller than colicins and share a number of physio-chemical properties with the bacteriocins produced by Gram-positive bacteria, including thermostability, resistance to some proteases, relative hydrophobicity, and resistance to extreme pH (Gillor et al. 2004; Severinov et al. 2007).

Presence of colicin-like substance has also been reported in Gram-positive bacteria and termed bacteriocins (Jacobs et al. 1953). In contrast to colicins, the bacteriocins produced by Gram-positive bacteria display a wider spectrum of inhibitory action against different species and a less solid host cell immunity to the homologous bacteriocins (Tagg et al. 1976).

As suggested by Klaenhammer (1988), 99% of all bacteria may make at least one bacteriocin. To date, a number of bacteriocin-producing bacteria (*Bacillus*, *Enterobacter*, *Lactobacilli*, *Salmonella*, *Shigella*, *Staphylococcus*, *Proteus*, *Pseudomonas*, etc.) and archaea have been reported (Fredericq 1957; Klaenhammer and Kullen 1999; Riley and Wertz 2002; Kaur et al. 2015; Sieiro et al. 2016).

8.3 Classification of Bacteriocins

8.3.1 Bacteriocins of Gram-Negative Bacteria

Bacteriocins produced by Gram-negative bacteria are classified into four categories, namely, colicins, colicin-like bacteriocins, microcins, and phage tail-like bacteriocins (Chavan and Riley 2007). Below is a brief description of each type of the mentioned bacteriocins.

8.3.1.1 Colicins

These antibacterial substances produced by *E. coli* are protease-sensitive, thermo-sensitive proteins with molecular size ranging from 25 to 90 KDa (Pugsley and Oudega 1987). Based on their mode of killing, two major types of colicins were defined by Gillor et al. (2004):

- Pore-former colicins: These type of colicins kill sensitive cells by forming pores in the cell membrane. Some examples include colicins A, B, E1, Ia, Ib, K, EI, and 5.
- Nuclease colicins: Nuclease colicins demonstrate killing action by acting as DNases, RNases, or tRNAses. Colicins E2, E3, E4, E5, E6, E7, E8, and E9 are examples of nuclease colicins.

8.3.1.2 Colicin-Like Bacteriocins

Proteinaceous bacteriocins produced by other Gram-negative species are termed colicin-like due to their structural and functional similarity to colicins. Like colicins, they can be pore formers (pyocin S5) and nucleases (pyocins S1 and S2). Some other examples include klebicins, S-pyocins, and alveicins produced by *Klebsiella* species, *Pseudomonas aeruginosa*, and *Hafnia alvei*, respectively (Michel-Briand and Baysse 2002).

8.3.1.3 Phage Tail-Like Bacteriocins

They are large structures resembling the tail of bacteriophages or defective phage particles. Some of the most studied phage tail-like bacteriocins are R and F pyocins produced by *P. aeruginosa* (Michel-Briand and Baysse 2002; Dingemans et al. 2016).

8.3.1.4 Microcins

Gram-negative bacteria produce much smaller peptides (<10 KDa) known as microcins. Gillor and his colleagues (2004) reported microcins to be chromosomally encoded. Microcins can be divided into two classes including:

- Posttranslationally modified microcins including microcins B17, C7, J25, and D93.
- Unmodified microcins like microcins E492, V, L, H47, and 24.

8.3.2 Bacteriocins of Gram-Positive Bacteria

Based on biochemical and genetic characteristics, the bacteriocins from Gram-positive bacteria have been classified into four distinct classes (Klaenhammer 1993). Some of these characteristics include molecular size, physical properties, chemical structure, and mode of actions. The four widely described classes of the Gram-positive bacteriocins are as follows:

8.3.2.1 Class I or Lantibiotics

Lantibiotics are very low-molecular-weight (<5 KDa, 19–38 amino acid), thermostable lanthionine peptides. Lantibotic-producing bacteria are well studied for their commercial use in the food industry for making dairy products like cheese, the main example being nisin. Another example is duramycin, which is used as a veterinary antibiotic, especially for chickens. Chatterjee and his colleagues (2005) defined three distinct types of class I lantibiotics based on their biosynthetic pathway and bioactivity.

- Type A: Relatively long, linear, and flexible cationic peptides, e.g., nisin, subtilin, bisin, epidermin, and gallidermin

- Type B: More globular, rigid peptides with no or negative net charge, e.g., mersacidin, cinnamycin, duramycin, and plantaricin C
- Type C: Also known as two-component peptides. Members of Type C require synergistic of two peptides for bioactivity. Some examples include haloduracin, avermipeptin, erythreapeptin, and griseopeptin

8.3.2.2 Class II

This class is composed of small (<10 KDa), thermostable, non-lanthionine peptides that are not posttranslationally modified (Heng et al. 2007). The bacteriocins from this class affect target cells similar to class I bacteriocins; however, their receptors seem to be proteins rather than lipids (Diep et al. 2007). The circular class II bacteriocins can be divided into five subclasses, corresponding to the four subclasses of unmodified LAB bacteriocins and one subclass of unmodified microcins:

- Subclass IIa: Pediocin and enterocins are the main representatives of subclass IIa (Cintas et al. 2001; Diez et al. 2012). Bacteriocins belonging to this class are 37–48 amino acid residues showing high specificity to *Listeria monocytogenes.*
- Subclass IIb: This subclass comprises of bacteriocins requiring combined activity of two peptides (heterodimeric). Garneau and his colleagues (2002) stated that these peptides have very low activity when employed individually. They further reported that the members of class IIb bacteriocins can be classified into two types: type E (enhanced) and type S (synergistic) peptides. Lactocin G is an example of subclass IIb bacteriocins.
- Subclass IIc: Bacteriocins belonging to this subclass have a cyclic structure owing to the covalent bondings between C and N terminals (Kawai et al. 2004). These circular bacteriocins display a broader spectrum of antimicrobial activity toward various Gram-positive bacteria, including many food-borne spoilage and pathogenic bacteria. Lactocin B, Enterocin AS-48, circularin A, and reutericin 6 are representatives of this subclass (Maqueda et al. 2008).
- Subclass IId: They are one-peptide, non-pediocin, linear, and leaderless bacteriocins. In contrast to the bacteriocins from other class, subclass IId bacteriocins do not share any common system for their killing mechanisms mainly due to fundamental diversity of their primary structure (Iwatani et al. 2011; Cotter et al. 2013). Lacticin Q belongs to this subclass.
- Subclass IIe: Microcins E492-like bacteriocins (formerly known as the class IIb microcins).

8.3.2.3 Class III

Large (>30 KDa), heat-labile, non-peptide bacteriocins having complex activity and protein structure are placed in class III. This class of bacteriocin differs from other bacteriocins in their action mechanism. Helveticin J and lysostaphin are examples of this class of bacteriocin.

8.3.2.4 Class IV

This class contains large (<10 KDa), complex cyclic peptides, combined with carbohydrate or lipid moieties (Oman et al. 2011; Stepper et al. 2011), e.g., Enterocin AS-48.

8.4 Mode of Action of Bacteriocins

The bacteriocins produced by Gram-positive and Gram-negative bacteria differ in their mode of actions. Majority of bacteriocins are bactericidal in their mode of action, while few are reported to be bacteriostatic, e.g., leuconocin S and leucocin A-UAL (Holzaphel 1997). A widely accepted hypothesis for the mode of action of bacteriocins is that these antimicrobial agents exert their actions on the sensitive cells in two steps (Tagg et al. 1976). In both, Gram-positive and Gram-negative bacteria, adsorption of bacteriocins to specific or nonspecific receptors on the cell surface is considered an essential step leading to the death of sensitive cells. Among Gram-negative bacteria, colicins are known to exert their bactericidal effects by inhibiting the cell wall synthesis, permeabilization of the target cell membrane, or inhibition of RNase or DNase activity (Gillor et al. 2004). The mode of action of bacteriocins in Gram-positive bacteria differs among its classes, as receptor molecule or a "docking molecule" found in the target bacterial cell membrane differs among different classes and subclasses.

The bacteriocins produced by Lactic Acid Bacteria (LAB) have been known to have both bactericidal and bacteriostatic effect on the sensitive cells (Sieiro et al. 2016). The bactericidal or bacteriostatic mode of action of these bacteriocins is often dependent on the aspects of assay system, including the arbitrary units, the buffer or broth, purity of the inhibitor, the sensitive indicator species, and used cell concentrations (DeVuyust and Vandamme 1994). Figure 8.1 depicts the killing mechanism of nisin belonging to class I lantibiotics.

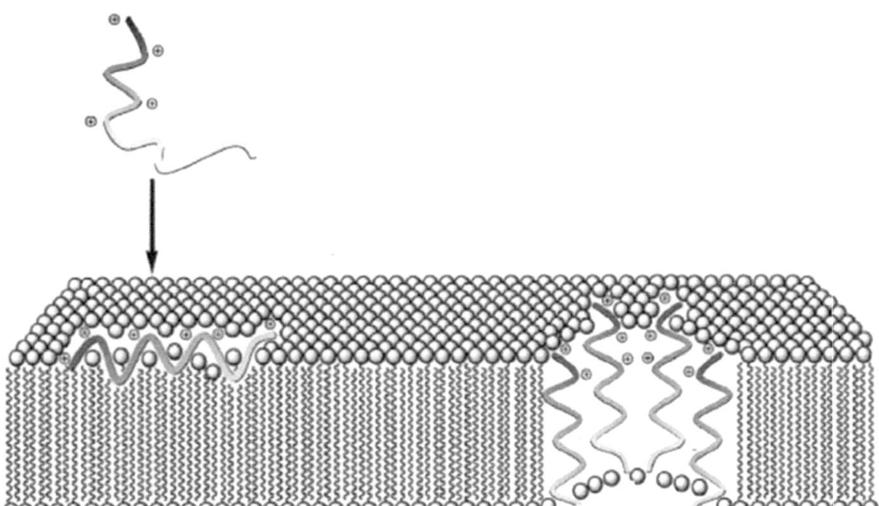

Fig. 8.1 Mode of action of bacteriocins by pore formation in the bacterial membrane (Ruiz-Larrea et al. 2007)

8.5 Genetics of Bacteriocins

The genes for bacteriocins are encoded on chromosomes, plasmids, and/or mobile elements such as transposons. Some examples of plasmid-associated bacteriocins include colicins, halocin H4, and lactocin RN78, while pyocins and lactocin NM24 and NM332 were reported to be chromosomally located (Michel-Briand and Baysse 2002; Chavan and Riley 2007; Mojgani et al. 2010). In some bacterial species like *Serratia marcescens*, the bacteriocin-producing genes are encoded both on plasmid and chromosome (Riley and Wertz 2002); while lacticin 481-producing genes are shown to reside on the transposon Tn5721 (Dufour et al. 2007).

In Gram-negative bacteria, the bacteriocin synthesis is under the control of three to four tightly linked genes, namely, toxin, immunity, transport, and lysis genes, typically located on one or two operons. Thus a functional operon harbors the (a) structural or activity gene, (b) a gene involved in specific immunity to the bacteriocin, (c) a gene that processes (removes leader sequences) and transports the bacteriocin across the membrane, and (d) lysis gene that encodes accessory protein required for the release of bacteriocin from the cell but whose specific role is yet not known (Riley and Gordon 1992). However, presence of all these genes is not ubiquitous especially in colicinogenic strains. Braun and his co-workers (1994) demonstrated that colicin B and M and colicin-like pyocin S3 lack the lysis genes involved in the secretion of bacteriocins.

In most of the class II bacteriocins, the genes encoding ABC transporter and the accessory proteins are on separate operons or near the operon which harbors the structural gene and immunity gene. Some class II bacteriocin structural genes are located adjacent to the promoter on an operon that also contains the immunity gene downstream, i.e., all genes within the same operon. Mostly the structural genes are transcribed in the same direction, although in some cases, as is the case with colicin B (Braun et al. 1994), the immunity gene is transcribed in the opposite orientation.

In Gram-positive bacteria, the gene clusters involved in bacteriocin synthesis might contain many more genes and have more complex transcriptional organization. The genetic organization in Gram-positive bacteria varies between the classes. Lactic Acid Bacteria are known to produce several bacteriocins which are encoded by a variety of bacteriocin genes scattered over the chromosomes and plasmids (Franz et al. 2007). *Carnobacterium piscicola* LV17 produces at least three bacteriocins which are encoded by three genes: two encoded on plasmids and one on the chromosome. Similarly, nisin has been shown to contain eight genes located on a polycistronic operon that are involved in nisin biosynthesis (Engelke et al. 1992). These multiple genes are located on the chromosome and are carried by a large conjugative transposon (Horn et al. 1999). The structural gene, *nis*A, is the first in the operon and is directly followed by three genes, *nis*B, *nis*T, and *nis*C, thought to be involved in the export of bacteriocin. Four additional genes that lie directly downstream are *nis*I, *nis*P, *nis*R, and *nis*K.

Enterocin, a bacteriocin produced by *Enterococcus* species, has also been known to be encoded by a number of structural enterocin genes. Many of the enterocin structural genes studied include *entA, entB, entP, entQ, bac31, entL50A*, and *entL50B* (Cintas et al. 2001; Sanchez-Hidalgo et al. 2003; Henning et al. 2015).

8.6 Resistance and Immunity to Bacteriocins

The immunity mechanism of the bacteriocin-producing strain to its own product is controlled by a special mechanism depending on a variety of bacteriocin-specific immunity proteins encoded by the related gene sequences in close proximity of other bacteriocin structural genes on the same operon (Fig. 8.2) (Cintas et al. 2001;

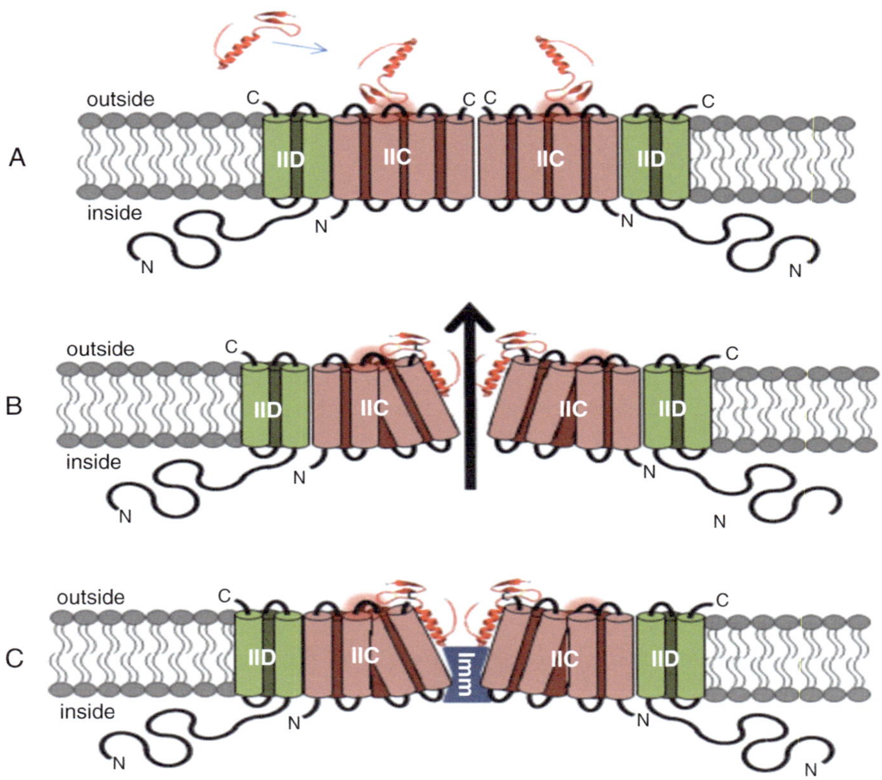

Fig. 8.2 Mode of action and immunity of Class IIa bacteriocins; (**a**) Class IIa bacteriocin specifically targets an extracellular loop of IIC, one of the two membrane-embedded components (IIC and IID, also called ManCD) of man-PTS; (**b**) the initial interaction leads to further interactions with some membrane helices of man-PTS, somehow causing the channel of the sugar permease to remain open, leading to leakage of solutes, destruction of membrane integrity, and eventually cell death; (**c**) in producer cells, the cognate immunity protein binds to IICD and locks the bacteriocin in a tight complex, thereby preventing the bacteriocin from opening the pore

Cleveland et al. 2001). The bacteriocin producer strains produce specific immunity proteins that prevent pore formation by the bacteriocins. However, a number of non-bacteriocin-producing LAB has also been reported to possess immunity genes. Meager information is available regarding the mode of action of these immunity proteins (Nes and Holo 2000). A number of observations have indicated that class IIa bacteriocin immunity proteins are free intracellular molecules which prevent bacteriocin action at the membrane site indirectly via a membrane bound protein (Ennahar et al. 2000). Immunity to nisin is thought to be mediated by the lipoprotein encoded by *nis*I, while *nis*R and *nis*K are involved in the regulation of nisin biosynthesis genes (Kuipers et al. 1993).

The bacteriocin resistance mechanisms are known to be completely different from the immunity. Several structural and physiological alterations are known to be responsible for making a cell resistant toward bacteriocin. According to reports, changes in the composition and structure of the cell wall and cellular membrane, alterations in the electrical potential, fluidity, membrane lipid composition, and load or cell wall thickness might be the factors related to the resistance of cells toward bacteriocins (Mantovani and Russel 2001). Additional factors might be that the specific receptors essential for bacteriocin action are either lacking or mutated on the cell surface of the resistant strain.

The resistance of *L. monocytogenes* to nisin has been attributed to the changes in fatty acid composition of cell membranes which reduces phospholipid concentrations and hinders pore formation, all preventing insertion of nisin molecules. The average frequency of nisin resistance may vary between 10^2 and 10^7 depending on the strain of *L. monocytogenes* (Gravesen et al. 2002).

8.7 Bacteriocin-Producing Rhizobacteria

Rhizosphere is densely populated by a diverse group of microorganisms among which the Gram-positive bacteria are dominant. The bacteria inhabiting the rhizosphere are usually termed as rhizobacteria, which on the basis of their effect on plant growth are grouped into (a) the beneficial bacteria responsible for plant growth and development and termed as plant growth-promoting rhizobacteria (PGPR), (b) the deleterious rhizobacteria responsible for plant disease, and (c) neutral group (Kloepper et al. 2004). The PGPR exert their beneficial effect either by providing hormones or by producing antagonistic substances like antibiotics and bacteriocins (Glick and Bashan 1997). A number of *Bacillus* and *Pseudomonas* spp. well known as PGPR are also potential bacteriocin producers (Podile and Kishore 2006).

Several bacteriocins isolated, identified, and characterized from *Bacillus* species include cerein7 and 8A produced by *B. cereus* (Oscariz and Pisabarro 2001; Bizani et al. 2005); Bac-GM17 from rhizosphere-derived *B. clausii* GM17; subtilisin H4, IH7, and Bac14B from *B. subtilis* strains (Compaore et al. 2013); and thricin17 and thricin Bn1 from *B. thuringiensis* strains (Gray et al. 2006). The growth-promoting function of thuricin 17, a bacteriocin produced by plant growth-promoting *B. thuringiensis*, has been reported earlier (Gray et al. 2006). Later, Lee and his colleagues (2009a, b) reported the plant growth-promoting effect of thuricin 17 after direct application onto leaves or roots of soybean plants. Additionally, Bac14B produced by *B. subtilis* 14B strain was isolated from the rhizosphere of healthy almond plant in Turkey (Hammami et al. 2009). Bac 14B showed significant antibacterial activity against *Agrobacterium tumefaciens*, the causal agent of crown gall disease.

Amylocyclin, a small peptide bacteriocin produced by *B. amyloliquefaciens* spp., is a circular bacteriocin with high antibacterial and antifungal activity (Scholz et al. 2014). Amylocyclin was found active against *Ralstonia solanacearum*, the causal agent of capsicum bacterial wilt, and *Xanthomonas campestris*, the causal agent of black rot disease in cruciferous plants (Hu et al. 2010).

The beneficial effect of Lactic Acid Bacteria (LAB) present in different ecological niches is not a hidden fact (Klaenhammer and Kullen 1999). This group of bacteria are considered GRAS (generally recognized as safe) and have found potential importance as probiotics (FAO/WHO 2001). In a study conducted by Anacarso et al. (2015), a number of bacteriocin-producing LAB were isolated from plants, flowers, and other vegetable matrices derived from house plants. In our previous studies, we were able to isolate bacteriocin producing *Lactobacillus* strains from green olives in Iran (Mojgani et al. 2009). Similarly, Kaur and his colleagues (2015) isolated and characterized bacteriocin-producing *Lactobacillus* species from rhizosphere soil. Although LAB are isolated from plants and plant rhizosphere, they are not considered part of the natural flora of the growing plants, indicating the role of insects in the spread of these organisms (Stirling and Whittenburg 1963).

Enterococci, member of LAB and common colonizer of gastrointestinal tract of man and animal, are also found in soil, plants, vegetables and water (Abriouel et al. 2010). A number of bacteriocin producer rhizosphere *Enterococcus* species have been reported earlier (Klibi et al. 2012). Bacteriocin-producing *Enterococcus* strains are considered to have an ecological advantage over non-bacteriocin producers, residing in the same ecosystem. In a report, Enterocin SE-K4 and mundticin KS were shown to be produced by *E. faecalis* and *E. mundtii* strains isolated from grass silage, respectively (Kawamoto et al. 2002). Furthermore, mundticin QU2 was also reported in an *E. mundtii* strain, isolated from soybeans, while Enterocin Xa and b were shown to be produced by *E. faecium*, isolated from sugar apples (Hu et al. 2010).

A large number of Gram-negative plant pathogens such as *Pseudomonas syringae*, *Pectobacterium* spp., and *Xanthomonas* spp. have been known for their bacteriocin-producing traits (Grinter et al. 2012). S-type pyocins and several high molecular mass phage tail-like bacteriocins have been reported in *P. syringae* and closely related plant-associated pseudomonads (Lavermicocca et al. 2002; Sisto et al. 2010; Ghequire et al. 2012). Putidacin are excreted by *P. putida* strain BW11MI, isolated from banana roots (Parret et al. 2003). Additionally, two lectin-like bacteriocins resembling putidacin in its spectrum of action were shown to be produced by a biocontrol strain *P. fluorescens* Pf 5 (Parret et al. 2005).

Several *Erwinia* spp. are known bacteriocin producers (Nguyen et al. 1999; Jabrane et al. 2002). Carotovoricin, a bacteriocin produced by *E. carotovora* subsp. *carotovora* (presently known as *Pectobacterium carotovora*), was the first ever bacteriocin described in *Erwinia* spp. (Hamon and Peron 1961). Later, Tovkach (1998) reported production of two types of bacteriocins including colicin-like small and macromolecular carotovoricin (MCTVs), respectively, by *E. carotovora*. Carotovoricin has been known to exert its antimicrobial action by self-assembling into cytotoxic phage tail-like fibers. Additionally, carotovoricin Er, carocin S1, and erwiniocin NA4 are some of the other bacteriocins produced by *Erwinia* species (Yamada et al. 2006; Chuang et al. 2007; Subramanian and Smith 2015). Several of these bacteriocins have been shown to exhibit anti-*Xanthomonas oryzae* activity which highlights their possible potential application in agriculture for the control of leaf blight in rice crop.

Phage tail-like bacteriocins have also been reported in *Ralstonia solanacearum* (reduces the development of bacterial wilt on tobacco), *Serratia plymithicum*, and *Rhizobium* (Holtsmark et al. 2008); while trifolitoxins, peptide bacteriocins produced by Gram-negative bacteria *Agrobacterium tumefaciens* and *R. leguminosarum*, have been reported by Scupham and Triplett (2006).

Minimum information is available regarding the bacteriocins produced by Gram-positive plant pathogens. One of the best studied bacteriocins from Gram-positive plant pathogens is ipomicin produced by *Streptomyces ipomoea*, a sweet potato pathogen (Zhang et al. 2003). In addition, a tomato pathogen, *Clavibacter michiganensis*, was shown to harbor a 14 KDa antimicrobial bacteriocin named michiganin A, which was active against the growth of potato pathogen *C. michiganensis* subsp. *sepedonices* (Holtsmark et al. 2008). Michiganin A resembles with other type B lantibiotics, produced by *Actinoplanes* and *Bacillus* species (Zimmermann and Jung 1997). Furthermore, Ahmad et al. (2014) reported a 25–35 KDa bacteriocin in *Lysinibacillus* jx416856, isolated from fruit and vegetable waste.

8.8 Bacteriocins as Biocontrol Agents

Bacteria and their metabolites that reduce the incidence or severity of plant diseases are often known as biocontrol agents, while those that exhibit antagonistic activity toward pathogen are defined as antagonists (Beattie 2006). The potential of bacteriocin producers and bacteriocin preparations to control major bacterial crop diseases is not a new concept and has been studied extensively in the last couple of decades. However, for effective biocontrol, the bacteriocins must be able to survive and grow under natural field conditions so as to be able to compete with the phytopathogens on long-term basis (Cabrefiga 2004). Based on observations, the rhizosphere-derived bacteriocin producers and the bacteriocins produced by phytopathogens are of great interest due to their possible contributions in determining the composition of microbial ecosystems, for example, in the rhizosphere, which reciprocally could affect the emergence and severity of plant disease outbreaks. Furthermore, such compounds might provide safe and natural tools for combating plant pathogens.

Some success in using bacteriocins as biocontrol agents of pre- and post-harvest crops has been reported. In a study conducted in Egypt, *In planta* biological control of potato brown rot disease caused by *Ralstonia solanacearum*, was carried out using active Biocine S2HA as biocontrol agents (Kabeil et al. 2008). The results of their findings indicated that treating infected tuber seeds with biocine S2HA increased potato yield significantly. Similarly, Lavermicocca and his colleagues (2002) showed that an uncharacterized bacteriocin produced by *P. syringae* pv. *ciccaronei* was able to reduce 60–80% olive knot disease caused by *P. syringae* pv. *savastanoi*. Sakthivel and Mew (1991) used non-pathogenic bacteriocin-producing mutants of *Xanthomonas oryzae* pv. *oryzae* to control bacterial blight of rice. A drawback of this study was that although the strains used were consistently able to survive as epiphytes up to 4-weeks post inoculation, the reduction in disease symptoms was highly variable, ranging from 31% to 99% in greenhouse tests. Similar inconsistencies were observed when using avirulent *Erwinia amylovora*, *E. herbicola*, or *Pseudomonas tabaci* to control fire blight (Mclntyre et al. 1973). Control was variable with respect to timing of treatment application relative to inoculation with the pathogen.

A number of other examples include attenuation of gall formation by bacteriocin (agrocin 84)-producing nonpathogenic *Agrobacterium radiobacter* strain 84. Despite its commercial success, the use of agrocin 84 encountered some problems, first of which is its restricted specificity limited to *A. tumefaciens* strains harboring agropine-specific enzymes. Under natural conditions, transfer of the agrocin plasmid into the pathogen presents a potential problem for biocontrol. In a field experiment in Greece, the use of strain K84 showed formation of some galls by agrocin-resistant *A. tumefaciens* strains (Stockwell et al. 1996). However, this problem was solved by a deletion of a 5.9 kb region of the agrocin plasmid which contained the genes necessary for plasmid mobilization. The modified strain has subsequently undergone substantial field testing and is marketed as the first genetically engineered organism to be used as a pesticide.

Bacteriocins have great potential for use as prophylactic treatment for seed or tuber-borne pathogens, prevention of secondary spread of pathogenic bacteria from infected plants, and protection of high-valued crops from bacterial plant pathogens. However, in order to reduce the chances of development of bacterial strains resistant to bacteriocins, it is recommended that at least two and preferably three serologically unrelated bacteriocins are used simultaneously.

8.9 Conclusion

The economic damages in the agriculture farming due to phytopathogens could be avoided by undertaking safe and efficacious strategies such as application of bacteriocins for combating the bacterial plant pathogens. One of the advantages of narrow-spectrum bacteriocins is that they could target specific pathogens without disturbing the wider microbial community.

It has become interestingly clear that bacteriocins have the potential to cover a very broad field of application including the food industry, the medical, and agriculture sectors. In agriculture industry, these natural and safe antimicrobials might play a dual role, as plant growth promotors and as antagonistic agents with disease suppression mechanisms.

Although certain limitations still exist in the use of these natural antimicrobial substances as biocontrol agents, they are of high agronomic importance. Investigations regarding the existing bacteriocins are essential as the possibility still exists that many of the bacteriocins that have been characterized to date may have additional, undiscovered functions. Once harnessed, the use of bacteriocins could not only reduce excessive use of antibiotics and overcome the problem of emerging multidrug-resistant pathogens in the health sector but would also minimize the use of chemical fertilizers, fungicides, and insecticides in the agriculture industry.

References

Abriouel H, Franz CM, Omar NB, Gálvez A (2010) Diversity and applications of *Bacillus bacteriocins*. FEMS Microbiol Rev 35:201–232. doi:10.1111/j.1574-6976.2010.00244.x

Ahmad V, Iqbal AN, Haseeb M, Khan MS (2014) Antimicrobial potential of bacteriocin producing *Lysinibacillus* jx 416856 against food borne pathogens, isolated from fruit and vegetable waste. Anaerobe 27:87–95

Anacarso I, Bassoli L, Sabia C, Condo C (2015) Isolation and identification of LAB from plants and other vegetable matrices and microbial recombination with *Enterococcus* spp. Amer Res Thoughts 1:1504–1515

Beattie GA (2006) Plant-associated bacteria: survey, molecular phylogeny, genomics and recent advances. In: Gnanamanickam SS (ed) Plant-associated bacteria. Springer, Dordrecht, pp 1–56

Bizani D, Motta AS, Morrissy JAC et al (2005) Antibacterial activity of cerein 8A, a bacteriocin-like peptide produced by *Bacillus cereus*. Int Microbiol 8:125–131

Braun V, Pilsl H, Gross P (1994) Colicins: structures, mode of action, transfer through membranes and evolution. Arch Microbiol 161:199–206

Cabrefiga J (2004) Fire blight (*Erwinia amylovora*) of rosaceous plants. Pathogen virulence and selection and characterization of biological control agents. Dissertation, University of Girona

Chatterjee C, Paul M, Xie L, van der Donk WA (2005) Biosynthesis and mode of action of lantibiotics. Chem Rev 105:633–684

Chavan MA, Riley MA (2007) Bacteriocins: molecular evolution of bacteriocins in gram negative bacteria. Springer, Heidelberg

Chuang DY, Chien YC, Wu HP (2007) Cloning and expression of the *Erwinia carotovora* subsp. *Carotovora* gene encoding the low-molecular weight bacteriocin carocin S1. J Bacteriol 189(2):620–626

Cintas LM, Casaus MP, Herranz C, Nes IF, Hernández PE (2001) Review: bacteriocins of lactic acid bacteria. Food Sci Tech Int 7:281–305

Cleveland J, Montville TJ, Nes IF, Chikindas ML (2001) Bacteriocins: safe, natural antimicrobials for food preservation. Int J Food Microbiol 71:1–20

Compaoré CS, Nielsen DS, Ouoba LII et al (2013) Co-production of surfactin and a novel bacteriocin by *Bacillus subtilis* subsp. *subtilis* H4 isolated from Bikalga, an African alkaline *Hibiscus sabdariffa* seed fermented condiment. Int J Food Microbiol 162:297–307

Cotter PD, Ross RP, Hill C (2013) Bacteriocins a viable alternative to antibiotics. Nat Rev Microbiol 11:95–105

De Vuyst L, Vandamme EJ (1994) Antimicrobial potential of lactic acid bacteria in bacteriocins of lactic acid bacteria: microbiology, genetics and applications. Blackie Academic and Professional, London, pp 91–142

Diep DB, Skaugen M, Salehian Z, Holo H, Nes IF (2007) Common mechanisms of target cell recognition and immunity for class II bacteriocins. Proc Natl Acad Sci U S A 104:2384–2389

Diez L, Rojo Benzares B, Zarazaga M et al (2012) Antimicrobial activity of pediocin PA-1 against *Oenococcus oeni* and other wine bacteria. Food Microbiol 31:167–172

Dingemans J, Ghequire MK, Craggs M et al (2016) Identification and functional analysis of a bacteriocin, pyocin S6 with ribonuclease activity from a *P. aeruginosa* cystic fibrosis clinical isolate. Microbiol Open 5:413–423

Dufour A, Hindré T, Haras D, Le Pennec JP (2007) The biology of lantibiotics from the lacticin 481 group is coming of age. FEMS Microbiol Lett 31:134–167

Engelke G, Gutowski-Eckel Z, Hammelman M, Entian KD (1992) Biosynthesis of the lantibiotic nisin: genomic organization and membrane localization of the nisb protein. Appl Environ Microbiol 58:3730–3743

Ennahar S, Deschamps N, Richard N (2000) Natural variation in susceptibility of *Listeria* strains to class IIa bacteriocins. Curr Microbiol 41:1–4

Franz CM, Van Belkum MJ, Holzapfel WH, Abriouel H, Gálvez A (2007) Diversity of enterococcal bacteriocins and their grouping in a new classification scheme. FEMS Microbiol Rev 31:293–310

Fredericq P (1946) Sur la pluralite des recepteurs d antibiose de *E. coli*. C R Soc Biol 140:1189–1191

Fredericq P (1957) Colicins. Annu Rev Microbiol 11:7–22

Garneau S, Martin NI, Vederas JC (2002) Two peptide bacteriocins produced by LAB. Biochimie 84:577–592

Ghequire MG, Li W, Proost P, De Mot R (2012) Plant lectin-like antibacterial protein from phytopathogens *P. syringae* and *Xanthomonas citri*. Environ Microbiol Rep 4:373–380

Gillor O, Benjamin CK, Margaret AR (2004) Colicins and Microcins: the next generation antimicrobials. Adv Appl Microbiol 54:129–146

Glick BR, Bashan Y (1997) Genetic manipulation of plant growth-promoting bacteria to enhance biocontrol of phytopathogens. Biotechnol Adv 15:353–378

Gravesen A, Axelsen AM, da Silva J, Hansen TB, Knochei S (2002) Frequency of bacteriocin resistance development and associated fitness costs in *L. monocytogenes*. Appl Environ Microbiol 68:756–764

Gray EJ, Lee KD, Souleimanov AM et al (2006) A novel bacteriocin, thuricin17, produced by PGPR strain *Bacillus thuringiensis* NEB17: isolation and classification. J Appl Microbiol 100:545–554. doi:10.1111/j.1365-2672.2006.02822.x

Grinter R, Milner J, Walker D (2012) Bacteriocins active against plant pathogenic bacteria. Biochem Soc Trans 40:1498–1502

Hammami I, Rhouma A, Jaouadi B, Rebai A, Nesme X (2009) Optimization and biochemical characterization of a bacteriocin from a newly isolated *Bacillus subtilis* strain14B for biocontrol of *Agrobacterium* spp. strains. Lett Appl Microbiol 48:253–260. doi:10.1111/j.1472-765X.2008.02524.x

Hamon Y, Peron Y (1961) Les proprieties antagonistes reciproques parmi les *Erwinia*. Discussion de la position taxonomique de ce genre. Compt Rend 253:913–915

Heng NK, Wescombe PA, Burton JP, Jack RW, Tagg JR (2007) The diversity of bacteriocins in Gram positive bacteria. In: Riley MA, Chavan M (eds) Bacteriocins; ecology and evolution. Springer, Berlin, pp 45–92

Henning C, Gautam D, Muriana P (2015) Identification of multiple bacteriocins in *Enterococcus* species using an *Enterococcus*-specific bacteriocin PCR Array. Microorganisms 3:1–16

Holtsmark I, Eijsink VG, Brurberg MB (2008) Bacteriocins from plant pathogenic bacteria. FEMS Microbiol 1(280):1–7

Holzaphel W (1997) Use of starter cultures in fermentation on a household scale. Food Control 8:241–258

Horn N, Martinez MI, Martinez JM et al (1999) Enhanced production of pediocin PA 1 and co-production of nisin and pediocin PA1 by *L. lactis*. Appl Environ Microbiol 65:4443–4450

Hu CB, Malaphan W, Zendo T, Nakayama J, Sonomoto K (2010) Enterocin X, a novel two-peptide bacteriocin from *Enterococcus faecium* KU-B5, has an antibacterial spectrum entirely different from those of its component peptides. Appl Environ Microbiol 76:4542–4545

Iwatani S, Zendo T, Sonomoto K (2011) Class IId or linear and non-pediocin-like bacteriocins. In: Drider D, Rebuffat S (eds) Prokaryotic antimicrobial peptides: from genes to applications. Springer, London, pp 237–253

Jabrane A, Sabri A, Compere P et al (2002) Characterization of Serracin P, a phage- tail like bacteriocin, and its activity against *E. amylovora,* the fire blight pathogen. Appl Environ Microbiol 68:5704–5710

Jacob F, Lwoff A, Siminovitch A, Wollman E (1953) Definition de quelue termes relatifs a la lysogenie. Ann Inst Pasteur (Paris) 84:222–224

Joint FAO/WHO (2001, October 1–4) Expert consultation on evaluation of health and nutritional properties of probiotics in food including powder milk with live lactic acid bacteria

Kabeil S, Amer MA, Matarand SM, El-Marsy MH (2008) In planta biological control of potato brown rot disease in Egypt. World J Agri Sci 4:803–810

Kaur S, Kaur HP, Kanghan (2015) Evaluation of antibacterial activity of bacteriocins produced from *Lactobacillus* spp. isolated from rhizosphere soil. Int J Pure Appl Biosci 3:136–142

Kawai Y, Kemperman R, Kok J, Saito T (2004) The circular bacteriocins gassericin A and circularin A. Curr Protein Pept Sci 5:393–398

Kawamoto S, Shima J, Sato R et al (2002) Biochemical and genetic characterization of mundticin KS, an antilisterial peptide produced by *E. mundtii* NFRI 7393. Appl Environ Microbiol 68:3830–3840

Klaenhammer TR (1988) Bacteriocins of lactic acid bacteria. Biochimie 70:337–349

Klaenhammer TR (1993) Genetics of bacteriocins produced by lactic acid bacteria. FEMS Microbiol Rev 12(1–3):39–85

Klaenhammer TR, Kullen MJ (1999) Selection and design of probiotics. Int J Food Microbiol 50:45–57

Klibi N, Slimen NB, Fhoula I et al (2012) Genotypic diversity, antibiotic resistance and bacteriocin production of *Enterococcus* isolated from rhizospheres. Microbes Environ 27:533–537

Kloepper JW, Ryu C-M, Zhang SA (2004) Induced systemic resistance and promotion of plant growth by *Bacillus* spp. Phytopathology 94:1259–1266

Kuipers OP, Beerthuyzen MM, Siezen RJ, Devos WM (1993) Characterization of the nisin gene-cluster *nisABTCIPR* of *Lactococcus lactis*–requirement of expression of the *nisA* and *nisI* genes for development of immunity. Eur J Biochem 216:281–291

Lavermicocca P, Lonigro SL, Valerio F, Evidente A, Visconti A (2002) Reduction of olive knot disease by a bacteriocin from *Pseudomonas syringae* pv. *ciccaronei*. Appl Environ Microbiol 68:1403–1407

Lee H, Churey JJ, Worobo RW (2009a) Biosynthesis and transcriptional analysis of thurincin H, a tandem repeated bacteriocin genetic locus, produced by *Bacillus thuringiensis* SF361. FEMS Microbiol Lett 299:205–213

Lee K, Gray EJ, Mabood F et al (2009b) The class IId bacteriocin thuricin-17 increases plant growth. Planta 229:747–755

Mantovani HC, Russell JB (2001) Nisin resistance of *Streptococcus bovis*. Appl Environ Microbiol 67:808–813

Maqueda M, Hidalgo MS, Fernandez M et al (2008) Genetic features of circular bacteriocins produced by Gram positive bacteria. FEMS Microbiol Rev 32:2–22

Mclntyre J, Kuc J, Williams EB (1973) Protection of pear against fire blight by bacteria and bacterial sonicates. Phytopathology 63:872–877

Michel-Briand Y, Baysse C (2002) The pyocins of *P. aeruginosa*. Biochimie 84:499–510

Mojgani N, Sabiri G, Ashtiani MP, Torshizi MK (2009) Characterization of bacteriocins produced by *L.brevis* NM 24 and *L. fermentum* NM332 isolated from green olives in Iran. Internet J Microbiol 6(2)

Mojgani N, Ameli M, Narjis V, Torshizi MA, Amirinia C (2010) Growth control of *Listeria monocytogenes* in experimental cheese samples by *Lactobacillus casei* RN 78 and its bacteriocin. Afr J Microbiol Res 4:1044–1050

Montesinos E (2007) Antimicrobial peptides and plant disease control. FEMS Microbiol Lett 270:1–11

Nes IF, Holo H (2000) Class II antimicrobial peptides from lactic acid bacteria. Biopolymers 55:50–61

Nguyen HA, Tomita T, Hirota M, Sato T, Kamio Y (1999) A simple purification method and morphology and component analyses for carotovoricin Er, a phage tail like bacteriocin from the plant pathogen *Erwinia carotovora* Er. Biosci Biotechnol Biochem 63:1360–1369
Oman TJ, Boettcher JM, Wang H et al (2011) Sublancin is not a lantibiotic but an S-linked glycopeptide. Nat Chem Biol 7:78–80
Oscariz JC, Pisabarro AG (2001) Classification and mode of action of membrane-active bacteriocins produced by gram-positive bacteria. Int Microbiol 4:13–19
Parret AH, Schoofs G, Proost P, de Mot R (2003) Plant lectin-like bacteriocin from a rhizosphere-colonizing *Pseudomonas* isolate. J Bacteriol 185:897–908
Parret AH, Temmerman K, De Mot R (2005) Novel lectin-like bacteriocins of biocontrol strain *Pseudomonas fluorescens* Pf-5. Appl Environ Microbiol 71:5197–5207
Podile AR, Kishore GK (2006) Plant growth-promoting rhizobacteria. In: Gnanamanickam SS (ed) Plant-associated bacteria. Springer, Dordrecht, pp 195–230
Pugsley A (1984) The ins and outs of colicins. Microbiol Sci 1(168–175):203–205. 09–819
Pugsley AP, Oudega B (1987) Methods of studying colicins and their plasmids. In: Hardy KG (ed) Plasmids, a practical approach. IRL Press, Oxford, pp 105–161
Riley MA, Gordon DM (1992) A survey of col plasmids in natural isolates of *E. coli* and an investigation into the stability of col plasmid lineages. J Gen Microbiol 138:1345–1352
Riley MA, Wertz JE (2002) Bacteriocins: evolution, ecology and application. Ann Rev Microbiol 56:117–137
Rogers LA, Whittier EO (1928) Limiting factors in lactic fermentation. J Bacteriol 16:211–229
Ruiz-Larrea F, Rojo-Bezares B, Sáenz Y, Navarro L, Díez L (2007) Bacteriocins for wine microbiological control and reduction of SO_2 levels. De la Vigne et du Vin:1–13
Sakthivel N, Mew TW (1991) Efficacy of bacteriocinogenic strains of *Xanthomonas oryzae* pv. *oryzae* on the incidence of bacterial blight disease of rice (*Oryza sativa* L.) Can J Microbiol 37:764–768
Sanchez-Hidalgo M, Maqueda M, Galvez A et al (2003) The genes coding for enterocin EJ97 production by E. *faecalis* EJ97 are located on a conjugative plasmid. Appl Environ Microbiol 69:1633–1641
Scholz R, Vater J, Budiharjo A et al (2014) Amylocyclicin, a novel circular bacteriocin produced by *Bacillus amyloliquefaciens* FZB42. J Bacteriol 196:1842–1852
Scupham AJ, Triplett EW (2006) Determination of the amino acid residues required for the activity of the anti-rhizobial peptide antibiotic trifolitoxin. J Appl Microbiol 100:500–507
Severinov K, Semenova AK, Kazakov T, Gelfand MS (2007) Low molecular weight post translationally modified microcins. Mol Micobiol 65:1380–1394
Sieiro AP, Lopez MM, Mu D, Kuipers OP (2016) Bacteriocins of LAB: extending the family. Appl Microbiol Biotechnol 100:2939–2951
Sisto A, Cipriani M, Morea M et al (2010) An rhs-like genetic element is involved in bacteriocin production by *P. savastanoi* pv *savastanoi*. Antonie Van Leeuwenhoek 98:505–517
Stepper J, Shashtri S, Loo TS et al (2011) Cysteine S glycosylation, a new post-translational modification found in glycopeptide bacteriocins. FEBS Lett 585:645–650
Stirling AC, Whittenburg R (1963) Sources of the LAB occurring in silage. J Appl Bacteriol 26:86–90
Stockwell VO, Kawalek MD, Moore LW, Loper JE (1996) Transfer of pAgK84 from the biocontrol agent *Agrobacterium radiobacter* K84 to A.*tumefaciens* under field conditions. Amer Phytopathol Soc 1120-03R
Subramanian S, Smith DL (2015) Bacteriocins from the rhizosphere microbiome – from an agriculture perspective. Front Plant Sci 6:909
Tagg JR, Dajani AS, Wannamaker LW (1976) Bacteriocins of gram positive bacteria. Bacteriol Rev 40:722–756
Tovkach FI (1998) Biological properties and classification of *Erwinia carotovora* bacteriocins. Mikrobiologia 67:636–642

WHO (1969) Specification for the identity and purity of food additives and their toxicological evolution. 12th report the joint FAO/WHO Expert Committee on Food Additives, WHO Tech. Rept. Series No 430. World Health Organization, Geneva

Yamada K, Hirota M, Niimi Y et al (2006) Nucleotide sequences and organization of the genes for carotovoricin (Ctv) from *Erwinia carotovora* indicate that Ctv evolved from the same ancestor as *Salmonella typhi* prophage. Biosci Biotechnol Biochem 70:2236–2247

Zhang X, Clark CA, Pettis GS (2003) Interstrain inhibition in the sweet potato pathogen *Streptomyces ipomoeae*: purification and characterization of a highly specific bacteriocin and cloning of its structural gene. Appl Environ Microbiol 69:2201–2208

Zhang J, Liu G, Shang N, Cheng W, Chen S (2009) Purification and partial amino acid sequence of pentocin 31–1, an anti-listeria bacteriocin produced by *Lactobacillus pentosus* 31–1. J Food Prot 72:2524–2529

Zimmermann N, Jung G (1997) The three-dimensional solution structure of the lantibiotic murein-biosynthesis inhibitor actagardine determined by NMR. Eur J Biochem 246:809–819

Role of Hydrolytic Enzymes of Rhizoflora in Biocontrol of Fungal Phytopathogens: An Overview

9

H.P. Jadhav, S.S. Shaikh, and R.Z. Sayyed

Abstract

Microbial community in the rhizosphere produces a variety of hydrolytic enzymes that are responsible for the degradation of various components of fungal pathogens. The extracellular hydrolytic enzymes excreted by soil rhizobia degrade cell wall components of plant pathogenic microbes. The enzymes of these types are able to breakdown glycosidic linkages present in the polysaccharide of the cell wall of phytopathogens. In this regard, plant growth-promoting rhizobacteria (PGPR) are known to colonize rhizosphere and enhance plant growth through different mechanisms that include (i) plant growth promotion and (ii) biological control of plant disease. Plant growth promotion mechanisms include mineralization of insoluble substances, production of plant growth hormones, biological nitrogen fixation, and promotion of root growth. Biocontrol mechanism involves competition, antibiosis, parasitism, induction of systemic acquired resistance (SAR), induction of systemic resistance (ISR), soil suppressiveness, and production of various antifungal metabolites; hydrolytic enzymes such as chitinase, glucanase, protease, and cellulase; and antibiotics such as 2,4-diacetyl phloroglucinol (DAPG), amphisin, oomycin A, hydrogen cyanide, phenazine, pyoluteorin, pyrrolnitrin, cyclic lipopeptides, oligomycin A, zwittermicin A, kanosamine, and xanthobaccin. Production of hydrolytic enzymes by PGPR is an important mechanism directed against phytopathogens for sustainable plant disease management. These enzymes break down the cell wall of fungal pathogens causing cell death. This review focuses on the different aspects of various hydrolytic enzymes produced by rhizoflora and their role in sustainable biocontrol of phytopathogens.

H.P. Jadhav • S.S. Shaikh • R.Z. Sayyed (✉)
Department of Microbiology, P.S.G.V.P. Mandal's Arts, Science & Commerce College, Shahada, Maharashtra 425 409, India
e-mail: sayyedrz@gmail.com

S. Mehnaz (ed.), *Rhizotrophs: Plant Growth Promotion to Bioremediation*, Microorganisms for Sustainability 2, DOI 10.1007/978-981-10-4862-3_9

9.1 Introduction

Soilborne phytopathogens are responsible for causing infection of roots, stems, leaves, and fruits. These phytopathogens occur in a broad spectrum of susceptible hosts under favorable environmental conditions. Such diseases are known to cause severe economic losses to variety of food crops and are commonly controlled by using synthetic fungicides or nonspecific chemical fungicides. These plant diseases are known to cause a loss of 30% crop yield posing economic hardship to producers (Sayyed et al. 2012; Shaikh and Sayyed 2015). All over the world, population is increasing tremendously every day and hence the agricultural practices to increase the yield. This need has compelled to use synthetic agrochemicals, but the chemical pesticides and fertilizers have caused even more destructive effects to the agricultural field because these chemicals are not eco-friendly. The present need of sustainable agricultural practices is focused on the safer alternatives to conventional agrochemicals (Pane et al. 2013). The intensive use of fungicides, to control plant pathogens and excessive use of chemical fertilizers to increase crop productivity, has severally imbalanced the agroecosystem (Logemann and Schell 1993). In this regard, PGPR have been seen as a greener approach to control plant pathogens and to promote plant growth (Sayyed and Chincholkar 2009; Sayyed et al. 2010, 2013, 2015; Sayyed and Patel 2011; Shaikh et al. 2014, 2016).

The mechanisms of plant growth promotion by PGPR include production of plant growth regulators, asymbiotic N_2 fixation, and solubilization of mineral phosphates and other nutrients (Sarvanakumar et al. 2007; Sayyed et al. 2007; Sharma et al. 2013), while biocontrol involves antagonistic action toward plant pathogens by production of siderophores, antibiotics, cyanide, and hydrolytic enzymes (Shaikh et al. 2014; Shaikh and Sayyed 2015). Antagonistic or biocontrol activity of PGPR is attributed to the production of different types of cell wall-lysing enzymes such as chitinase, protease/elastase, cellulase, and β-1,3 glucanase.

9.2 Plant Growth-Promoting Rhizobacteria (PGPR)

Rhizospheric bacteria, having plant growth-promoting ability by colonizing the plant roots, are known as PGPR (Kloepper and Schroth 1978). PGPR are potentially useful in stimulating plant growth and increasing crop yields (Sayyed et al. 2010). Thus the rhizosphere of crop plants is a promising source of PGPR (Lucas et al. 2001 and Barriuso et al. 2005). PGPR can be differentiated into two categories on the basis of their relationship with the plants: symbiotic rhizobacteria and free-living rhizobacteria (Khan 2005; Freitas et al. 2007). Worldwide literature clearly states that the use of PGPR in agriculture is increased tremendously, and significant increase in growth and yield of agronomically important crops has been obtained (Asghar et al. 2002; Vessey 2003; Gray and Smith 2005; Silva et al. 2006; Figueiredo et al. 2008; Araujo 2008). The plant growth-promoting ability of some bacteria is highly specific to certain plant species, cultivar, and genotype (Bashan 1998; Gupta et al. 2000; Lucy et al. 2004). PGPR not only provide essential nutrients for plant growth promotion, but they are also important in biocontrol of pathogen; they improve the health of soil in the long term and, hence, are potentially important in reducing the use of chemical fertilizers and chemical pesticides (Lugtenberg and Kamilova 2009). However, the better understanding of mechanisms of plant growth promotion and the biocontrol is vital aspect for the better utilization of PGPR in agriculture. The knowledge of structure and diversity of rhizosphere microbial consortium with respect to their complexity; natural selection; interpopulational relations like symbiosis, parasitism, mutualism, or competence; and succession is equally important in this aspect (Barriuso et al. 2008).

9.3 Fungal Plant Diseases

The vast range of phytopathogens causes various types of diseases by infecting the whole or a specific part of the plants. Their effect ranges from mild symptoms to catastrophes in which huge plantations of food crops are destroyed and hence causes loss of yield. Catastrophic plant disease exerts the current deficit of food supply in which at least 800 million people are not properly fed. The strengths of phytopathogens like their populations are variable in time and space, and genotype increases the difficulties to control them (Strange and Scott 2005). The continuous use of fungicides has developed the resistance which causes the loss in productivity. The biological controls have been found more promising than chemical fertilizers, discussed in Sect. 9.5.

The worldwide reporting shows that not all but various fungal species are found to be pathogenic to the plants and their products. Some of the plants affected by phytopathogenic diseases are listed in Table 9.1.

9.4 Composition of Fungal Cell Wall

The cell wall of fungal or any pathogen is meant for protection of its internal constituent from various environmental factors. The structure of fungal cell wall is unique and is therefore an excellent target for the development of antifungal metabolites. The structure and biosynthesis of various antifungal metabolites have been reviewed. These studies have clearly demonstrated that fungal cell walls are mainly composed of chitin, glucans, mannans, and glycoproteins (Bowman and Free 2006).

The fungal cell walls contain fibrillar materials attached to sugars, proteins, lipids, and a variety of polysaccharides (Fig. 9.1). These fibrillar materials are inert. The functional components of cell wall are needed for nutrient transport, extracellular degradation of non-permeable substrates, communication, and modifications of cell wall structure.

About, 80% of the cell wall of fungi is made up of polysaccharides. The fibrillar structure is built on chitin, chitosan, ß-glucans, and a variety of heteropolysaccharides (Table 9.2). These fibers are encompassed in a complex gel-like matrix. Proteins in the form of glycoprotein are present in small amount, i.e., 20%. All

Table 9.1 List of phytopathogens infecting plants with various diseases

No.	Disease	Target plant or plant part	Phytopathogen	References
1	Brown patch	Patches of brown and yellow color appear on the lawn in irregular shapes	*Rhizoctonia solani*	Giesler and Yuen (1998)
2	Cankers	Woody plants	*Gibberella circinata* (*Fusarium circinatum*)	Wingfield et al. (2002)
3	Damping off	All types of plants	*Pythium* and *Fusarium*	Mao et al. (1997)
4	Powdery mildew	Grains, alfalfa, onions, cucumbers	*Uncinula necator*	Doster and Schnathorst (1985)
5	Ergot	Rye, barley, wheat, and other grasses	*Claviceps purpurea*	Giesbert et al. (2008)
6	Root rots	All types of plants	*Phytophthora* sp.	Thomas et al. (2003)
7	Rusts	Wheat, oats, barley, rye	*Puccinia*	Uchida et al. (2006)
8	Scab	Wheat, rye, barley, potatoes	*Fusarium graminearum*	O'Donnell et al. (2000)
9	Seed decay	All types of plants	*Phomopsis*	Li et al. (2015)
10	Smuts	Oats, barley, grasses, corn, wheat	*Ustilaginomycetes*	Müller (2015)
11	Soft rots, dry rots	Potatoes, onions, carrots, fleshy organs, etc.	*Syncephalastrum racemosum*, *Fusarium* sp.	Misra (2016) Heltoft (2016)
12	Wilts	Potatoes, alfalfa, trees	*Fusarium oxysporum*	Pietro et al. (2003)

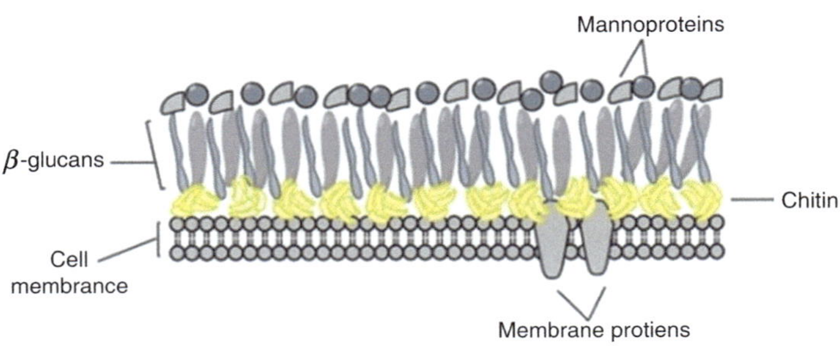

Fig. 9.1 Typical structure of fungal cell wall (Adapted from Vega and Kalkuma 2011)

proteins are not generally the structural components. Lipids are present only in small amount. Proteins and lipids regulate movement of water and protect the fungal cell from desiccation (Cox and Hooley 2009).

Table 9.2 Fungal cell wall-forming polymers

No.	Classification	Fibrous polymers	Gel-like polymers
1	Basidiomycota	Chitin β-(1-3), β-(1-6) glucan	Xylomannoproteins α (1-3) glucan
2	Ascomycota	Chitin β-(1-3), β-(1-6) glucan	Galactomannoproteins α (1-3) glucan
3	Zygomycota	Chitin chitosan	Polyglucuronic acid, glucuronomannoproteins, polyphosphate
4	Chytridiomycota	Chitin glucan	Glucan

Adapted from Gooday (1995)

9.5 Biocontrol Through Hydrolytic Enzymes

It has been studied that many rhizobacteria/biocontrol agents (BCAs) synthesize extracellular hydrolytic enzymes that are involved in hydrolysis of fungal cell wall components such as chitin, proteins, cellulose, hemicellulose, and DNA; these hydrolytic enzymes have the potential of inhibiting phytopathogens (Pal and Gardener 2006).

9.5.1 Hydrolytic Enzymes

The term biocontrol/biological control denotes the direct or indirect manipulation of microbes for reducing plant disease (Baker and Cook 1974; Maloy 1993). Among the wide genetic biodiversity of prokaryotes, PGPR play crucial role in the biocontrol of plant diseases and in improvement of crop productivity through various mechanisms (Fernando et al. 2005). Biotic agents like harmful insects, parasitic weeds, and phytopathogens are among the major causes of serious loss and damage to agricultural crop and products. This needs to be controlled to sustain the quality and quantity of agriculture products. Currently numerous strategies are employed to combat this problem (Bargabus et al. 2002; Benhamou 2004; Kloepper et al. 2004, Islam et al. 2005; Chisholm et al. 2006; Heydari 2007; Heydari et al. 2007). A natural, safe, and productive option for the control of these pathogens is the use of BCAs. BCAs include the number of microbial genera from rhizosphere including PGPR. Consequently, to improve biocontrol strategies by manipulating soil environment, the study of mechanism of biocontrol of plants diseases through the interaction between BCAs and pathogens is the key to create successful biocontrol conditions (Fravel 1998). The biocontrol of plant disease includes the secretion of microbial metabolite which controls the diseases by acting on or by inhibiting the growth of phytopathogens.

Hydrolytic enzymes (chitinase, glucanase, protease, and cellulase) produced by PGPR are responsible for the lysis of phytopathogens through hyperparasitism. The antagonistic properties of hydrolytic enzymes against various phytopathogens play a major role in biocontrol (Kim et al. 2003; Shaikh and Sayyed 2015). BCAs producing hydrolytic enzymes are used in biocontrol of phytopathogens thereby improving plant growth. These attributes make PGPR an effective BCA (Garbeva et al. 2004; Ran et al. 2005). The cell wall of most of the phytopathogenic fungi (except oomycetes) is made up of chitin ($C_8H_{13}O_5N)_n$, which is an unbranched, long-chain polymer of glucose derivatives, composed of β-1,4-linked units of the amino sugar N-acetyl-D-glucosamine (NAG).

The biocontrol activity of BCAs/PGPR can be achieved through the following mechanisms:

(a) **Niche competition** – this excludes the growth of phytopathogens from soil or host tissue.
(b) **Mycoparasitism** – leading to the lysis of fungal pathogen.
(c) **Production of antibiotics** – that interfere with the metabolism of phytopathogen.
(d) **Production of hydrolytic enzymes** – that degrade the cell wall of phytopathogens (Sayyed et al. 2013).

9.5.1.1 Cell Wall Lysis

Hydrolytic enzymes are capable of breaking down glycosidic bonds in chitin. Thus, they play a vital role in the biological control of many plant diseases by degrading the cell walls of phytopathogens.

It affects fungal growth by its lytic action on cell walls, hyphal tips, and germ tubes (Kim et al. 2003) and partial swelling in the hyphae and at the hyphal tip leading to hyphal curling or bursting of the hyphal tip (Fig. 9.2; Someya et al. 2000). Among the huge population of hydrolytic enzymes, chitinase, glucanase, protease, and cellulase are of major interest due to their ability to degrade and lyse fungal cell wall, and thus hydrolytic enzymes are employed in biocontrol of fungal phytopathogens (Mabood et al. 2014). Cell wall-degrading enzymes of rhizobacteria damage the structural integrity of the cell wall of phytopathogen (Budi et al. 2000). Felse and Panda (1999) reported the control of *Sclerotium rolfsii* and *F. oxysporum* through the cell wall degradation on beans.

9.5.1.2 Mycoparasitism

The other concept regarding the inhibition of phytopathogens is mycoparasitism that directly attacks which is defined as a direct attack on a fungal thallus leading to its lysis (Chet et al. 1997). According to Barnett and Binder (1973), mycoparasites play an important role in biocontrol. Mycoparasitism can be divided into two types: necrotrophic and biotrophic. Necrotrophic mycoparasites are those that kill the host cells before or just after invasion and use the released nutrients. These mycoparasites are more aggressive and destructive than biotrophs. They have a broad host range and are relatively unspecialized in their mode of parasitism. The antagonistic activity of necrotrophs is due to the production of antibiotics, toxins, or hydrolytic enzymes (Manocha and Sahai 1993). In biotrophic parasitism, the development of the parasite is favored by a living rather than a dead host structure (Chet et al. 1997). Biotrophic mycoparasites have a more restricted host range and in many cases

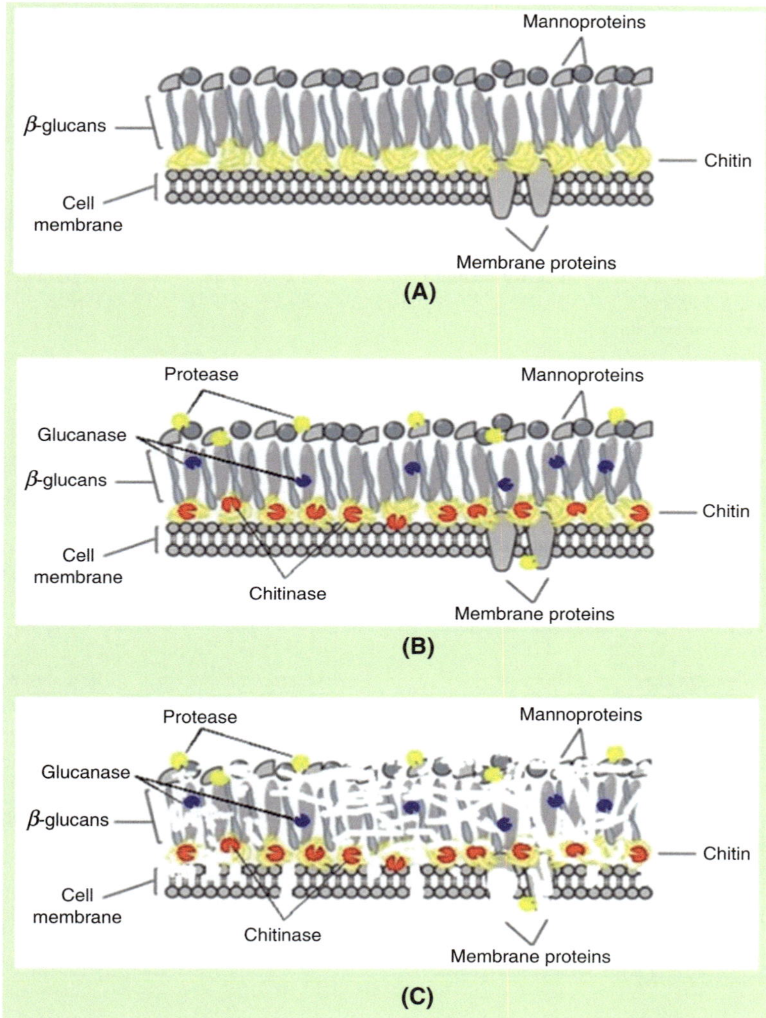

Fig. 9.2 Mechanism of fungal cell wall hydrolysis. (**a**) Typical structure of fungal cell wall. (**b**) Hydrolytic enzymes (chitinase, glucanase, and protease) acting on chitin, β-glucan, and proteins. (**c**) Fungal cell wall losing integrity after hydrolysis

Table 9.3 List of microorganisms showing hydrolytic activity

No.	Microbes showing hydrolytic activity	Hyd. enz. produced	Target phytopathogen	References
1	*S. marcescens*	Chitinase	*R. solani* and *F. oxysporum*	Someya et al. (2000)
2	*B. subtilis* NPU 001	Chitinase	*F. oxysporum*	Chang et al. (2010)
3	*S. plymuthica* C48	Chitinase	*Botrytis cinerea*	Frankowski et al. (2001)
4	*Paenibacillus* sp. strain 300 and *Streptomyces* sp. strain 385	β -1,3-glucanase	*F. oxysporum*	Singh et al. (1999)
5	*Bacillus subtilis* YJ1	Cellulase	–	Li-Jung et al. (2010)
6	*Cellulomonas* sp. ASN2	Cellulase	–	Muhammad et al. (2012)
7	*Bacillus coagulans*	Carboxymethyl cellulase and polygalacturonase	–	Odeniyi et al. (2009)
8	*Bacillus cereus*, *Bacillus subtilis*, *Bacillus thuringiensis*	Cellulase	–	Basavaraj et al. (2014)
9	*P. aeruginosa* PGPR2	Protease	*Macrophomina* sp., *Rhizoctonia* sp., and *Fusarium* sp.	Illakkiam et al. (2013)
10	*Bacillus subtilis* PE-11	Alkaline protease	–	Adinarayana et al. (2003)
11	Paenibacillus and Streptomyces	–	*F. oxysporum*	Compant et al. (2005)
12	*B. cepacia*	–	*R. solani*, *P. ultimum*, and *S. rolfsii*	
13	*P. fluorescens LRB3W1* and *S. marcescens* B2	–	*F. oxysporum*	Someya et al. (2007)

produce specialized structures (haustoria) to absorb nutrients from their host (Manocha and Sahai 1993). Rhizobacteria capable of producing hydrolytic enzymes and inhibiting phytopathogens are listed in Table 9.3.

9.5.2 Chitinases in Biocontrol of Phytopathogenic Fungi

Chitinase [EC 3.2.1.14] plays a vital role in the biocontrol of many plant diseases by lysing fungal cell wall through degradation of chitin polymer present in the cell walls of fungal phytopathogens. The enzyme can either be used directly in the biocontrol on microorganisms or indirectly by using purified proteins or through manipulation of genes coding for chitinase (Kim et al. 2003). Chitinases have been reported from various microorganisms, such as insects, crustaceans, yeasts, and fungi, and also organisms that do not contain chitin, such as bacteria, higher plants, and vertebrates (Kramer et al. 1997). Chitinase was isolated, purified, and characterized in 1992 (Cruz et al. 1992). Chitinase produced by rhizobacteria exhibits antagonism in vitro against fungi (Gay et al. 1992; Fridlender et al. 1993). Schlumbaum et al. (1986) and Skujins et al. (1965) demonstrated the inhibition of fungal growth by chitinases of *Streptomycetes*. The importance of chitinase activity was further demonstrated by the loss of biocontrol efficacy in *Serratia marcescens* chitinase mutants in which the *chiA* gene had been inactivated (Jones et al. 1986). The potential BCAs can be produced by cloning *chiA* gene into rhizosphere competent model organisms. Oppenheim and Chet (1992) cloned the *chiA* gene of *S. marcescens* into *E.coli* for the control of *S. rolfsii* and *R. solani* and found *E.coli* to be better in reducing disease incidence. Likewise the chitinase genes from *S. marcescens* were expressed in *Pseudomonas* and the plant symbiont *Rhizobium meliloti* to control the pathogens *F. oxysporum* var. *redolens* and *Gaeumannomyces graminis* var. *tritici* (Sundheim 1992). The antifungal activity of the transgenic *Rhizobium* during symbiosis on alfalfa roots was verified by lysis of *R. solani* hyphal tips treated with cell-free nodule extracts (Sitrit et al. 1993).

The fungal spp. *Trichoderma* and *Gliocladium virens* have been studied more extensively (Cook 1993; Chet et al. 1997). Weindling (1932) reported the potential of *Trichoderma* species as BCAs. The chitinase of *T. harzianum* was used as a means of biocontrol of phytopathogens such as *Rhizoctonia solani* (Chet and Hornby 1990). Several species of *Trichoderma* have been tested as BCAs; among them *T. harzianum* was found to be more effective and can be used to control the number of economically important soilborne phytopathogens (Chet 1987). Using genetic modification technology, Lorito (1998) cloned the tobacco and potato with gene encoding endochitinase from *T. harzianum* (P1) and reported the high level and broad spectrum of resistance against a number of phytopathogens.

9.5.2.1 Mode of Action of Chitinase

Chitinases are chitin-degrading enzymes which play an important role in biological control and plant defense mechanisms against phytopathogens. Chitin is the second most abundant polymer in nature, an unbranched homopolymer of 1,4-β-linked *N*-acetyl-D-glucosamine (GlcNAc) after cellulose. It is abundant as a structural polymer in most fungi and insects, including those that are agricultural pests (Havukkala 1991).

On the basis of mode of action, chitinase is divided into three types:

(A) **β-1,4-*N*-acetyl-glucosaminidases** (EC 3.2.1.30) split the chitin polymer into GlcNAc monomers in an exo-type pattern.
(B) **Endochitinases** (EC 3.2.1.14) cleave randomly at internal sites over the entire length of the chitin microfibril.

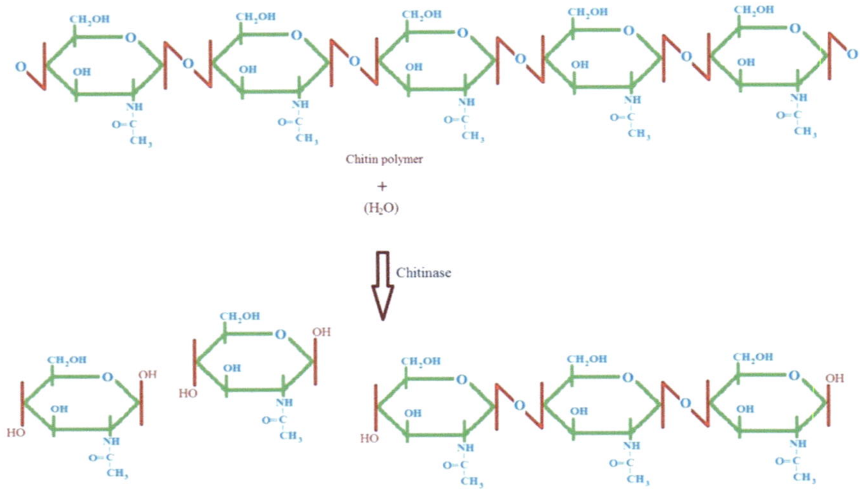

Fig. 9.3 Chitinolysis of 1,4-β-linked *N*-acetyl-D-glucosamine (GlcNAc)

(C) **Exochitinases** (EC 3.2.1.14) catalyze the progressive release of diacetylchitobiose in a stepwise manner such that no monosaccharides or oligosaccharides are formed (Fig. 9.3) (Harman et al. 1993; Manocha and Sahai 1993).

An extracellular chitinase of *Myrothecium verrucaria* inhibits germination and germ tube elongation of the groundnut rust fungus *Puccinia arachidis*. Similarly, *Acremonium obclavatum* produces and secretes a chitinase in vitro which inhibits germination of uredospores of the peanut rust (Manocha and Balasubramanian 1994).

9.5.2.2 Molecular Characterization of Chitinase

Cruz et al. (1992) reported the purification and characterization of three chitinases from *T. harzianum*; the isozymes' mol. wt. were 37, 33, and 42 kDa, respectively. Only the purified 42 kDa chitinase hydrolyzed *B. cinerea* purified cell walls in vitro, but this effect was heightened in the presence of either of the other two isoenzymes. According to Haran et al. (1995), the chitinolytic system of *T. harzianum* was more complex, consisting of six distinct enzymes. The system is apparently composed of two ß-(1,4)-N-acetylglucosaminidases of 102 and 73 kDa, respectively, and four endochitinases of 52, 42, 33, and 31 kDa, respectively. Among these, the 42 kDa endochitinase was found more effective because of its ability to hydrolyze *B. cinerea* cell walls in vitro. The 1,4-β-*N*-acetyl-glucosaminidases of 72 kDa have been purified from *T. harzianum* strain (Lorito et al. 1994). Haran et al. (1995) reported the chitinase isolated from respective *T. harzianum* had different molecular weights: 73 kDa heat-stable glucosaminidase (CHIT 73), isolated from *T. harzianum* strain TM, an endochitinase of 52 kDa (CHIT 52), an endochitinase of 42 kDa (CHIT 42), the endochitinases produced by the other strains of *T. harzianum* which are of 33 kDa (CHIT 33) and 31 kDa (CHIT 31), and two endochitinases, having molecular weights of 37 kDa and 33 kDa, which were expressed by *T. harzianum* strain CECT 2413.

9.5.3 Proteases in Biocontrol of Phytopathogenic Fungi

Proteases [E.C. 3.4.24] play a significant role in cell wall lysis of phytopathogenic fungi, since chitin and/or fibrils of β-glucan are embedded into the protein matrix. Thus proteolytic activity is prerequisite to lyse whole fungal cells (Elad and Kapat 1999). Proteases are wide spread in nature; microbes are the preferred source of these enzymes due to their fast growth and easy cultivation and the ease in genetic manipulation to get the enzyme with desired properties for specific applications (Anwar and Saleemuddin 1998; Beg and Gupta 2003). *Bacillus* sp. produces extracellular proteases; several *Bacillus* species like *Bacillus cereus*, *Bacillus stearothermophilus*, *Bacillus mojavensis*, *Bacillus megaterium*, and *Bacillus subtilis* are known to produce protease (Sookkheo et al. 2000; Beg and Gupta 2003; Banik and Prakash 2004; Gerze et al. 2005). Bacterial proteases are generally extracellular, easily produced in greater amounts, and active under various environmental conditions.

Proteases purified from *Bacillus* have significant activity, stability, broad substrate specificity, short period of fermentation, simple downstream purification, and low-cost production process (Maurer 2004; Haddar et al. 2009). Extracellular proteases of *Trichoderma* sp. also play a significant role in the lysis of cell walls of phytopathogenic fungi. Some of the proteases produced by *Trichoderma* sp. are involved in inactivating extracellular enzymes of phytopathogenic fungi (Elad and Kapat 1999). The protease enzymes break down major proteins into peptide chains and/or their constituent amino acids of phytopathogens and thereby destroy their capacity to act on plant cells.

9.5.3.1 Mode of Action of Protease

Proteins are degraded by a hydrolysis that involves cutting of one or more peptide bonds by addition of water to liberate peptide or amino acids. Enzymes that hydrolyze the proteins are called proteases. Each protease recognizes the chemical structures of certain specific amino acids and then catalyzes the breaking of the peptide bond (Fig. 9.4).

9.5.3.2 Molecular Characteristics of Protease

The recent studies by Asker et al. (2013) reported the molecular weight of the purified proteases P1 and P2 as 28 and 25 kDa, respectively. The purified P1 and P2 were rich in aspartic acid and serine and relatively have higher amounts of alanine, leucine, glycine, valine, threonine valine, and glutamic acid. Gessesse et al. (2003) purified an alkaline protease of 24 kDa from *Bacillus pseudofirmus* AL-89. Adinarayana et al. (2003) purified an alkaline protease of 15 kDa from *B. subtilis* PE-11. A halotolerant alkaline protease of 28 kDa was purified from *Bacillus*

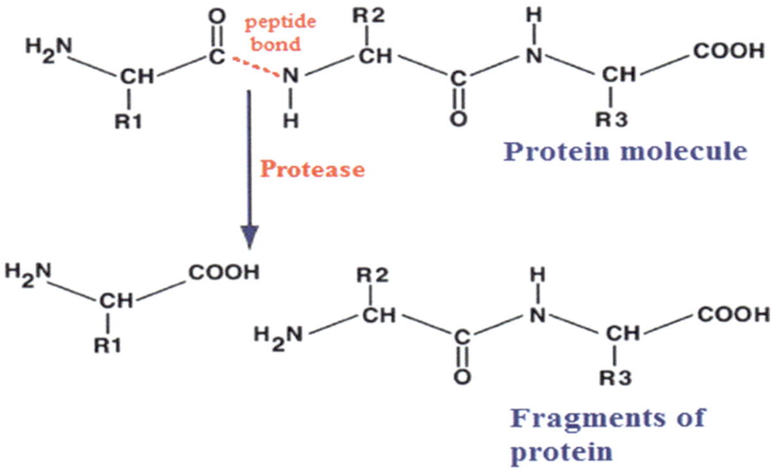

Fig. 9.4 Schematic representation of proteolysis (Modified from Donohue and Osna 2003)

clausii I-52 using a combination of Diaion HPA75, phenyl-Sepharose, and DEAE-Sepharose column chromatography (Joo and Chang 2005). Gupta et al. (2005) purified an alkaline protease from *B. pseudofirmus* to tenfold purity with an 85% yield using a single-step method with a phenyl-Sepharose 6 fast-flow column. The apparent molecular weight of this protease was 29 kDa. Sareen and Mishra (2008) purified a 55 kDa alkaline protease from *Bacillus licheniformis* RSP-09-37.

9.5.4 Cellulase in Biocontrol of Phytopathogenic Fungi

Cellulases [EC 3.2.1.4] catalyze the hydrolysis of 1,4-β-D-glycosidic linkages in cellulose and play a significant role in nature by recycling this polysaccharide. Cellulose is a linear polymer of β-D-glucose units linked through 1,4-β-linkages with a degree of polymerization ranging from 2,000 to 25,000 (Kuhad et al. 1997). Cellulose chains form numerous intra- and intermolecular hydrogen bonds, which account for the formation of rigid, insoluble, crystalline microfibrils. Cellulose is structurally heterogeneous having both amorphous and crystalline regions. Resistance to microbial degradation depends on the degree of crystallinity, and highly crystalline regions are more resistant to enzymatic hydrolysis. Cellulases belong to a class of enzymes that catalyze the hydrolysis of cellulose and are produced chiefly by fungi, bacteria, and protozoa as well as other organisms like plants and animals. The cellulolytic enzymes are inducible since they can be synthesized by microorganisms during their growth on cellulosic materials (Lee and Koo 2001).

9.5.4.1 Mode of Action of Cellulase

Complete degradation of cellulose involves a complex interaction between different cellulolytic enzymes. It has been widely accepted that three types of enzymes including cellulose/endoglucanases [EC 3.2.1.4], exo-cellobiohydrolase/exo-glucanases [EC 3.2.1.91], and β-glucosidases [EC 3.2.1.21] act synergistically to convert cellulose into β-glucose (Lynd et al. 2002). Cellulases are a mixture of endo-1,4-β-glucanase enzymes and exo-1,4-β-glucanase enzymes. Endo-1,4-β-glucanase cleaves the internal bonds, while exo-1,4-β-glucanase cleaves two to four units from the ends of cellulose strands and cellobiase, which cleaves the disaccharide cellobiose into two glucose moieties (Fig. 9.5).

9.5.4.2 Molecular Characterization of Cellulase

Hurst et al. (1977) reported the cellulase of molecular weight of 26,000 on the basis of amino acid composition and PAGE analysis. Carboxymethyl cellulase produced by *B. pumilus* EB3 was having the range of a molecular weight from 30 to 65 kDa (Ariffin et al. 2006). Li-Jung et al. (2010) reported the strain *Bacillus subtilis* YJ1 producing cellulase; they purified and characterized cellulase, having a molecular mass of 32.5 kDa.

9.5.5 Glucanases in Biocontrol of Phytopathogenic Fungi

β-1,3-Glucanases [EC 3.1.1.6] are widely spread in bacteria, fungi, and higher plants (Simmons 1994). This enzyme has interesting and important physiological roles and practical applications in the degradation of cell wall in fungi, yeasts, and higher plants (Pang et al. 2004). These enzymes are classified as either exo- or endo-β-1,3-glucanases (β-1,3-glucan glucanohydrolase). Fridlender et al. (1993) reported the hydrolytic inhibition of *Rhizoctonia solani*, *Sclerotium rolfsii*, and *Pythium ultimum* by β-1,3-glucanases of *Bacillus cepacia*. Singh et al. (1999) reported two strains of *Paenibacillus* and *Streptomyces* sp. which produce β-1,3-glucanases that inhibited the growth of *F. oxysporum*. Vazquez et al. (1998) reported the seven β-1,3-glucanases produced by *T. harzianum* strain under diverse growth conditions.

9.5.5.1 Mode of Action of Glucanase

β-1,3(1,6)-Glucans are major components in cell wall of yeasts and fungi. The cell wall polysaccharide glucan is consisted of predominantly β-1,3-linked backbone having some branches via β-1,6-linkages, 4,6,8,9. Glucanase causes degradation of cell wall and further penetration into the host mycelium (Fridlender et al. 1993). These enzymes can hydrolyze the substrate by two possible mechanisms: (a) exo-1,3-glucanases (EC 3.2.1.58) hydrolyze the substrate by sequentially cleaving glucose residues from the nonreducing end and (b) endo-1,3-glucanases (EC3.2.1.39) cleave linkages at random sites along the polysaccharide chain, releasing smaller oligosaccharides (Noronha and Ulhoa 1996).

9.5.5.2 Molecular Characteristics of Glucanase

Cruz et al. (1992) and Noronha and Ulhoa (1996) have reported two 1,3-glucanases having molecular weights of 78 and 36 kDa, respectively, purified from the supernatants of *T. harzianum* grown in minimal medium, supplemented with chitin as

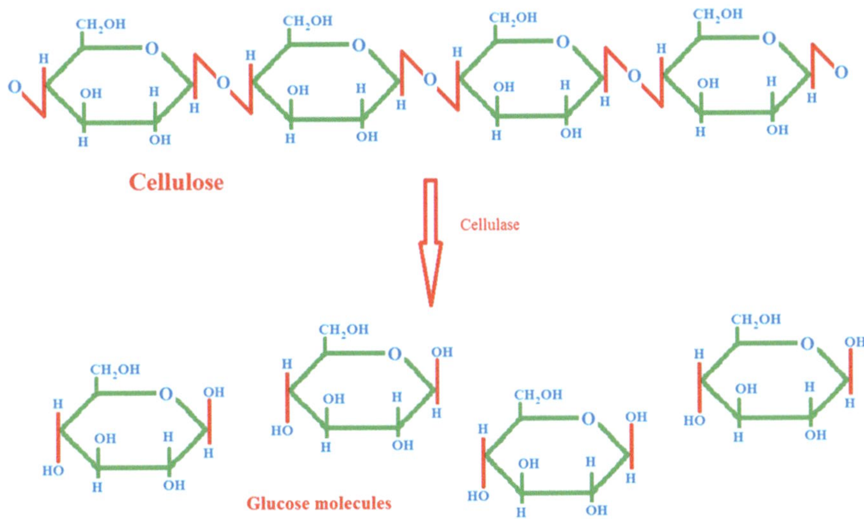

Fig. 9.5 Structure of cellulose and its digestion to glucose

carbon source. The characterization of these enzymes revealed that they are endo-1,3-glucanases, as confirmed by the production of oligosaccharides rather than glucose from the laminarin. Noronha and Ulhoa (2000) purified and characterized the 29 kDa extracellular-1,3-glucanase produced by *T. harzianum*, grown on chitin-containing medium. Maria et al. (2003) report the two purified 83.1 kDa extracellular exo-β-1, 3-glucanases produced by *T. asperellum*.

9.6 Conclusion

In concern with the current scenario toward chemical pesticides and fertilizers, and their huge consumption, there is a prominence/focus on utilization of microbial inoculants and organic inputs for its application in agricultural field. Hence, the potential of rhizobacteria in crop protection by producing different defensive antifungal metabolites like antibiotics, hydrolytic enzymes, and other metabolites is hoped to provide sustainable and eco-friendly plant disease control. Application of these rhizobacteria in agricultural field in the form of formulated product will give the greener and eco-friendly approach for the sustainable agriculture to combat the fungal diseases. Application of efficient rhizobacterial strain secreting various hydrolytic enzymes will help to reduce the liberal use and doses of agrochemicals which is the most important prospect in rhizobacterial/PGPR research. Commercial production of these organisms will have sustained release of antifungal metabolites in the environment, and these metabolites do not develop the resistance to target organism as in chemical pesticides.

Application of single or consortium of these organisms has shown the promising prospect in the field of biocontrol and plant growth promotion. These microbes can successfully utilize their potential for agricultural integrated plant disease management (IPDM) strategies. Study of hydrolytic enzymes of rhizobacteria will help in manipulating the bacterial community with biological control and plant growth promotion ability in rhizospheric zone of different sites. So these rhizobacteria will be the key determinant in plant health and productivity with sustainability.

References

Adinarayana K, Ellaiah P, Prasad DS (2003) Purification and partial characterization of thermostable serine alkaline protease from a newly isolated *Bacillus subtilis* PE-11. AAPS Pharm Sci Tech 4(4):440–448

Anwar A, Saleemuddin M (1998) Alkaline proteases – a review. Bioresour Technol 64(3):175–183

Araujo FF (2008) Inoculacao de sementes com *Bacillus subtilis*, formulado com farinha de ostras e desenvolvimento de milho, soja e algodao. Cienc Agrotec 32:456–462

Ariffin H, Abdullah N, Umi Kalsom MS, Shirai Y, Hassan MA (2006) Production and characterisation of cellulase by *Bacillus Pumilus* EB3. Int J Eng Technol 3(1):47–53

Asghar HN, Zahir ZA, Arshad M, Khalig A (2002) Plant growth regulating substances in the rhizosphere: microbial production and functions. Adv Agron 62:146–151

Asker MM, Mahmoud MG, El Shebwy K, el Aziz MSA (2013) Purification and characterization of two thermostable protease fractions from *Bacillus megaterium*. J Genet Eng Biotechnol 11:103–109

Baker KF, Cook RJ (1974) Biological control of plant pathogens. WH Freeman and Company, San Francisco

Banik RM, Prakash M (2004) Laundry detergent compatibility of the alkaline protease from *Bacillus cereus*. Microbiol Res 159:135–140

Bargabus RL, Zidack NK, Sherwood JE, Jacobsen BJ (2002) Characterization of systemic resistance in sugar beet elicited by a non-pathogenic, phyllosphere-colonizing *Bacillus mycoides*, biological control agent. Physiol Mol Plant Pathol 61:289–298

Barnett HL, Binder FL (1973) The fungal host-parasite relationship. Annu Rev Phytopathol 11:273–292

Barriuso J, Pereyra MT, Lucas García JA et al (2005) Screening for putative PGPR to improve establishment of the symbiosis *Lactarius deliciosus-Pinus* sp. Microbial Ecol 50:82–89

Barriuso J, Ramos B, Lucas JA et al (2008) Ecology, genetic diversity and screening strategies of plant growth promoting rhizobacteria (PGPR). In: Ahmad I, Pichtel J, Hayat S (eds) Plant-bacteria interactions: strategies and techniques to promote plant growth, Wiley-VCH Verlag GmbH & Co. KGaA, Weinheim ISBN:978-3-527-31901-5

Basavaraj I, Patagundi CT, Shivasharan KBB (2014) Isolation and characterization of cellulase producing bacteria from soil. Int J Curr Micro Appl Sci 3(5):59–69

Bashan Y (1998) Inoculants of plant growth promoting bacteria for use in agriculture. Biotechnol Adv 16:729–770

Beg QK, Gupta R (2003) Purification and characterization of an oxidation-stable, thiol-dependent serine alkaline protease from *Bacillus mojavensis*. Enz Microb Technol 32:294–304

Benhamou N (2004) Potential of the mycoparasite, *Verticillium lecanii,* to protect citrus fruit-against *Penicillium digitatum*, the casual agent of green mold: a comparison with-the effect of chitosan. Phytopathology 94:693–705

Bowman SM, Free SJ (2006) The structure and synthesis of the fungal cell wall. Bioessays 28:799–808

Budi SW, van Tuinen D, Arnould C et al (2000) Hydrolytic enzyme activity of *Paenibacillus sp.* strain B2 and effects of the antagonistic bacterium on cell integrity of two soilborne pathogenic bacteria. Appl Soil Ecol 15:191–199

Chang WT, Chen M, Wang SL (2010) An antifungal chitinase produced by *Bacillus subtilis* using chitin waste as a carbon source. World J Microbiol Biotechnol 26:945–950

Chet I (1987) *Trichoderma*-application, mode of action and potential as a biocontrol agent of soil-borne plant pathogenic fungi. In: Chet I (ed) Innovative approaches to plant disease control. Wiley, New York, pp 137–160

Chet I, Hornby D (1990) Biological control of soil-borne plant pathogens with fungal antagonists in combination with soil treatments. In: Hornby D (ed) Biological control of soil-borne plant pathogens. Cab International, Wallingford, pp 15–25

Chet I, Inbar J, Hadar I (1997) Fungal antagonists and mycoparasites. In: Kubicek CP, Druzhinina IS (eds) The Mycota IV: environmental and microbial relationships. Springer-Verlag, Berlin, pp 165–184

Chisholm ST, Coaker G, Day B, Staskawicz BJ (2006) Host-microbe interactions: shaping the evolution of the plant immune response. Cell 124:803–814

Compant S, Duffy B, Nowak J, Clément C, Barka EA (2005) Use of plant growth-promoting bacteria for biocontrol of plant diseases: principles, mechanisms of action, and future prospects. Appl Environ Microbiol 71:4951–4959

Cook RJ (1993) Making greater use of introduced microorganisms for biological control of plant pathogens. Annu Rev Phytopathol 31:53–80

Cox PW, Hooley P (2009) Hydrophobins: new prospects for biotechnology. Fungal Biol Rev 23:40–47

De La Cruz J, Rey M, Lorca JM et al (1992) Isolation and characterization of three chitinases from *Trichoderma harzianum*. Eur J Biochem 206:859–867

Donohue TM, Osna NA (2003) Intracellular proteolytic systems in alcohol-induced tissue injury. Alcohol Res Health 27:317–324

Doster MA, Schnathorst WC (1985) Effects of leaf maturity and cultivar resistance on development of the powdery mildew fungus on grapevines. Phytopathology 75:318–321

Elad Y, Kapat A (1999) The role of *Trichoderma harzianum* protease in the biocontrol of *Botrytis cinerea*. Eur J Plant Pathol 105:177–189

Felse AP, Panda T (1999) Production of microbial chitinases. Bioprocess Eng 23:127–134

Fernando WGD, Nakkeeran S, Zhang Y (2005) Biosynthesis of antibiotics by PGPR and its relation in biocontrol of plant diseases. In: Siddiqui ZA (ed) PGPR: biocontrol and biofertilization. Springer, Dordrecht, pp 67–109

Figueiredo MVB, Burity HA, Martinez CR, Chanway CP (2008) Alleviation of water stress effects in common bean (*Phaseolus vulgaris* L.) by co-inoculation *Paenibacillus x Rhizobium tropici*. Appl Soil Ecol 140:182–188

Frankowski J, Lorito M, Scala F, Schmidt R, Berg G, Bahl H (2001) Purification and properties of two chitinolytic enzymes of *Serratia plymuthica* HRO-C48. Arch Microbiol 176:421–426

Fravel DR (1998) Role of antibiosis in the biocontrol of plant diseases. Annu Rev Phytopathol 26:75–91

Freitas ADS, Vieira CL, Santos CERS, Stamford NP, Lyra MCCP (2007) Caracterizacao de rizobios isolados de Jacatupe cultivado em solo salino no Estado de Pernanbuco, Brasil. Bragantia 66(3):497–504

Fridlender M, Inbar J, Chet I (1993) Biological control of soilborne plant pathogens by a ß-1, 3-glucanase producing *Pseudomonas cepacia*. Soil Biol Biochem 25:1211–1221

Garbeva P, van Veen JA, van Elas JD (2004) Assessment of the diversity, and antagonism towards *Rhizoctonia solani* AG3, of *Pseudomonas* species in soil from different agricultural regimes. FEMS Microbiol Ecol 47:51–64

Gay PA, Saikurnar KV, Cleveland TE, Tuzun S (1992) Antagonistic effect of chitino1ytic bacteria against toxin producing fungi. Phytopathology 82:1074

Gerze A, Omay D, Guvenilir Y (2005) Partial purification and characterization of protease enzyme from *Bacillus subtilis megatherium*. In: Davison BH, Evans B, Finkelstein M, Mcmillan JD (eds) Twenty-sixth symposium on biotechnology for fuels and chemicals. Humana Press, New York, pp 335–345

Gessesse A, Hatti-Kaul R, Gashe BA, Mattiasson BO (2003) Novel alkaline proteases from alkaliphilic bacteria grown on chicken feather. Enz Microb Technol 32:519–524

Giesbert S, Schuerg T, Scheele S, Tudzynski P (2008) The NADPH oxidase Cpnox1 is required for full pathogenicity of the ergot fungus *Claviceps purpurea*. Mol Plant Pathol 9:317–327

Giesler LJ, Yuen GY (1998) Evaluation of *Stenotrophomonas maltophilia* strain C3 for biocontrol of brown patch disease. Crop Prot 17:509–513

Gooday GW (1995) Cell walls. In: Gow NA, Gadd GM (eds) Growing fungus. Springer Science & Business Media, Dordrecht, pp 43–62

Gray EJ, Smith DL (2005) Intracellular and extracellular PGPR: commonalities and distinctions in the plant bacterium signaling processes. Soil Biol Biochem 37:395–412

Gupta A, Gopal M, Tilak KV (2000) Mechanism of plant growth promotion by rhizobacteria. Indian J Exp Biol 38:856–862

Gupta A, Roy I, Patel RK, Singh SP, Khare SK, Gupta MN (2005) One-step purification and characterization of an alkaline protease from haloalkaliphilic *Bacillus* sp. J Chromat A 1075:103–108

Haddar A, Agrebi R, Bougatef A, Hmidet N, Sellami-Kamoun A, Nasri M (2009) Two detergent stable alkaline serine-proteases from *Bacillus mojavensis* A21: purification, characterization and potential application as a laundry detergent additive. Bioresour Technol 100:3366–3373

Haran S, Schickler H, Oppenheim A, Chet I (1995) New components of the chitinolytic system of *Trichoderma harzianum*. Mycol Res 99:441–446

Harman GE, Hayes CK, Lorito M et al (1993) Chitinolytic enzymes of *Trichoderma harzianum* : purification of chitobiosidase and endochitinast. Phytopathology 83:313–318

Havukkala I (1991) Chitinolytic enzymes and plant pests. In: Ilag LL, Raymundo AK (eds) Biotechnology in the Philippines towards the year 2000, proceedings of the Second Asia-Pacific Biotechnology Congress. SEARCA, University of the Philippines, Los Banos, pp 127–140
Heltoft P, Brierley JL, Lees AK, Sullivan L, Lynott J, Hermansen A (2016) The relationship between soil inoculum and the development of *Fusarium* dry rot in potato cultivars Asterix and Saturna. Eur J Plant Pathol 146:1–4. doi:10.1007/s10658-016-0946-2
Heydari A (2007) Biological control of turfgrass fungal diseases. In: Pessarakli M (ed) Turfgrass management and physiology. CRC Press, Boca Raton
Heydari A, Misaghi IJ, Balestra GM (2007) Pre-emergence herbicides influence the efficacy of fungicides in controlling cotton seedling damping off in the field. Int J Agric Res 2:1049–1053
Hurst PL, Nielsen J, Sullivan PA, Shepherd MG (1977) Purification and properties of a cellulase from *Aspergillus niger*. Biochem J 165:33–41
Illakkiam D, Anuj NL, Ponraj P, Shankar M, Rajendhran J, Gunasekaran P (2013) Proteolytic enzyme mediated antagonistic potential of *Pseudomonas aeruginosa* against *Macrophomina phaseolina*. Indian J Exp Biol 51:1024–1031
Islam MT, Yasuyuki H, Abhinandan D, Toshiaki I, Santoshi T (2005) Suppression of damping-off disease in host plants by the rhizoplane bacterium *Lysobacter* sp. strain SB-K88 is-linked to plant colonization and antibiosis against soilborne *peronosporomycetes*. Appl Environ Microbiol 71:3786–3796
Jones IDG, Grady KL, Suslow TV, Bedbrook IR (1986) Isolation and characterization of genes encoding two chitinase enzymes from *Serratia marcescens*. EMBO J 5:467–473
Joo HS, Chang CS (2005) Oxidant and SDS-stable alkaline protease from a halo-tolerant *Bacillus clausii* I-52: enhanced production and simple purification. J Appl Microbiol 98:491–497
Khan AG (2005) Role of soil microbes in the rhizospheres of plants growing on trace metal contaminated soils in phytoremediation. J Trace Elem Med Biol 18:355–364
Kim KJ, Yang YJ, Kim JG (2003) Purification and characterization of chitinase from *Streptomyces* sp. M-20. J Biochem Mol Biol 36:185–189
Kloepper JW, Schroth MN (1978) Plant growth-promoting rhizobacteria on radishes. Proceedings of the fourth international conference on plant pathogen bacteria, vol 2 INRA, pp 879–882
Kloepper JW, Ryu CM, Zhang S (2004) Induce systemic resistance and promotion of growth by *Bacillus sp*. Phytopathology 94:1259–1266
Kramer KJ, Muthukrishnan S, Johnson L, White F (1997) Chitinases for insect control. In: Carozzi N, Koziel M (eds) Advances in insect control: the role of transgenic plants. Taylor and Francis Publishers, Bristol, pp 185–193
Kuhad RC, Singh A, Eriksson KEL (1997) Microorganisms and enzymes involved in the degradation of plant fiber cell walls. Adv Biochem Eng Biotechnol 57:45–125
Lee SM, Koo YM (2001) Pilot-scale production of cellulase using T. reesei rut C-30 in fed-batch mode. Microb Biotechnol 11:229–233
Li S, Rupe J, Chen P, Shannon G, Wrather A, Boykin D (2015) Evaluation of diverse soybean germplasm for resistance to *Phomopsis* seed decay. Plant Dis 99:1517–1525
Li-Jung Y, Hsin-Hung L, Zheng-Rong X (2010) Purification and characterization of a cellulase from *Bacillus Subtilis* YJ1. J Mar Sci Technol 18:466–471
Logemann J, Schell J (1993) The impact of biotechnology on plant breeding, or how to combine increases in agricultural productivity with an improved protection of the environment. In: Chet I (ed) Biotechnology in plant disease control. Wiley-Liss, New York, pp 1–14
Lorito M (1998) Chitinolytic enzymes and their genes. In: Kubicek CP, Harman GE (eds) *Trichoderma* and *Gliocladium*: enzymes, biological control and commercial applications, vol 2. Taylor and Francis, London, pp 73–99
Lorito M, Hayes CK, Di Pietro A, Woo SL, Harman GE (1994) Purification, characterization, and synergistic activity of a glucan 1, 3-β-glucosidase and an N-acetyl-β-glucosaminidase from *Trichoderma harzianum*. Phytopathology 84:398–405
Lucas García J, Probanza A, Ramos B, Gutiérrez Manero FJ (2001) Genetic variability of rhizobacteria from wild populations of *Lupinus* species based on PCR-RAPDs. J Plant Nutr Soil Sci 164:1–7

Lucy M, Reed E, Glick BR (2004) Applications of free living plant growth-promoting rhizobacteria. Antonie Van Leeuwenhoek: Int J Gen Mol Microbiol 86:1–25

Lugtenberg B, Kamilova F (2009) Plant-growth-promoting Rhizobacteria. Annu Rev Microbiol 63:541–556

Lynd LR, Weimer PJ, van Zyl WH, Pretorius IS (2002) Microbial cellulose utilization: fundamentals and biotechnology. Microbiol Mol Biol Rev 66:506–577

Mabood F, Zhou X, Smith DL (2014) Microbial signaling and plant growth promotion. Can J Plant Sci 94:1051–1063

Maloy OC (1993) Plant disease control: principles and practice. Wiley, New York

Manocha MS, Balasubramanian R (1994) Fungal chitinases: their properties and roles in morphogenesis, mycoparasitism and control of pathogenic fungi. In: Petrini O, Ouel-lette GB (eds) Host wall alterations by parasitic fungi. APS Press, St. Paul, pp 81–90

Manocha MS, Sahai AS (1993) Mechanisms of recognition in necrotrophic and biotrophic mycoparasites. Can J Microbiol 39:269–275

Mao W, Lewis JA, Hebbar PK, Lumsden RD (1997) Seed treatment with a fungal or a bacterial antagonist for reducing corn damping-off caused by species of *Pythium* and *Fusarium*. Plant Dis 81:450–454

Maria Teresa FB, Adilson LL, Cirano JU (2003) Purification and characterization of an exo-β -1,3-glucanase produced by *Trichoderma asperellum*. FEMS Microbiol Lett 219:81–85

Maurer KH (2004) Detergent proteases. Curr Opin Biotechnol 15:330–334

Misra AK, Garg N, Yadav KK (2016) First report of shell soft rot of Bael (*Aegle marmelos*) caused by *Syncephalastrum racemosum* in North India. Plant Dis PDIS-12 100:1779

Muhammad I, Safdar A, Syed Q, Nadeem M (2012) Isolation and screening of cellulolytic bacteria from soil and optimization of cellulase production and activity. Turk J Biochem 37:287–293

Müller J (2015) Rarer downy mildews, rusts and smuts of the Czech and Slovak Republics. Czech Mycol 67:69–83

Noronha EF, Ulhoa CJ (1996) Purification and characterization of an endo-β-1,3-glucanase from *Trichoderma harzianum*. Can J Microbiol 42:1039–1044

Noronha EF, Ulhoa CJ (2000) Characterization of a 29-kDa β-1,3-glucanase from *Trichoderma harzianum*. FEMS Microbiol Lett 183:119–123

O'Donnell K, Kistler HC, Tacke BK, Casper HH (2000) Gene genealogies reveal global phylogeographic structure and reproductive isolation among lineages of *Fusarium graminearum*, the fungus causing wheat scab. Proc Nat Acad Sci 97:7905–7910

Odeniyi OA, Onilude AA, Ayodele MA (2009) Production characteristics and properties of cellulase/polygalacturonase by a *Bacillus coagulans* strain from a fermenting palm fruit industrial residue. Afr J Microbiol Res 3:407–417

Oppenheim AB, Chet I (1992) Cloned chitinases in fungal plant pathogen control strategies. Trends Biotechnol 10:392–394

Pal KK, Gardener BM (2006) Biological control of plant pathogens. Plant Health Instr 1–25, doi: 10.1094/PHI-A-2006-1117-02

Pane C, Piccolo A, Spaccini R, Celano G, Villecco D, Zaccardelli M (2013) Agricultural waste-based composts exhibiting suppressivity to diseases caused by the phytopathogenic soil-borne fungi *Rhizoctonia solani* and *Sclerotinia minor*. Appl Soil Ecol 65:43–51

Pang Z, Otaka K, Suzuki Y, Goto K, Ohnishi M (2004) Purification and characterization of an endo-1,3-β-glucanase from *Arthrobacter* sp. J Biol Macromol 4:57–66

Pietro AD, Madrid MP, Caracuel Z, Delgado-Jarana J, Roncero MIG (2003) *Fusarium oxysporum*: exploring the molecular arsenal of a vascular wilt fungus. Mol Plant Pathol 4:315–325

Ran LX, Liu CI, Wu GJ, van Loon LC, Bakker PAHM (2005) Suppression of bacterial wilt in *Eucalyptus urophylla* by *fluorescent Pseudomonas* spp. in China. Biol Control 32:111–120

Sareen R, Mishra P (2008) Purification and characterization of organic solvent stable protease from *Bacillus licheniformis* RSP-09-37. Appl Microbiol Biotechnol 79:399–405

Sarvanakumar D, Vijaykumar C, Kumar N, Samiyappan R (2007) ACC deaminase from *Pseudomonas fluorescens* mediate saline resistance in groundnut plants. J Appl Microbiol 102:1283–1292

Sayyed RZ, Chincholkar SB (2009) Siderophore producing *A. faecalis* more biocontrol potential vis-a`-vis chemical fungicide. Curr Microbiol 58:47–51

Sayyed RZ, Patel PR (2011) Biocontrol potential of siderophore producing heavy metal resistant *Alcaligenes* sp. and *Pseudomonas* sp. vis-a`-vis organophosphorus fungicide. Indian J Microbiol 51:266–272

Sayyed RZ, Patel DC, Patel PR (2007) Plant growth promoting potential of P solubilizing *Pseudomonas* sp. occurring in acidic soil of Jalgaon. Asian J Microbiol Biotechnol Environ Sci 4:925–928

Sayyed RZ, Gangurde NS, Patel PR, Josh SA, Chincholkar SB (2010) Siderophore production by *Alcaligenes faecalis* and its application for plant growth promotion by *Arachis hypogaea*. Indian J Biotechnol 9:302–307

Sayyed RZ, Reddy MS, Kumar KV et al (2012) Potential of plant growth-promoting rhizobacteria for sustainable agriculture. In: Maheshwari DK (ed) Bacteria in agrobiology: plant probiotics. Springer, Berlin/Heidelberg, pp 287–313

Sayyed RZ, Chincholkar SB, Reddy MS, Gangurde NS, Patel PR (2013) Siderophore producing PGPR for crop nutrition and phytopathogen suppression. In: Maheshwari DK (ed) Bacteria in agrobiology: disease management. Springer, Berlin/Heidelberg, pp 449–471

Sayyed RZ, Patel PR, Shaikh SS (2015) Plant growth promotion and root colonization by EPS producing *Enterobacter* sp. RZS5 under heavy metal contaminated soil. Indian J Exp Biol 53:116–123

Schlumbaum A, Mauch F, Vögeli U, Boller T (1986) Plant chitinases are potent inhibitors of fungal growth. Nature 324:365–367

Shaikh SS, Sayyed RZ (2015) Role of plant growth promoting rhizobacteria and their formulation in biocontrol of plant diseases. In: Arora NK (ed) Plant microbes symbiosis: applied facets. Springer, New Delhi, pp 337–351. doi:10.1007/978-81-322-2068-8_18

Shaikh SS, Patel PR, Patel SS, Nikam SD, Rane TU, Sayyed RZ (2014) Production of biocontrol traits by banana field fluorescent pseudomonads and their comparison with chemical fungicides. Indian J Exp Biol 52:917–920

Shaikh SS, Sayyed RZ, Reddy MS (2016) Plant growth promoting rhizobacteria: a sustainable approach to agro-ecosystem. In: Hakeem KR et al (eds) Plant, soil and microbes – interactions and implications in crop science. Springer international publishing AG, Cham, pp 181–201. doi:10.1007/978-3-319-27455-3_10

Sharma SB, Sayyed RZ, Trivedi MH, Gobi TA (2013) Phosphate solubilizing microbes: sustainable approach for managing phosphorus deficiency in agricultural soils. Springer Plus 2:587

Silva VN, Silva LESF, Figueiredo MVB (2006) Atuacao de rizobios com rizobacterias promotoras de crescimento em plantas na cultura do caupi (*Vigna unguiculata* L. Walp). Acta Sci Agron 28:407–412

Simmons CR (1994) The physiology and molecular biology of plant 1-3-β-D-glucanase and 1,3;1-4-β-D-glucanase. Crit Rev Plant Sci 13:325–387

Singh PP, Shin YC, Park CS, Chung YR (1999) Biological control of *Fusarium* wilt of cucumber by chitinolytic bacteria. Phytopathology 89:92–99

Sitrit Y, Barak Z, Kapulnik Y, Oppenheim AB, Chet I (1993) Expression of a *Serratia marcescens* chitinase gene in *Rhizobium meliloti* during symbiosis on alfalfa roots. Mol Plant-Microbe Interact 6:293–298

Skujins JJ, Potgieter HJ, Alexander M (1965) Dissolution of fungal cell walls by a *Streptomycete* chitinase and ß-(1,3)-glucanase. Arch Biochem Biophys 111:358–364

Someya N, Kataoka N, Komagata T, Hirayae K, Hibi T, Akutsu K (2000) Biological control of cyclamen soil borne diseases by *Serratia marcescens* strain B2. Plant Dis 84:334–340

Someya N, Tsuchiya K, Yoshida T, Noguchi MT, Akutsu K, Sawada H (2007) Fungal cell wall degrading enzyme-producing bacterium enhances the biocontrol efficacy of antibiotic-producing bacterium against cabbage yellows/Ein Zellwand zersetzendes Bakterium steigert die antagonistische Wirkung eines Antibiotika bildenden Bakteriums gegenüber der Fusarium-Kohlwelke. J Plant Dis Protec, 108–112

Sookkheo B, Sinchaikul S, Phutrakul S, Chen ST (2000) Purification and characterization of the highly thermostable proteases from *Bacillus stearothermophilus* TLS33. Protein Expr Purif 20:142–151
Strange RN, Scott PR (2005) Plant disease: a threat to global food security. Phytopathology 43
Sundheim L (1992) Effect of chitinase encoding genes in biocontrol *Pseudomonas* spp. In: Tjamos EC, Papavizas GC, Cook RJ (eds) Biological control of plant diseases: progress and challenges for the future. Plenum, New York, pp 331–333
Thomas JUNG, Nechwatal J, Cooke DE et al (2003) *Phytophthora pseudosyringae* sp. nov., a new species causing root and collar rot of deciduous tree species in Europe. Mycol Res 107:772–789
Uchida J, Zhong S, Killgore E (2006) First report of a rust disease on ohia caused by *Puccinia psidii* in Hawaii. Plant Dis 90:524–524
Vazquez-Garcidueiias S, Leal-Morales CA, Herrera-Estrella A (1998) Analysis of the ß-l,3-g1ucanolytic system of the biocontrol agent *Trichoderma harzianum*. Appl Environ Microbiol 64:1442–1446
Vega K, Kalkum M (2011) Chitin, chitinase responses, and invasive fungal infections. Int J Microbiol, Vol 2012
Vessey JK (2003) Plant growth promoting rhizobacteria as biofertilizers. Plant Soil 255:571–586
Weindling R (1932) *Trichoderma lignorum* as a parasite of other fungi. Phytopathology 22:837–845
Wingfield MJ, Jacobs A, Coutinho TA, Ahumada R, Wingfield BD (2002) First report of the pitch canker fungus, *Fusarium circinatum*, on pines in Chile. Plant Pathol 51:397–397

Role of Quorum Sensing Signals of Rhizobacteria for Plant Growth Promotion

10

Anton Hartmann and Michael Rothballer

Abstract

Signaling events between rhizosphere microbes and plants substantially contribute to establish different qualities of microbe-plant interactions from beneficial cooperativity to pathogenicity. In addition to the pathogen-associated molecular patterns (PAMPs), like exo- and lipopolysaccharides or flagellins, which are effectively recognized by the plants' innate immune system, various secondary metabolites, such as antibiotics or the so-called autoinducers involved in the quorum sensing response of bacteria, are additional modulators of plants' perception of associated microbes. In Gram-negative bacteria, *N*-acyl homoserine lactones (AHLs) are the major quorum sensing autoinducing molecules, which have a central role in the differentiation of specific phenotypes of sessile cells, living in root-attached microcolonies or biofilm consortia. AHLs turned out to have profound effects on plant development and/or defense priming and development of systemic resistance against pathogens. AHLs have different structural modifications (e.g., short or long hydrocarbon chain residues). While the hydrophilic ones can be taken up by plants, the lipophilic stay in the roots. Different modes of plant growth promotion by these AHL types in various plants are summarized in this chapter. We hypothesize, that in the absence of pathogenic patterns, AHLs support a beneficial to symbiotic interaction with plants. In cases when plant pathogens use AHLs for virulence development, AHLs reinforce plant's defense. Alternatively, AHL degradation activities of certain rhizosphere bacteria can be used to suppress the pathogenic attack. To foster beneficial interactions of rhizotrophs with plants, consortia of bacteria using the same autoinducers could be developed.

A. Hartmann (✉) • M. Rothballer
Department Environmental Sciences, Research Unit Microbe-Plant Interactions, Helmholtz Zentrum München, German Research Center for Environmental Health (GmbH), Ingolstaedter Landstrasse 1, 85764 Neuherberg, Germany
e-mail: ahartmanndr@gmail.com; rothballer@helmholtz-muenchen.de

S. Mehnaz (ed.), *Rhizotrophs: Plant Growth Promotion to Bioremediation*, Microorganisms for Sustainability 2, DOI 10.1007/978-981-10-4862-3_10

10.1 Introduction

The plant microbiome – the community of microbes, associated with plant surfaces and tissues – is currently in the focus of plant biotechnology, since molecular characterization technologies, so-called omic approaches, are providing almost unbiased insights into the diversity in the community structure and its activities (Turner et al. 2013). Plant-colonizing microbes exert great influences on plant health and development. This had been already recognized more than hundred years ago by the pioneers of rhizosphere microbiology like Lorenz Hiltner (Hartmann et al. 2008). Since the genetic capacity of the microbes associated with all higher organisms is by far higher than and complementary to the hosts' genetic repertoire, the importance of prokaryote-eukaryote interactions has led to the metagenome or holobiont view of higher organisms and their associated microbiomes (Zilber-Rosenberg and Rosenberg 2008). The hypothesis was created that these assemblages/symbioses are the true, evolutionary superior life strategies for accommodating with rapidly changing and challenging environmental conditions. There is much to discover regarding prokaryotic diversity and their function within the plant microbiome, since many microbes are difficult to isolate and to grow in pure culture. Microbe-plant interactions cover a wide spectrum, from pathogenic to beneficial and even symbiotic interactions in plant and animal/human hosts (Berg et al. 2005; Mendes et al. 2013). Plant growth promotion by rhizosphere-associated, root-colonizing microbes is a well-documented phenomenon (Dessaux et al. 2010). It can be considered as a symbiotic and synergistic microbe-plant interaction, although no particular symbiotic organs are visible. Benefits of these associations can be observed particularly when the plant is challenged by limiting nutrient supply, by abiotic stresses like hypersaline conditions or lack of water, or when attacked by pathogens (Raaijmakers et al. 2009).

While the discovery and description of the microbial diversity within the plant microbiome has made substantial progress, only comparably little details are understood about signaling mechanisms involved in microbe-host interactions. Several groups of signaling molecules are known, like a multitude of bacterial volatile substances (Ryu 2015) or bacterial-synthesized phytohormones (Spaepen 2015). The autoinducing small molecules (*N*-acyl-L-homoserine lactones; Fig. 10.1) of the quorum sensing response in Gram-negative bacteria are only recently discovered signals between bacteria and plants. They are in the focus of this chapter.

10.2 Bacteria as Social Organisms

Bacteria rarely live as a single dispersed organism; mostly they live in communities and colonize surfaces. The ability of bacterial populations to form biofilms is often under the quorum sensing (QS) control. Quorum sensing describes the phenomenon that bacteria produce and perceive signal molecules to coordinate their behavior in a population-dependent manner; thus QS is considered a social trait (Parsek and Greenberg 2005). Biofilm formation is essential to colonize surfaces of minerals and higher organisms, such as roots of plants. Biofilms are highly structured in which the metabolic activities are distributed between the different bacteria of the

N-acyl homoserine lactones

General formula: N-$C_n$oyl-L-HSL (n= 4-14)

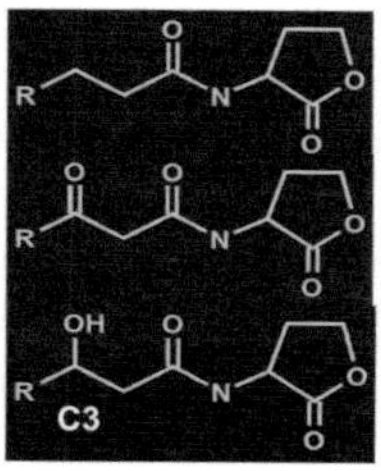

example:

N-hexanoyl-L-HSL	C6-HSL
N-3-oxo-hexanoyl-L-HSL	Oxo-C6-HSL
N-3-hydroxy-hexanoyl-L-HSL	OH-C6-HSL

Fig. 10.1 Structure of *N*-acyl homoserine lactones (AHLs), the autoinducer molecule of the quorum sensing (QS) response in many Gram-negative bacteria. *HSL* homoserine lactone residue, *R* acyl chain residue, ranging from 4 to 14 carbon atoms. Further variations: hydroxyl group or carbonyl group at C3-atom and/or double bond in the acyl chain R

consortium. This requires a high degree of coordination and is quite similar to the situation in multicellular organisms. In bacteria, several types of autoinducers (AIs) are known. While in Gram-negative bacteria, frequently *N*-acyl-L-homoserine lactones (AHL) and occasionally 4-hydroxyl 2-alkyl quinolones (HAQ) are found; Gram-positive bacteria mainly use cyclic oligopeptides. A furanon derivative, borate complex (the so-called AI-2), seems to act as a more general AI in a variety of Gram-negative and Gram-positive bacteria. In addition, other molecules with unknown structures might be involved in unidentified QS systems. Usually, the signaling molecules are produced in an autoinducing process: it consists of a very low constitutive biosynthesis and release of the autoinducer (AI) and a receptor protein (R), which senses the overall concentration of the AI within the cells. Beyond a certain threshold level of AI, the AI/R complex binds to the promoter of the biosynthesis gene I, which is then greatly stimulated; also other AI/R-regulated promoters are activated. This regulatory cascade finally leads to the expression of a whole set of genes and operons depending on the QS activation which drastically alters the overall behavior of the population.

The QS mechanism of Gram-negative bacteria using *N*-acyl homoserine lactones (AHLs) is established since some time (Fuqua and Greenberg 2002), and until now more than 200 species are known to produce different AHLs (Kimura 2014). However, there are bacteria which do not produce AHLs but harbor so-called LuxR solos (Patel et al. 2013). It could be demonstrated that these receptors evolved in some cases to respond to different molecules including signals from plant origin (Patel et al. 2013). *N*-Acyl-L-homoserine lactones are known to have a wide variety of chemical structures (Fig. 10.1). They can carry a hydroxy- or oxo-function at the C3 atom of the hydrocarbon side chain and occur as quite hydrophilic molecules with a length of the CH chain from C4-, C6-, and C8-HSLs (hydrophilic) to C12- and C14-HSLs (hydrophobic). QS regulation is of high ecophysiological and evolutionary importance because it optimizes gene expression under changing environmental conditions. This led to the suggestion of the QS autoinducing regulation as an “efficiency sensing” mechanism (Hense et al. 2007). Important physiological traits, like expression of virulence, chemotaxis and swarming, siderophore, antibiotic, and exoenzyme production, are regulated by quorum sensing (Eberl 1999). In developing microcolonies and biofilms on root surfaces, the production of AHL substances has been demonstrated using AHL biosensor bacteria (Steidle et al. 2001; Gantner et al. 2006; Fekete et al. 2007), which indicate the presence of AHLs by the activation of AHL regulated promoters fused to a gene for fluorescence protein synthesis, like the green fluorescence gene (GFP). These colonies and biofilms of Gram-negative bacteria are the source for the AHL signaling molecules to interact with plant roots. Recently, it was discovered that these AHLs are also active as direct signal to the plant, stimulating different plant responses (see below). This may be regarded as a consequence of the coevolution of plants with the omnipresent microbiota. Since Gram-negative bacterial pathogens also coordinate their colonization and virulence using these signals, it is highly advantageous for plants to recognize and perceive these signals. However, the AI signals also play important roles in the coordination mechanisms of interactions of plant hosts with beneficial bacteria within holobionts (see below).

10.3 Perception of AHL Molecules by Plants

The first study reporting evidences for the impact of bacterial AHLs on plant physiological activities was published by Joseph and Phillips (2003). They measured the influence of several water-soluble compounds taken up by roots with the natural transpiration stream on stomatal conductance and transpiration of bean (*Phaseolus vulgaris* L.) plants. In these experiments, 10 nM homoserine lactones as well as the respective homoserine had similar stimulating effects. Shortly afterward, the group of Ulrike Mathesius at the University of Brisbane and our research unit at the Helmholtz Zentrum München communicated detailed plant responses toward different AHLs applied to roots. Using a differential proteome analysis approach, Mathesius et al. (2003) revealed that defense- and stress-related proteins of *Medicago truncatula* to be regulated in their expression by the addition of 3-oxo-C12-L-homoserine lactone. Proteins associated with flavonoid metabolism, several regulatory proteins (including protein degradation and synthesis), and an auxin-responsive promoter were within this AHL regulon. The first demonstration of C6 and C8 AHL-induced systemic resistance development was reported for tomato plants by Hartmann et al. (2004) and Schuhegger et al. (2006). Upon the addition of C6- and C8-HSL-producing *Serratia liquefaciens* MG1 to roots of the tomato variety MicroTom, growing in regular soil, the plant developed increased resistance in the leaves against the attack by the fungal pathogen *Alternaria alternata*. In a clean and axenic quartz sand-based hydroponic system, also the addition of pure C6- and C8-HSLs to the rooting solution of tomato seedlings caused induction of the expression of the pathogen defense genes PR1 and chitinase in the leaves. This supported the primary finding of a systemic induction due to inoculation with AHL-producing *Serratia liquefaciens*, especially because an AHL-deficient mutant was unable to prevent the pathogen attack like the wild type. It was very remarkable that salicylic acid was increased in AHL-treated tomato plants, which is known as a transmitter of induced systemic resistance (Schuhegger et al. 2006). When the same hydrophilic autoinducers C6- or C8-HSL were added to *Arabidopsis thaliana*, no induced resistance but an increased auxin/cytokinin ratio and an increased root growth response were observed (von Rad et al. 2008). It could be demonstrated that a diversity of phytohormone-regulated genes was up- or downregulated after addition of C6- or C8-HSL but no defense-regulated genes were upregulated (von Rad et al. 2008).

The root stimulatory effect of C6- and C8-HSLs was corroborated by Liu et al. (2012), who found that *Arabidopsis* mutants in the G-protein receptors GCR1 and GAP1 lost this effect, while constructs with increased levels of the G-protein-related receptor showed increased root stimulation. In contrast, when oxo- or hydroxyl-C14-HSL was added to the rooting solution of *Arabidopsis thaliana* or barley seedlings, a clear induction of systemic resistance responses was found (Schikora et al. 2011). Based on these observations, two contrasting effects of AHL molecules with different hydrophilicities (due to short versus long hydrocarbon chains) were hypothesized for *Arabidopsis thaliana* (Hartmann and Schikora 2012; Schenk et al. 2012) (Fig. 10.2). However, the impact of AHLs on plant growth and pathogen defense in different plant species seems to be quite diverse (Hartmann et al. 2014). For example, it was found for *Medicago truncatula* that the long-chain 3-oxo-C14-HSL produced by *Sinorhizobium meliloti* enhanced the nodulation in roots (Veliz-Vallejos et al. 2014). Very striking was the observation that only 3-oxo-C14-HSL, the predominant AHL of symbiotic bacterium *S. meliloti*, increased the number of roots but no other AHLs. In mung bean plants, an induction of the growth of adventitious roots was stimulated specifically only by 3-oxo-C10-HSL but failed to be increased by unsubstituted C10-HSL or C12-HSL (Bai et al. 2012). Palmer and coworkers (2014) proposed that a change in transpiration rate induced by the hydrolysis of AHLs to L-homoserines increased stomatal opening fostering water and mineral flow through the plant, which has been already described by Joseph and Philipps (2003).

10.4 Uptake and Physiological Interactions of AHL with Plants

The strong impact of the addition of AHLs to the rooting solution of plants could be either caused by the initiation of a systemic signal (or signals) which would be triggered after the perception of AHLs on the root surface or whether they are taken up by the roots and transported to the shoots and then act locally at the plant tissues. Using highly resolving chemical analysis equipment, like a 12 Tesla Fourier Transform Ion Cyclotron Resonance Mass Spectrometer (FTICR-MS), which can resolve the molecular weights of small molecules with highest accuracy, the identification of different AHL molecules could be achieved even within plant tissues (Fekete et al. 2007; Götz et al. 2007). The identification and quantification was further supported and confirmed by ultrahigh performing liquid chromatography coupled with MS detection (Fekete et al. 2010; Buddrus-Schiemann et al. 2014; Rothballer et al. 2016). When different short and long carbon chain AHLs were applied to the rooting solution of barley plants, the C4- to C8-HSLs were clearly identified in the roots and even in the shoots. Only traces of C10-HSL and no AHLs

N-acyl homoserine lactone effects on Arabidopsis

	C6-	C8-	C10-	C12-	C14-HSL
Root Growth	++	++	ND	-	-
Resistance (systemic and local)	-	-	-	+	+
AHL-uptake	++	+	(+)	-	-
Membrane receptor	+	+	ND	not	known
Calcium accumulation	+	+	ND	ND	ND
Signaling cascade	-	-	ND	+	+

Fig. 10.2 Synopsis: effects of short and long *N*-acyl chain HSL in *Arabidopsis thaliana*. *HSL* homoserine lactone, *ND* not determined

with longer carbon chains were detectable in shoots (Fig. 10.2). Obviously, their hydrophobicity prevented the transport. In contrast to barley plants, in the leguminous yam beans (*Pachyrhizus erosus*), no uptake of AHLs was detectable (Götz et al. 2007). Obviously, these plants had highly active AHL lactonases, which effectively destroyed the signal molecule. When the AHL transport in barley was measured using AHL-specific monoclonal antibodies for short carbon chain HSLs (Chen et al. 2010a), the uptake into the shoots could be quantified using a quantitative enzyme-linked immunosorbent assay (ELISA) (Chen et al. 2010b), when 60 μM of C6-HSL was added to roots. The quantification gave a concentration of 30 μM of antigen in the phloem liquid of cut barley stalks. To prove that the transported AHL molecule in the plant sap still is in the active form, a C6-HSL-specific biosensor bacterium was applied to this plant liquid, resulting in the quantification of a similar concentration of authentic C6-HSL (Fekete et al. 2010; Sieper et al. 2013). Finally, radioactively labeled C8- and C10-HSLs were applied by Sieper et al. (2013) to derive further details about the AHL transport in barley plants. Since the uptake was completely blocked in the presence of the energy poison iodic acetate, an ATP-dependent uptake of these AHLs into the vascular system was suggested. The further transport along the barley roots could be blocked by vanadate, which argues for a symplasmic transport of the majority of AHLs in the vascular system. The transport rate of the AHL molecules was quantified in excised roots in a Pitman chamber, using radiolabeled ^{3}H-AHLs. The transport rate along the root was higher in the more hydrophilic AHLs, favoring short-chain AHLs (Sieper et al. 2013). Autoradiography clearly showed that the highest concentration of ^{3}H-AHL was concentrated in the vascular system of roots. In *Arabidopsis*, the short-chain AHL, C6-HSL, was also taken up into the shoot after application to roots, while the long-chain AHL, oxo-C14-HSL, was not transported (Schikora et al. 2011) (Fig. 10.2).

In barley, which had taken up C6-, C8-, and C10-HSLs, specific detoxification enzymes, like glutathione S-transferase and dihydro-ascorbate reductase, were affected by AHLs. When the most hydrophilic compound C6-HSL was applied to roots, the largest influence on the leaf-located enzyme was found (Götz-Rösch et al. 2015). In yam beans, which have a high lactonase activity, no influences of AHL application on foliar enzyme activities were found. Since these short-term effects on enzyme activities were not accompanied by a concomitant increase of transcription level of the respective genes, it was hypothesized that the AHL compounds exert a direct modulation of certain enzyme activities (Götz-Rösch et al. 2015).

An influence of bacterial AHLs and the plant reproduction was found in the study of Singh et al. (2015) with the green macroalga *Ulva* and the red macroalga *Gracilaria*. The short-chain C4- and C6-HSLs produced by bacterial biofilms colonizing the algae stimulated the release of carpospores (Singh et al. 2015). Interestingly, the protein patterns of both the cystocarps and the cystocarp-bearing plantlets treated with AHLs had protein patterns different from the pattern in control algae. Another very interesting finding was the influence of bacterial QS signals on herbivore defense of plants, since the jasmonic acid (JA)-mediated herbivore resistance was reduced in tobacco plants by AHLs (Heidel et al. 2010).

10.5 AHLs and Priming/Induction of Induced Resistance

In addition to the induction of a fast and highly sensitive innate immune response to microbial pathogens, which are recognized by their microbe-associated molecular patterns (MAMPs), also other additional mechanisms, like AHL signals, may be integrated into a network of interactive perceptions of environmental signals to further specify and strengthen the plant defense response. At situations of biotic or abiotic stresses, plants can further induce steps toward defense stimulation with a sensibilization mechanism, called priming (Jung et al. 2005). A diversity of metabolites which induce priming also include a diversity of volatile substances, known to be produced by a number of biocontrol bacteria, which can prepare the plant for an upcoming pathogen attack (Ryu et al. 2003; Ryu 2015).

Using a set of *Arabidopsis thaliana* mutants in the signaling chain and downstream response effects, the specific effects of AHL compounds on the induction of resistance responses could be revealed (Schikora et al. 2011). When the major innate immune elicitor protein flg22 was added to the model plant *Arabidopsis thaliana*, the phosphorylation of mitogen-activated protein kinases MPK 3 and MPK 6 and an activation of the transcription factors WRKY 18, WRKY 22, and WRKY 29 occurred. In the presence of C12- and C14-HSL compounds in the rooting solution, MPK 3 and MPK 6 phosphorylation was modified and the expression of WRKY transcription factors was increased. In the presence of long carbon chain AHLs, the expression of pathogenesis-related proteins and accumulation of H_2O_2 was increased (Fig. 10.3). Barley plants also showed an enhanced production of ROS. In the presence of C12- and C14-HSLs, an increased level of hypersensitive response (HR) occurred after infection with *Blumeria graminis*.

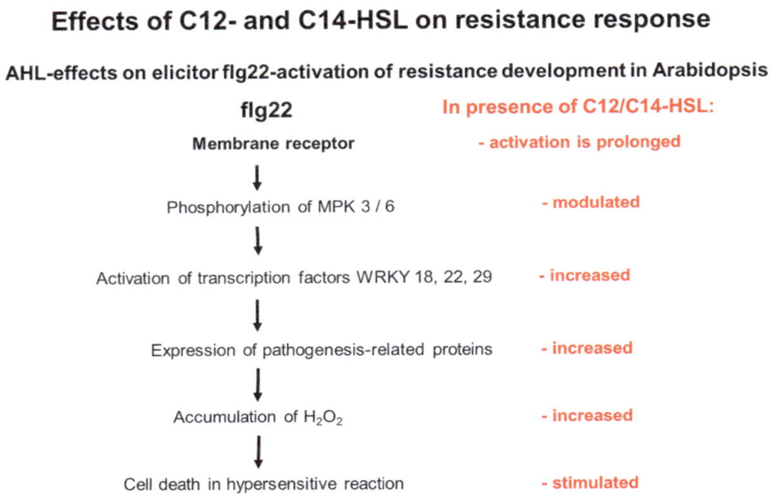

Fig. 10.3 Effects of C12- and C14-HSL on resistance response in *Arabidopsis thaliana* (according to Schikora et al. 2011) Flg22, flagellin as elicitor of microbe-associated molecular pattern (MAMP) response; MPK 3 and MPK 6, mitogen-associated protein kinases (MAPKs)

Also using the *Arabidopsis* model, further details of the molecular mechanism of AHL-induced priming were revealed by Schenk and Schikora (2015). Using the pathogen *Pseudomonas syringae* pathovar *tomato* (*Pst*), the treatment with oxo-C14-HSL resulted in increased pathogen resistance in accordance with SA- and oxylipin-dependent signaling (Schenk et al. 2014). Another AHL resistance phenotype could be the increased amount of closed stomata in response to *Pst* challenge in oxo-C14-HSL-pretreated plants (Schenk et al. 2014). This AHL-induced stomata defense response seems similar to the flg22-induced stomata closure, described as RES oxylipin and SA-dependent by Montillet et al. (2013). Another very striking similarity to other cellular signaling and response cascades was the demonstration of the increase of intracellular levels of calcium in *Arabidopsis* root cells (Song et al. 2011) and the involvement of calmodulin in the primary root elongation response to 3-oxoC6-HSL in *Arabidopsis* (Zhao et al. 2015). The occurrence of priming events as part of the response patterns of plants toward AHLs was recently reviewed in detail by Schikora et al. (2016).

10.6 Role of AHLs in the Integrated Plant Perception of Rhizobacteria

In most experiments about the effects of AHL autoinducers on plants, commercially available AHLs of different structures were applied to roots of plant seedlings in a clearly defined axenic hydroponic system. This approach has the great advantage that disturbances with the effects of other organic compounds from the soil solution or other components of the inoculated bacterium are avoided, which certainly also affect the response of the plant. However, the more realistic approach to investigate the role of AHLs and to learn about the true integrated role of AHLs of a certain beneficial rhizosphere bacterium on its perception by plants is to apply wild-type rhizosphere bacteria and compare their effects on plants with an AHL-deficient mutant. In the case of development of systemic resistance, against plant pathogens, induced by certain rhizosphere bacteria, specific AHL deletion mutants or phenotypic variants devoid of AHL production were already used and clearly showed the importance of the AHL production to achieve this trait (Schuhegger et al. 2006; Pang et al. 2009). However, in a general beneficial endophytic bacterium, the influence of AHL production on the detailed transcriptional response by plants was not investigated. Therefore, the Gram-negative bacterium *Acidovorax radicis* N35, characterized as a plant beneficial endophytic bacterium in wheat and barley (Li et al. 2011), was studied in detail in this respect. This bacterium produces only 3-OH-C10-HSL and according to its genome analysis harbors only one AHL-biosynthesis (*araI*) gene (Fekete et al. 2007). Therefore, an *araI* mutant was constructed, devoid of any AHL production (Han et al. 2016). When the wild type and the *araI* mutant, labeled differently with GFP and YFP, were compared for their colonization abilities, the mutant was less competitive when both strains were inoculated in a 1:1 mixture to barley roots. The wild type showed a biofilm-like colonization pattern, while the mutant occurred at the root surface only as dispersed single cells. However, both bacteria could finally exert a comparable plant growth-promoting effect (Han et al. 2016). When the transcription profile of the barley seedlings was tested by RNA sequencing, the wild-type bacteria showed several priming effects and only very weak induction of early defense responses, while the *araI* mutant caused severely increased expression of flavonoid biosynthesis genes among other defense genes. This was corroborated by directed qPCR quantification and the accumulation of the flavonoids, lutonarin and saponarin, and related compounds in barley leaves, which had been inoculated with the *araI* mutant (Han et al. 2016). These flavonoids were not or much less found after inoculation with the AHL-producing *A. radicis* wild type. Therefore, the production of AHL has pronounced implications on the perception by the host plant and contributes to the better establishment of the endophytic bacterium. AHL-producing bacteria may be used as co-inoculants to better pave the way for other inoculants. However, the compatibility of these bacteria in colonization has to be checked for each combination.

10.7 Possible Future Use of Quorum Sensing Mechanisms in Sustainable Agriculture

The ongoing and even increased application of chemicals in industrialized agriculture and the possible consequences on food quality are key arguments toward changes to more sustainable agricultural practices. Among biology-based plant protection methods, the use of biologicals or biocontrol agents is increasing in agriculture, but their application by far did not reach its full potential. Today, several products based on bacterial inocula mainly consisting of several strains of different *Bacillus* spp., *Pseudomonas* spp., or *Serratia* spp. are successfully used by farmers. The application of bacteria, producing specific AHLs, could enhance the beneficial effects of other rhizosphere bacteria, especially bacterial inocula, and enlarge the impact to plant species usually not associated with the particular strain (Zarkani et al. 2013; Hernández-Reyes et al. 2014). Furthermore, the potential of AHLs or AHL-producing bacteria which are able to prime or induce several immune responses could open new possibilities in the prevention of pathogen infections also in field crops. The possibility to interfere with bacterial QS mechanism, via mimicry or enzymatic degradation of QS molecules by the plant or rhizosphere microbes, provides additional possible strategies to compete with pathogen attack, since such approaches lower the virulence by pathogenic bacteria. A variety of rhizosphere bacteria have quorum-quenching activities, like many bacilli (Dong et al. 2007), which harbor efficient AHL lactonases and are therefore good candidates for practical pathogen control in the field. QS mechanisms are therefore in the center of new strategies against different infectious diseases (LaSarre and Federle 2013; Kusari et al. 2015).

Acknowledgments This chapter is dedicated to Dr. Michael Schmid, who unexpectedly died in June 2016. His contributions as commission head of the Research Unit Microbe-Plant Interactions and as expert on molecular microbial taxonomy and ecology are greatly appreciated. In addition, the fruitful cooperation with Burkhard Hense and Philippe Schmitt-Kopplin (Helmholtz Zentrum München), Leo Eberl (University of Zürich), Karl-Heinz Kogel (University of Giessen), and Adam Schikora (presently at Julius-Kühn Institute Braunschweig) over the years to discover more and more details about the perception of AHLs by plants is deeply acknowledged.

References

Bai X, Todd CD, Desikan R, Yang Y, Hu X (2012) *N*-3-oxo-decanoyl-L-homoserine-lactone activates auxin-induced adventitious root formation via hydrogen peroxide- and nitric oxide-dependent cyclic GMP signaling in mung bean. Plant Physiol 158:725–736

Berg G, Eberl L, Hartmann A (2005) The rhizosphere as a reservoir for opportunistic human pathogenic bacteria. Environ Microbiol 7:1673–1685

Buddrus-Schiemann K, Rieger M, Mühlbauer M et al (2014) Analysis of *N*-acyl-homoserine lactone dynamics in continuous culture of *Pseudomonas putida* IsoF using ELISA and UPLC/qTOF-MS-related measurements and mathematical models. Anal Bioanal Chem 406:6373–6383

Chen X, Kremmer E, Gouzy MF et al (2010a) Development and characterization of rat monoclonal antibodies for *N*-acylated homoserine lactones. Anal Bioanal Chem 398:2655–2667

Chen X, Buddrus-Schiemann K, Rothballer M, Krämer P, Hartmann A (2010b) Detection of quorum sensing molecules in *Burkholderia cepacia* culture supernatants with enzyme-linked immunosorbent assays. Anal Bioanal Chem 398:2669–2676

Dessaux Y, Hinsinger P, Lemanceau P (eds) (2010) Rhizosphere: achievements and challenges. Springer-Science-Press, Berlin

Dong YH, Wang LH, Zhang LH (2007) Quorum-quenching microbial infections: mechanisms and implications. Philos T Roy Soc B 362:1201–1211

Eberl L (1999) *N*-acyl-L-homoserine lactone-mediated gene regulation in Gram-negative bacteria. Syst Appl Microbiol 22:493–506

Fekete A, Frommberger M, Rothballer M et al (2007) Identification of bacterial *N*-acyl homoserine lactones (AHL) using ultra performance liquid chromatography (UPLC™), ultrahigh resolution mass spectrometry and in situ biosensor constructs. Anal Bioanal Chem 387:455–467

Fekete A, Rothballer M, Hartmann A, Schmitt-Kopplin P (2010) Identification of bacterial autoinducers. In: Kraemer R, Jung K (eds) Bacterial signalling. Wiley Publishers, Weinheim, pp 95–111
Fuqua C, Greenberg EP (2002) Listening in on bacteria: acyl homoserine lactone signalling. Nat Rev Mol Cell Biol 3:685–695
Gantner S, Schmid M, Dürr C et al (2006) In situ spatial scale of calling distances and population density-dependent *N*-acyl homoserine lactone mediated communication by rhizobacteria colonized on plant roots. FEMS Microbiol Ecol 56:188–194
Götz C, Fekete A, Gebefügi I et al (2007) Uptake, degradation and chiral discrimination of *N*-acyl--D/L-homoserine lactones by barley (*Hordeum vulgare*) and yam bean (*Pachyrhizus erosus*) plants. Anal Bioanal Chem 389:1447–1457
Götz-Rösch C, Sieper T, Fekete A, Schmitt-Kopplin P, Hartmann A, Schröder P (2015) Influence of bacterial N-acyl-homoserine lactones on growth parameters, pigments, antioxidative capacities and the xenobiotic phase II detoxification enzymes in barley and yam bean. Front Plant Sci 6:205
Han, S, Li D, Trost E et al (2016) Systemic response of barley to the 3-hydroxy-C10-homoserine lactone producing plant beneficial endophyte *Acidovorax radicis* N35. Front Plant Sci 7:1868
Hartmann A, Schikora A (2012) Quorum sensing of bacteria and trans-kingdom interactions of N-acyl homoserine lactones with eukaryotes. J Chem Ecol 38:704–713
Hartmann A, Gantner S, Schuhegger R et al (2004) *N*-acyl homoserine lactones of rhizosphere bacteria trigger systemic resistance in tomato plants. In: Lugtenberg B, Tikhonovich I, Provorov N (eds) Biology of plant-microbe interactions. APS Press, St. Paul, pp 554–556
Hartmann A, Rothballer M, Schmid M (2008) Lorenz Hiltner, a pioneer in rhizosphere microbial ecology and soil bacteriology research. Plant Soil 312:7–14
Hartmann A, Rothballer M, Hense BA, Schröder P (2014) Bacterial quorum sensing compounds are important modulators of microbe-plant interactions. Front Plant Sci 5:131
Heidel A, Barazani O, Baldwin I (2010) Interaction between herbivore defense and microbial signaling: bacterial quorum-sensing compounds weaken JA-mediated herbivore resistance in *Nicotiana attenuata*. Chemoecology 20:149–154
Hense BA, Kuttler C, Mueller J, Rothballer M, Hartmann A, Kreft JU (2007) Does efficiency sensing unify diffusion and quorum sensing? Nat Rev Microbiol 5:230–239
Hernández-Reyes C, Schenk ST, Neumann C, Kogel KH, Schikora A (2014) N-acyl-homoserine lactones-producing bacteria protect plants against plant and human pathogens. Microb Biotechnol. doi:10.1111/1751-7915.12177
Joseph CM, Phillips DA (2003) Metabolites from soil bacteria affect plant water relations. Plant Physiol Biochem 41:189–192
Kimura N (2014) Metagenomic approaches to understanding phylogenetic diversity in quorum sensing. Virulence 5:433–442
Kusari P, Kusari S, Spiteller M, Kayser O (2015) Implications of endophyte-plant crosstalk in light of quorum responses for plant biotechnology. Appl Microbiol Biol 99:5383–5390
LaSarre B, Federle MJ (2013) Exploiting quorum sensing to confuse bacterial pathogens. Microbiol Mol Biol Rev 77:73–111
Li D, Rothballer M, Schmid M, Esperschütz J, Hartmann A (2011) *Acidovorax radicis* sp. nov., a rhizosphere bacterium isolated from wheat roots. Int J Syst Evol Microbiol 61:2589–2594
Liu F, Bian Z, Jia Z, Zhao Q, Song S (2012) The GCR1 and GPA1 participate in promotion of *Arabidopsis* primary root elongation induced by *N*-acyl-homoserine lactones, the bacterial quorum-sensing signals. Mol Plant-Microbe Interact 25:677–683
Mathesius U, Mulders S, Gao M et al (2003) Extensive and specific responses of a eukaryote to bacterial quorum-sensing signals. Proc Natl Acad Sci U S A 100:1444–1449
Mendes R, Garbeva P, Raaijmakers JM (2013) The rhizosphere microbiome. Significance of plant beneficial, plant pathogenic, and human pathogenic microorganisms. FEMS Microbiol Rev 37:634–663

Montillet JL, Leonhardet N, Mondy S et al (2013) An abscisic acid-independent oxylipin pathway controls stomatal closure and immune defense in *Arabidopsis*. PLoS Biol 11(3):e1001513. doi:10.1371/journal.pbio.1001513

Pang YD, Liu XG, Ma YX, Chernin L, Berg G, Gao KX (2009) Induction of systemic resistance, root colonisation and biocontrol activities of the rhizospheric strain of *Serratia plymuthica* are dependent on *N*-acyl homoserine lactones. Eur J Plant Pathol 124:261–268

Parsek MR, Greenberg EP (2005) Sociomicrobiology: the connections between quorum sensing and biofilms. Trends Microbiol 13:27–33

Patel HK, Suarez-Moreno ZR, Degrassi G, Subramoni S, Gonzalez JF, Venturi V (2013) Bacterial LuxR solos have evolved to respond to different molecules including signals from plants. Front Plant Sci 4:447

Raaijmakers JM, Paulitz TC, Steinberg C, Alabouvette C, Moenne-Loccoz Y (2009) The rhizosphere: a playground and battlefield for soilborne pathogens and beneficial microorganisms. Plant Soil 321:341–361

Rothballer M, Uhl J, Kunze J, Schmitt-Kopplin P, Hartmann A (2016) Detection of the bacterial quorum sensing molecules *N*-acyl homoserine lactones (HSL) and N-acyl-homoserines (HS) with the enzyme-linked immunosorbent assay (ELISA) and via ultrahigh performance liquid chromatrography coupled to mass spectrometry (UPLC-MS). In: Leoni L, Ramioni G (eds) Quorum sensing: methods and protokolls, Springer series methods in molecular biology. Springer International Publishing, Cham. (in press)

Ryu CM (2015) Bacterial volatiles as airborne signals for plants and bacteria. In: Lugtenberg B (ed) Principles of plant-microbe interactions. Springer International Publishing, Cham, pp 53–64

Ryu CM, Farag MA, Hu CH et al (2003) Bacterial volatiles promote growth in *Arabidopsis*. Proc Natl Acad Sci U S A 100:4927–4932

Schenk ST, Schikora A (2015) AHL-priming functions via oxylipin and salicylic acid. Front Plant Sci 5:784

Schenk ST, Stein E, Kogel KH, Schikora A (2012) *Arabidopsis* growth and defense are modulated by bacterial quorum sensing molecules. Plant Signal Behav 7:178–181

Schenk ST, Hernandez-Reyes C, Samans B et al (2014) *N*-acyl-homoserine lactone primes plants for cell wall reinforcement and induces resistance to bacterial pathogens via the salicylic acid/oxylipin pathway. Plant Cell 26:2708–2723

Schikora A, Schenk ST, Stein E, Molitor A, Zuccaro A, Kogel KH (2011) *N*-acyl-homoserine lactone confers resistance towards biotrophic and hemibiotrophic pathogens via altered activation of AtMPK6. Plant Physiol 157:1407–1418

Schikora ST, Schenk ST, Hartmann A (2016) Beneficial effects of bacteria-plant communication based on quorum sensing molecules of the *N*-acyl-homoserine lactone group. Plant Mol Biol 90:605–612

Schuhegger R, Ihring A, Gantner S et al (2006) Induction of systemic resistance in tomato by *N*-acyl-L-homoserine lactone-producing rhizosphere bacteria. Plant Cell Environ 29:909–918

Sieper T, Forczek S, Matucha M, Kramer P, Hartmann A, Schröder P (2013) N-acyl-homoserine lactone uptake and systemic transport in barley rest upon active parts of the plant. New Phytol 201:545–555

Singh RP, Baghel RS, Reddy CR, Jha B (2015) Effect of quorum sensing signals produced by seaweed-associated bacteria on carpospore liberation from *Gracilaria dura*. Front Plant Sci 6:117

Song S, Jia Z, Xu J, Zhang Z, Bian Z (2011) N-butyryl-homoserine lactone, a bacterial quorum-sensing signaling molecule, induces intracellular calcium elevation in *Arabidopsis* root cells. Biochem Biophys Res Commun 414:355–360

Spaepen S (2015) Plant hormones produced by microbes. In: Lugtenberg B (ed) Principles of plant-microbe interactions. Springer, Berlin, pp 247–256

Steidle A, Sigl K, Schuhegger R et al (2001) Visualization of *N*-acyl-L-homoserine lactone-mediated cell-cell communication between bacteria colonizing the tomato rhizosphere. Appl Environ Microbiol 67:5761–5770

Turner TR, James EK, Poole PS (2013) The plant microbiome. Genome Biol 14:209

Veliz-Vallejos DF, van Noorden GE, Yuan M, Mathesius U (2014) A *Sinorhizobium meliloti*-specific *N*-acyl-L-homoserine lactone quorum-sensing signal increases nodule numbers in *Medicago truncatula* independent of autoregulation. Front Plant Sci 5:551

von Rad U, Klein I, Dobrev PI et al (2008) Response of *Arabidopsis thaliana* to *N*-hexanoyl-L-homoserine-lactone, a bacterial quorum sensing molecule produced in the rhizosphere. Planta 229:73–85

Zarkani AA, Stein E, Rohrich CR et al (2013) Homoserine lactones influence the reaction of plants to rhizobia. Int J Mol Sci 14:17122–17146

Zhao Q, Zhang C, Jia Z, Huang Y, Li H, Song S (2015) Involvement of calmodulin in regulation of primary root elongation by N-3-oxo-hexanoyl homoserine lactone in *Arabidopsis thaliana*. Front Plant Sci 5:807

Zilber-Rosenberg I, Rosenberg E (2008) Role of microorganisms in the evolution of animals and plants: the hologenome theory of evolution. FEMS Microbiol Rev 32:732–735

Tripartite Interaction Among Root-Associated Beneficial Microbes Under Stress

11

Dilfuza Egamberdieva, Stephan Wirth,
and Elsayed Fathi Abd-Allah

Abstract

Microbes living on the root surface and the inner plant tissues such as rhizobacteria, arbuscular mycorrhizal fungi, endophytic bacteria, and rhizobia mutually cooperate with each other and induce a beneficial effect on plant growth. They colonize the root system and play an important role in stimulation of plant growth, stress tolerance, and nutrient acquisition of plants by altering the root system and altering physiological processes of plants. PGPR-rhizobia-AMF tripartite symbiosis improves plant growth under stress through induction of osmoregulation, hormonal balance, increase in nutrient acquisition, improving physiochemical activity, compositions, tissue water content, and alters metabolic interactions among the partners. These are coordinately involved in the adaptation of plants to abiotic stress through several mechanisms which include production of phytohormones, ACC deaminase enzyme, and mitigation of oxidative damage by improving enzymatic and nonenzymatic antioxidant defense system, modulation of phytohormones, and induction of acquired systemic tolerance. Mutualistic relationship of microbial symbionts could be an approach to increase plant stress tolerance to various abiotic stress factors.

D. Egamberdieva (✉) • S. Wirth
Institute of Landscape Biogeochemistry, Leibniz Centre for Agricultural Landscape Research (ZALF), Eberswalder str. 84, 15374 Müncheberg, Germany
e-mail: egamberdieva@yahoo.com; swirth@zalf.de

E.F. Abd-Allah
Department of Plant Production, College of Food and Agricultural Sciences, King Saud University, P.O. Box. 2460, Riyadh 11451, Saudi Arabia
e-mail: eabdallah@ksu.edu.sa

S. Mehnaz (ed.), *Rhizotrophs: Plant Growth Promotion to Bioremediation*, Microorganisms for Sustainability 2, DOI 10.1007/978-981-10-4862-3_11

11.1 Introduction

11.1.1 Root-Associated Microbes

The microbial associations in the rhizosphere are diverse and are mostly mediated by plant root exudates which contain soluble sugars, amino acids, phenolic compounds, and polysaccharides (Lugtenberg and Kamilova 2009). Root-associated microbes are categorized as rhizosphere colonizing bacteria, endophytic bacteria living in plant tissue, rhizobia living inside root nodule cells, and mycorrhizal fungi colonizing the living root tissue (Berg et al. 2013, Berg and Martinez 2015; Cho et al. 2015). Microbes colonizing rhizosphere and living in plant tissue cooperate with each other and produce various biological active metabolites, which resulted in improved root growth, higher stress tolerance, and the modulation of plant defense mechanisms (Egamberdieva et al. 2016). In addition, root-associated microbes significantly alter soil's physiochemical properties and play and important role in ecosystem functioning and nutrient cycling. Several microbes associated with plants such as arbuscular mycorrhizal fungi, endophytic bacteria, and rhizobia have been shown to have mutualistic interaction and impact on plant growth, nutrient acquisition, and disease suppression (Ghosh et al. 2014; Hou et al. 2015). The microbial community structure and activities depend on the composition of the root exudates released by the plants (Tamilarasi et al. 2008). The bioactive secondary metabolites synthesized by plants can also strongly affect plant-associated microbial communities and their physiological functions (rev. in Köberl et al. 2014; Chaparro et al. 2014). Moreover, plants rely on their microbiome for specific traits and activities, including growth promotion, nutrient acquisition, induced systemic resistance, and tolerance to abiotic stress factors (Malfanova et al. 2011; Sessitsch et al. 2013; Berg et al. 2014). Numerous studies have shown that abiotic stresses reduce the plant growth by the antagonistic relationship of sodium, and this may cause inhibition of available nutrients to plants and also affect root-associated microbes (Ahmad 2010; Egamberdieva et al. 2010; Porcel et al. 2012). Numerous studies have shown that the symbiotic relationship between legumes, and their rhizobia are susceptible to abiotic factors such as salinity, drought, and soil temperature, which can cause a failure in the infection and nodulation process (Rabie et al. 2005; Mensah and Ihenyen 2009; Egamberdieva et al. 2013, 2015a).

Root-associated microbes, including mycorrhizal fungi, closely live in the rhizosphere/internal plant tissue and facilitate important physiological processes, related to improved nutrient uptake and plant stress tolerance (Berg et al. 2013; Ahanger et al. 2014; Abd-Allah et al. 2015a). The plants inoculated with PGPR produce more root hairs and take up N, P, K, and microelements more efficiently from the soil. The metabolites synthesized by microbes may be used as a source of nutrients for the survival and proliferation of partner organism (Yadegari and Rahmani 2010; Sessitsch et al. 2013; Berg et al. 2014; Egamberdieva et al. 2015a, b).

It has been proposed that root-associated plant beneficial bacteria living in a free or an endophytic lifestyle may directly or indirectly contribute to the infection and colonization processes of the *Rhizobium*-host association (Egamberdieva et al.

2015b). They have the potential of promoting the stress tolerance, protecting plants from various plant pathogens (Malfanova et al. 2011), and some of them has shown the ability to form nodules and fix N_2 in legumes (Salwani et al. 2012).

The arbuscular mycorrhizal fungi (AMF) establish symbiotic association with many plant species, and cooperation between host plant and fungi shows positive effect on plant growth and development, through improving nutrient availability to plant and resistance to various abiotic stresses (Ahmad et al. 2015; Abd-Allah et al. 2015a).

Leguminous plants form important symbiotic relationships with nitrogen-fixing rhizobia and play an important role in the nitrogen cycle (Egamberdieva et al. 2013; Santi et al. 2013). The colonization of rhizobia in the root system of leguminous plants is essential for the establishment of the legume – *Rhizobium* symbiosis (Gulash et al. 1984; Hashem et al. 2015). The synergism between symbiotic rhizobia, AMF, and other rhizobacteria has been shown to stimulate plant root system and nutrient uptake and improved plant tolerance to various stress factors (Prakamhang et al. 2015). An understanding of the mutualistic interactions of microbes in the plant root is necessary for their potential effect on the tolerance of plant to abiotic stresses and for the improvement of crop management practices under extreme soil conditions. In this review, the tripartite interaction of endophytic bacteria, AMF, and rhizobia and their impact on plant-growth and fitness in the hostile environment are discussed.

11.1.1.1 Plant Growth-Promoting Rhizobacteria (PGPR)

The plant growth-promoting rhizobacteria (PGPR) can colonize the root surface of plant and live endophytically (Berg et al. 2015). They belong to various genera, e.g., *Achromobacter*, *Alcaligenes*, *Azospirillum*, *Arthrobacter*, *Azotobacter*, *Bacillus*, *Brevibacillus*, *Burkholderia*, *Cellulosimicrobium*, *Cellulomonas*, *Enterobacter*, *Flavobacterium*, *Paenibacillus*, *Pseudomonas*, *Stenotrophomonas*, and *Serratia* (Egamberdiyeva and Hoflich 2003; Guo et al. 2015). These can stimulate the plant height and root system, improve nutrient acquisition, and yield under various ecological conditions. The plant biomass of chickpea was stimulated (27%) by *Pseudomonas* spp. Inoculants significantly compared to control plants (Goswami et al. 2013). Other studies reported that PGPR enhanced the symbiotic performance of rhizobia with host plants, stimulated number of nodules, dry weight of nodules, and nitrogen fixation (Ahmad et al. 2013; Egamberdieva et al. 2015a). PGPR can suppress both, bacterial and fungal pathogens, that caused diseases of various plants under natural and saline soil conditions (Lucas et al. 2009; Berg et al. 2010).

Several studies reported the improvement of salt stress tolerance of plants for goat's rue, licorice, tomato, wheat, bean, and lettuce (Pliego et al. 2011; Nadeem et al. 2014; Egamberdieva et al. 2015a, b). An ameliorative effect of *Bacillus* species on plant growth under stress conditions has been extensively reviewed by Arora et al. (2012). The *Pseudomonas* species improved plant growth, nutrient uptake of agricultural important crop plants, and also phytochemical constitutes of medicinal and aromatic plants (Egamberdiyeva 2005; Mahalakshmi and Reetha 2009; Mishra et al. 2010). Several reports indicate the presence of the genus *Enterobacter* in the rhizosphere of wheat, rice, and sugarcane and their positive effect on plant growth, nutrient uptake, and reduced disease incidence in plants (Kämpfer et al. 2005; Egamberdieva et al. 2008; Hassen et al. 2007; Zakria et al. 2008). The growth, nutrient acquisition of soybean under salt stress condition was increased by *Stenotrophomonas rhizophila* (Egamberdieva et al. 2015a). The similar observation was reported for soybean, when plants were inoculated with *Azospirillum* sp., the root and shoot biomass, as well as nodulation, was enhanced as compared to non-inoculated control plants (Aung et al. 2013).

The strains of *Pseudomonas* that showed antagonistic activity against soilborne pathogens, such as *Sclerotium rolfsii*, *Fusarium oxysporum*, and *Rhizoctonia solani*, were able to suppress soybean root diseases caused by fungal pathogens (Susilowati et al. 2011). *Macrophomina phaseolina* cause the charcoal root rot of soybean, and plant inoculated with antagonistic bacterial strains *P. agglomerans* and *Bacillus* sp. showed reduced disease incidence (Vasebi et al. 2013).

11.1.1.2 Endophytic Bacteria

The endophytic bacteria live in the plant tissue and form mutualistic relationship with host (Naz and Bano 2012; Zhu et al. 2012). The genera of *Bacillus*, *Enterobacter*, and *Pseudomonas* were found in plant tissue of *Aloe vera* (Akinsanya et al. 2015), *Andropogon gerardii* (Rosenzweig et al. 2013), strawberry (*Fragaria*) (Pereira et al. 2012), cucumber (*Cucumis sativis*), chickpea (*Cicer arietinum* L.) (Saini et al. 2014), and soybean (*Glycine max* L.) (Egamberdieva et al. 2016). They are able to stimulate plant growth, stress tolerance, fix inorganic nitrogen, and protect plants from various pathogens (Berg et al. 2013). Arun et al. (2012) isolated endophytic bacteria from *Cassia occidentalis* and observed an increase in the plant growth of mung bean in pot experiments.

Several endophytic bacteria belonging to genera *Pseudomonas*, *Variovorax*, *Rhizobium*, *Caulobacter*, *Bacillus*, and *Paenibacillus* were found in the lavender roots (Pereira et al. 2016). These bacterial strains showed multiple PGP traits and stimulated plant growth of lavender. The biological control of *Verticillium* wilt disease of cotton by endophytic bacteria, *B. subtilis* and *B. megaterium*, isolated from the medical plant *Sophora alopecuroides*, was reported by Lin et al. (2013). The endophytic strains belonging to genera *Sphingomonas* sp., *Bacillus* sp., and *Methylobacterium* sp. were found in tomato, which showed significant increase in the shoot and root biomass and photosythetic pigments as compared to control plants (Khan et al. 2016). Recently, endophytic bacterium was isolated from chickpea nodules and identified as *Serratia marcescens* (Zaheer et al. 2016). The inoculation of seeds with *S. marcescens* resulted in 30.85% increase in the grain yield of chickpea under nutrient-deficient soil condition. The inoculation of the thal tree (*Acacia gerrardii*) with endophytic *B. subtilis* enhanced the synthesis of osmoprotectants and modulated the antioxidant enzyme system in such a way that it alleviated the oxidative damage caused by salt stress (Hashem et al. 2016). In another study, *B. subtilis*, which produces IAA and ACC deaminase enzyme, increased the stress tolerance of *Trigonella* plants to drought, increased proline content, and lipid peroxidation (Barnawal et al. 2013). Mohamed and Gomaa (2012) also observed an improved salt stress tolerance, plant biomass in radish by *B. subtilis*. The strain was also able to alter physiological properties of plants such as proline, total free amino acids, and nutrient (N, P, K) as compared to uninoculated control plants. The phosphate-solubilizing bacteria (PSB) *Pseudomonas* increased rhizobia chickpea symbioses, plant biomass, and yield of chickpea (Messele and Pant 2012). The inoculation of green gram (*Vigna radiata* L. Wilczek) with PSB showed an increase in P availability to plants (Vikram and Hamzehzarghani 2008).

11.1.1.3 Rhizobia

Leguminous plants form symbiotic associations with rhizobia, belonging to the genera *Rhizobium*, *Sinorhizobium*, *Bradyrhizobium*, *Azorhizobium*, and *Mesorhizobium* (Deaker et al. 2004). The symbiotic association of rhizobia with host plant is considered as the most efficient system for biological nitrogen fixation (BNF) (Molla et al. 2001). Kumar et al. (2011) isolated and identified *Ensifer meliloti* and *Rhizobium leguminosarum*, from root nodules of fenugreek (*Trigonella foenum-graecum*). The strains were able to increase nodule number, plant biomass, and grain yield of fenugreek and showed antagonistic activity against *F. oxysporum* (Kumar et al. 2011). Several rhizobia have also shown to confer increased resistance of plants against plant pathogens (Avis et al. 2008).

In the rhizosphere, competition of microbes for nutrients and niches are high; thus rhizobia must have the ability to effectively colonize root hairs to form nodules (Laranjo et al. 2014). It has been demonstrated that the colonization and infection of root hairs by rhizobial cells are sensitive to environmental stresses (Räsänen et al. 2003). For example, salt stress inhibited colonization ability of *R. galegae* sv. *officinalis* in the rhizosphere of goat's rue (*Galega officinalis*) (Egamberdieva et al. 2013).

The growth of chickpea and nodulation was completely depressed under saline soil condition (Egamberdieva et al. 2014). In such stressed condition, salt-tolerant *Mesorhizobium* strains significantly increased shoot and root dry matter and nodule number of chickpea by 20% under arid saline soil condition. The selection of salinity-tolerant chickpea cultivar with their symbiont is supported by higher root-to-shoot ratio, yield, and improved nodulation and N_2 fixation (Tejera et al. 2006). Mhadhbi et al. (2004) and Sadiki and Rabih (2001) also observed an increased plant growth, nodule number, protein composition, and yield of chickpea under stress condition that depends on the rhizobia association and plant genotype. Inoculation of common bean with *Rhizobium tropici* CIAT 899 and *Rhizobium gallicum* 8a3 improved osmotic stress tolerance of a bean cultivar, which is sensitive to drought stress (Sassi-Aydi et al. 2012).

11.1.1.4 Arbuscular Mycorrhizal Fungi

Most of the terrestrial plants live in symbiosis with arbuscular mycorrhizal fungi (AMF), and they play an important role in mitigating stress-triggered damaging effects in plants (Abd-Allah et al. 2015b). Through colonization of root system, AMF induced several changes in morphological, physiological and nutritional status of host plant (Hameed et al. 2014). AMF inoculation improved development of root which resulted in better absorption of water and essential nutrients such as nitrogen, phosphorus, and potassium from soil by plants (Aroca et al. 2013;

Abd-Allah et al. 2015b). The inoculation of plants with AMF mitigated salt stress for sunflower (*Helianthus annuus* L.) (Abd-Allah et al. 2015a) and lettuce (Aroca et al. 2013). The plants, *Trifolium alexandrinum* L. and *Trifolium resupinatum* L., inoculated with AMF showed higher plant biomass and nodule number (Zarea et al. 2011). The improved nutrient acquisition of olive (*Olea europaea* L.) (Porras-Soriano et al. 2009) and wheat (*Triticum aestivum* L.) (Talaat and Shawky 2013) by AMF inoculation was also reported.

The enhanced chlorophyll content due to AMF inoculation under NaCl-stressed conditions was reported by Aroca et al. (2013) for lettuce, Alqarawi et al. (2014) for *Ephedra aphylla*, and Abd-Allah et al. (2015b) for *Sesbania sesban*. *Trichoderma harzianum* was also able to increase plant biomass and improve some physiological properties of *Brassica juncea*, under salt stress condition (Ahmad et al. 2015).

In another study, sweet potato was inoculated with AMF, and the tuber numbers, fresh weight, proline and sugar accumulation, and plant tolerance to drought were improved (Yooyongwech et al. 2016). However, an inhibited AMF colonization in plant roots by salt stress has been reported by Alqarawi et al. (2014) for *Ephedra aphylla* and Hashem et al. (2015) for *Vigna unguiculata*. AMF inoculated plants showed well-maintained activities of photosynthetic pigments as compared to stressed counterparts (Aroca et al. 2013).

11.1.2 Mutualistic Interactions Between Plants and Microbes

In the plant rhizosphere, synergism of microbes was observed, whereas such interactions resulted in improved plant growth and nutrient acquisition. The positive effect of combined inoculation of PGPR and rhizobia/AMF were reported for soybean (Egamberdieva et al. 2015a), goats rue (Egamberdieva et al. 2013), chickpea (Rokhzadi et al. 2008), and faba bean (Fatnassi et al. 2015). The combined inoculation of *Galega orientalis* with *Pseudomonas* spp. and *R. galegae sv. orientalis* affected positively on plant growth, nodule formation, and nitrogen uptake of goats rue as compared to rhizobial strain alone (Egamberdieva et al. 2010). Singh et al. (2014) observed an increased plant biomass and improvement in some plant physiological properties, such as total phenolic and flavonoid content, and free radical and hydroxyl radical scavenging activities of chickpea inoculated with *Pseudomonas*, *Mesorhizobium*, and *Trichoderma*. The endophytes which effectively colonize plant tissues could be more beneficial in co-inoculation with rhizobia under various growth conditions (Panjebashi et al. 2012; Berg et al. 2013). A positive effect of combined microbial inoculation on plant growth and tolerance to abiotic stresses has been also extensively reviewed by Nadeem et al. (2014). The mutualistic interaction of PGPR and AMF is believed to perform as essential bio-ameliorators of stress through enhancing root system architecture, producing biological active compounds and regulating nutritional and hormonal balance (Ruiz-Lozano et al. 2012; Abd-Allah et al. 2015a, b; Egamberdieva et al. 2015a, b).

11.1.2.1 PGPR and Rhizobia

The positive effects of co-inoculation of leguminous plants with PGPR and rhizobia under abiotic stress conditions were reported previously (Egamberdieva et al. 2015a; Fatnassi et al. 2015). Under saline conditions, the inoculation of *Pseudomonas* with rhizobia enhanced nodule number, root and shoot biomass, and nutrient acquisition of pigeon pea (Tilak et al. 2006) and mung bean (Ahmad et al. 2013). In another study, the salt tolerance of *Galega officinalis* was improved when the plant was inoculated in combination with two strains, *R galegae sv. officinalis* and *P. extremorientalis* TSAU20 (Egamberdieva et al. 2013).

Increased salt concentration inhibits colonization of legume roots by rhizobia and the infection process where rhizobia enter the root or root hair (Zahran 1999; Egamberdieva et al. 2015b). It has been observed that *Mesorhizobium* spp. colonization in the rhizosphere of *Glycyrrhiza uralensis* was decreased by 95% at 75 mM NaCl, from 11.1×10^3 to 0.65×10^3 CFU cm^{-1} of root tip (Egamberdieva et al. 2015b). Co-inoculation of *Mesorhizobium* spp. with *P. extremorientalis* TSAU20 increased the number of mesorhizobial cells colonizing *G. uralensis* roots under salt stress condition. It has been proposed that *Pseudomonas*, or other PGPR strains, have endophytic life style and may directly or indirectly assist the infection and colonization processes of the *Rhizobium*-host association (Egamberdieva et al. 2015b). The soybean, treated with *B. japonicum* and endophytic bacteria *Stenotrophomonas rhizophila*, and grown under 75 mM NaCl condition, showed significantly higher nodule number, root and shoot biomass, and N and P uptake compared to *B. japonicum* alone. This result indicates that *B. japonicum* form synergestic cooperation with *S. rhizophila*, and it may enhance nodulation and plant growth (Egamberdieva et al. 2015a). Alavi et al. (2013) observed glucosylglycerol production by *S. rhizophila* in response to root exudates. This compound is well-known osmoprotectant that has ability to protect plant and their associated microbes from abiotis stresses.

The plant growth and biological yield of chickpea were increased by combined inoculation of *Pseudomonas* and *Mesorhizobium* strains (Panjebashi et al. 2012). The chickpea plants inoculated with *Serratia proteamaculans* (J119) and *Mesorhizobium ciceri* (S14) showed an improved plant biomass and nodule numbers of chickpea (Shahzad et al. 2010).

Co-inoculation of faba bean with *Pseudomonas* and *Rhizobium* decreased copper uptake up to 80% in the roots of 1 mM, Cu-treated plants as compared to non-inoculated control. Combined inoculation also increased the dry weights of plant as compared with Cu-treated and uninoculated plants (Fatnassi et al. 2015).

Mutualistic interaction between *R. leguminosarum* and *Pseudomonas aeruginosa* resulted in increased nodule number, dry weight, and plant biomass and yield of chickpea (*Cicer arietinum* L.) (Yadav and Verma 2014). The possible mechanism of stimulation caused by combined inoculation was explained as improved acquisition of P and Fe and production of phytohormone (IAA) and antifungal compounds by *Pseudomonas* strain (Yadav and Verma 2014).

There are several mechanisms of plant growth and stress tolerance improvement by PGPR, which include production of exopolysaccarides, plant growth regulators, ACC deaminase, competition for nutrient and niches, and modulation of antioxidant

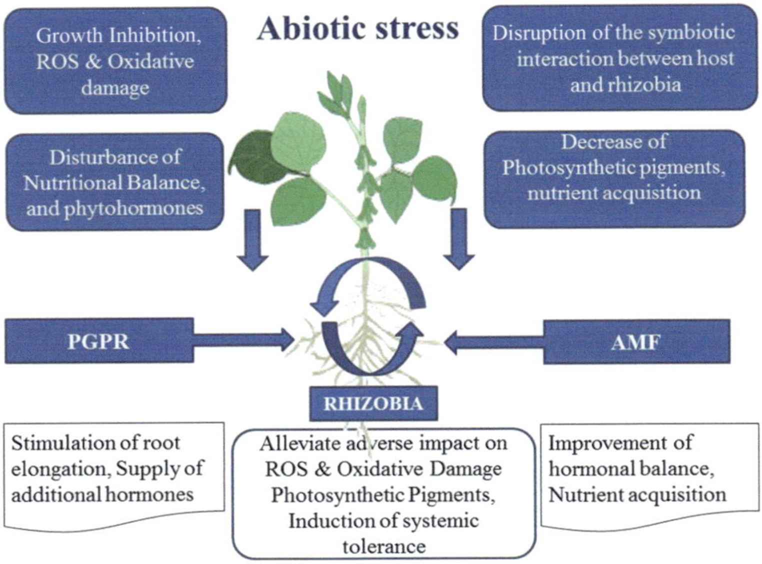

Fig. 11.1 Interactions between PGPR, rhizobia and arbuscular mycorrhizal fungi (AMF), and plant under abiotic stress

enzymes and others (Fig. 11.1; Naz and Bano 2012; Zhu et al. 2012). For example, PGPR with ACC-deaminase activity increased root length, number and length of lateral roots, and root biomass of chickpea under salt stress, through inhibiting the extra ethylene synthesis (Shaharoona et al. 2006). The strain of *Enterobacter hormaechei* producing IAA under saline condition was able to stimulate the root and shoot growth of tomato compared to untreated control (Egamberdieva et al. 2014). It was reported that synthesis of plant growth regulators inhibited by salt and drought stresses (Debez et al. 2001). The endopyhtes which are able to produce phytohormones may replace hormones into plant tissue and enhance root system (Malfanova et al. 2011; Berg et al. 2013).

The cell wall degrading enzymes play a major role in erosion of root epidermal cell walls of host plant, and it is related to primary host infection in the *Rhizobium*-legume symbiosis (Mateos et al. 2001). PGPR strains produced cell wall-degrading enzymes such as cellulase, glucanase, and pectinase, which have been proposed to locally degrade the root-hair cell wall (Sindhu and Dadarwal 2001). For example, *Pseudomonas* strain which produced cellulase, pectinase, and chitinase, improved the number of nodules of chickpea inoculated with *Mesorhizobium ciceri* (Egamberdieva et al. 2014).

11.1.2.2 AMF and Rhizobia

Synergistic interactions among AMF, *Rhizobium*, and legumes were reported to increase plant growth and nutrient uptake, in *Vicia faba* (Yinsuo et al. 2004), *Medicago truncatula* (de Varennes and Goss 2007), and *Medicago sativa* (Ardakani et al. 2009). The combined inoculation of legumes with AMF and rhizobia also showed an improved salt stress tolerance, plant growth, nodulation, and nitrogen fixation of soybean under salinity condition (Rasaei et al. 2012). Zarea et al. (2011) demonstrated that AMF inoculation of *Trifolium alexandrinum* L. and *Trifolium resupinatum* L. resulted in enhanced growth associated with increased nitrogenase activity as well as other attributes like nodule growth and number.

The modulation of endogenous levels of growth regulators by microbes is important for several physiological and biochemical key functions like stomatal closure, growth regulation, and maintenance of developmental events. The inoculation of plants with AMF promotes the synthesis of plant growth regulators such as indole acetic acid (IAA) and indole butyric acid (IBA) (Waqas et al. 2012). Abd-Alla et al. (2013) isolated and identified salt-tolerant *Rhizobium leguminosarum* bv. *viciae* from saline soil and applied to faba bean in combination with AMF (*Acaulospora laevis*, *Glomus geosporum*, *Glomus mosseae*, *Scutellospora armeniaca*). They observed an increase in nodule formation, leg-hemoglobin content of nodule, and plant biomass of plant under salt stress condition. Similar observation was reported by Ndoye et al. (2015), whereas AMF (*Glomus fasciculatum*, *Rhizophagus irregularis*, *Glomus verriculosum*) and rhizobia enhanced nodulation, the colonization of AMF, and also soil phosphatase activities of *A. senegal* (Ndoye et al. 2015). Barnawal et al. (2013) reported that the combination of ACC deaminase enzyme producing *B. subtilis*, *Ensifier meliloti*, and AMF acted synergistically to induce protective mechanisms against drought stress in *Trigonella* plants and improved plant weight by 56%. The combined inoculation also improved chlorophyll concentration, proline content, AMF colonization in the root, and nutrient asquisition. Dual inoculation of AMF and *Sinorhizobium meliloti* stimulated nodule formation and plant biomass of alfalfa. In addition, some physiological properties such as mineral element concentrations and proline content, were also increased (Ashrafi et al. 2014).

11.1.2.3 AMF and PGPR

Recent studies have shown that the colonization of plant roots with AMF and PGPR increases plants' tolerance to salinity through induction of osmoregulation and modulation of the impact of salt stress (Aroca et al. 2013; Latef et al. 2016; Sofo et al. 2016). Inoculation by AMF and *Bacillus subtilis* increased the tissue water content and the nutrient uptake and caused hormonal balance in the treated plants resulting in optimal activity of metabolic processes to meet the growing needs of host plants for coping with the stress (Hashem et al. 2016). The synergistic interactions of AMF and *B. subtilis* were evidenced by the fact that *B. subtilis* supported significant increase in AMF colonization of plants in saline conditions. The production of plant growth hormones by *B. subtilis* play direct and indirect role in promotion of mycorrhizal colonization. On the other hand, *B. subtilis* may have more essential role in management of ROS (Gill and Tuteja 2010) which in turn tends to decrease the affinity between AMF and host plant (Adesemoye et al. 2009). The dual inoculation of plants with AMF and endophytic *B. subtilis* also results in improved AMF colonization in the root system. Abiotic stress tolerance and induced systemic resistance by root-associated microbes include a variety of mechanisms such as production of phytohormones, osmoprotectants, exopolysaccharides, antifungal compounds, and regulation of plant hormone balance and antioxidant enzymes (Upadhyay et al. 2011). AMF results in considerable increase in the uptake of mineral nutrients by plants which ultimately promote synthesis of metabolically important metabolites and enzymes (Yuan et al. 2010). Among these essential plant metabolites, plant hormones have an intriguing role in plant growth maintenance. Auxins play a major role in signaling events between arbuscular mycorrhizal fungi (AMF) and host plants (Fernandez et al. 2014) and also in the development of nodule vasculature in leguminous plants (Mathesius 2008).

Accumulation of higher content of phenolic compounds like lignins, tannins, and fibers is another important strategy for avoiding the stress-induced changes. They have important role in plant physiology as their antioxidant property, which are involved in eliciting the proper response in plants during biotic and abiotic factors (Tomar and Agarwal 2013; Ahanger et al. 2015). Improved phenol and tannin content supports better growth and also mediates the radical scavenging. In our target plant, species inoculated with AMF- and *B. subtilis*-enhanced accumulation of phenols and tannin was observed as reflected in enhanced membrane stability in such plants. Plants inoculated with AMF showed lower lipid peroxidation and enhanced antioxidant enzyme activities in cowpea (Hashem et al. 2015) as systemic resistance tools against salt stress. The inoculation of plants with *Glomus etunicatum* increased the absorption of Mg^{+2} and inhibited Na^{+} transport, which improved chlorophyll biosynthesis (Zhu et al. 2010).

Accumulation of organic solutes including proline, sugars, and glycine betaine is one of the important tolerance strategy adapted by plants during stressful conditions (Ahanger et al. 2014; Hashem et al. 2014). AMF inoculation of plants enhances the accumulation of osmolytes which results in the maintenance of tissue water content and stimulates proline, glycine betaine contents. In *Ephedra aphylla* (Alqarawi et al. 2014), *Sesbania sesban* (Abd-Allah et al. 2015b) and *Solanum lycopersicum* (Hashem et al. 2015), accumulation of proline leads to salinity stress amelioration through better extraction of water from the soil solution by its active role in osmotic adjustment. The AMF- and *Bacillus subtilis*-inoculated plants showed increased content of osmoprotectants such as glycine, betaine, and proline.

It has been reported that abiotic stress induces excessive production of reactive oxygen species (ROS) which result in the loss of membrane integrity and desiccation (Hameed et al. 2014). Plants inoculated with AMF and endophytic bacteria showed improved stability of lipids from the oxidative degradation of toxic reactive oxygen species (Alqarawi et al. 2014; Hashem et al. 2014). Recently, Ahmad et al. (2015) observed increased production of free radicals like H_2O_2 in salt-stressed plants and reduced damaging effect by microbial inoculation. Abd-Allah et al. (2015a) demonstrated that AMF colonization provides protection to membrane lipids from the oxidative stress. Up-regulation of antioxidant system and scavenging of ROS goes hand in hand and is often correlated with stress tolerance (Alqarawi et al. 2014). AMF inoculation induced a significant increment in the activities of antioxidant enzyme activities under normal as well as salt-stressed condition, and these results are in confirmation with the results of Alqarawi et al. (2014).

Lipids have vital roles in the tolerance to several physiological stressors in plants such as drought and salinity (Singh et al. 2002). The higher salt concentration inhibits neutral lipids and phospholipids in plant tissue (Kerkeb et al. 2001). The PGPR strains colonizing root system may increase lipid concentrations in plants compared to control plants under saline soil conditions. The percentage of oleic acid (C18:1), linoleic (C18:2) and linolenic (C18:3) acids, in Indian brassica was increased by *B. subtilis* (Hashem et al. 2015). The inoculation of canola (*Brassica napus* L.) with *Azospirillum* strains significantly increased oleic acid (C18:1) and linolenic acid (C18:3) content (Nosheen et al. 2013).

The synergistic interaction of endophytic bacteria *Bacillus subtilis* and host plant *Robinia pseudoacacia* L. was explained as colonization ability of *Bacillus* in plant tissue similar to the rhizobia and formed bacteroids inside plant cortical cells (Huang et al. 2011). The *B. subtilis* strain was able to produce cellulase that could be one mechanism that strain help rhizobia enter into target root hair cells to form nodules (Sindhu and Dadarwal 2001). The combination of such cellulase producing endophytic strains can increase nodule formation by rhizobia and enhance nitrogenase activity.

The inoculation of endophytes also enhances nutrient uptake by plants under abiotic stress condition. It is known that salt stress impedes the acquisition of mineral elements by plants including nitrogen, which affects the nitrogen metabolism potential (Näsholm et al. 2009). In that condition, microbes such as AMF and endophytic bacteria colonizing inner part of plant roots may produce various biological active compunds and help plants to resist osmotic stress and improve plant nutrient uptake (Ahanger et al. 2015).

11.2 Conclusion

The studies mentioned above indicate that PGPR-rhizobia-AMF tripartite symbiosis leads to marked changes in the growth pattern of plants, improving physio-biochemical activity and compositions (Fig. 11.1). Mutualistic relationship of microbes in the plant root brings benefits to the plant through an increase in nutrient acquisition, alters metabolic interactions among the partners, alleviates salt stress and improves symbiotic performance of legumes. The microbes associated with plants and live in plant tissue are coordinately involved in the plant growth stimulation and resistance to various abiotic stresses. The mechanisms involved in such interactions include production of plant growth regulators, exopolysaccarides, osmolytes, ACC deaminase enzyme, enhancement of antioxidant defense system, and induction of acquired systemic tolerance. Mutualistic relationship of microbial symbionts could be an approach to increase the tolerance of plants to various abiotic stress factors.

References

Abd-Alla MH, El-Enany AWE, Nafady NA, Khalaf DM, Morsy FM (2013) Synergistic interaction of *Rhizobium leguminosarum* bv. viciae and arbuscular mycorrhizal fungi as a plant growth promoting biofertilizers for faba bean (*Vicia faba* L.) in alkaline soil. Microbiol Res 169(1):49–58

Abd-Allah EF, Hashem A, Alqarawi AA, Alwhibi MS (2015a) Alleviation of adverse impact of salt in *Phaseolus vulgaris* l. by arbuscular mycorrhizalfungi. Pak J Bot 47(3):1167–1176

Abd-Allah EF, Hashem A, Alqarawi AA, Bahkali AH, Alwhibi MS (2015b) Enhancing growth performance and systemic acquired resistance of medicinal plant *Sesbania sesban* (L.) Merr, using arbuscular mycorrhizal fungi, under salt stress. Saudi J Bio Sci 22:274–283

Adesemoye AO, Torbert HA, Kloepper JW (2009) Plant growth-promoting rhizobacteria allow reduced application rates of chemical fertilizers. Microb Ecol 58:921–929

Ahanger MA, Hashem A, Abd-Allah EF, Ahmad P (2014) Arbuscular mycorrhiza in crop improvement under environmental stress. In: Ahmad P, Rasool P (eds) Emerging technologies and management of crop stress tolerance, vol 2. Elsevier, New York, pp 69–95

Ahanger MA, Agarwal RM, Tomar NS, Shrivastava M (2015) Potassium induces positive changes in nitrogen metabolism and antioxidant system of oat (*Avena sativa* L cultivar Kent). J Plant Interact 10(1):211–223

Ahmad P (2010) Growth and antioxidant responses in mustard (*Brassica juncea* L.) plants subjected to combined effect of gibberellic acid and salinity. Arch Agron Soil Sci 56(5):575–588

Ahmad M, Zahir ZA, Khalid M, Nazli F, Arshad M (2013) Efficacy of *Rhizobium* and *Pseudomonas* strains to improve physiology, ionic balance and quality of mung bean under salt-affected conditions on farmer's fields. Plant Physiol Biochem 63:170–176

Ahmad P, Hashem A, Abd-Allah EF et al (2015) Role of *Trichoderma harzianum* in mitigating NaCl stress in Indian mustard (*Brassica juncea* L) through antioxidative defense system. Front Plant Sci 6:868

Akinsanya MA, Goh JK, Lim SP, Tinga ASY (2015) Metagenomics study of endophytic bacteria in *Aloe vera* using next-generation technology. Genomics Data 6:159–163

Alavi P, Starcher MR, Zachow C, Müller H, Berg G (2013) Root-microbe systems: the effect and mode of interaction of stress protecting agent (SPA) *Stenotrophomonas rhizophila* DSM14405T. Front Plant Sci 4:141

Alqarawi AA, Abd-Allah EF, Hashem A (2014) Alleviation of salt-induced adverse impact via mycorrhizal fungi in *Ephedra aphylla Forssk.* J Plant Interact 9:802–810

Ardakani MR, Pietsch G, Moghaddam A, Raza A, Friedel JK (2009) Response of root properties to tripartite symbiosis between lucerne (*Medicago sativa* L.), rhizobia and mycorrhiza under dry organic farming conditions. Am J Agric Biol Sci 4:266–277

Arora NK, Tewari S, Singh S, Lal N, Maheshwari DK (2012) PGPR for protection of plant health under saline conditions. In: Maheshwari DK (ed) Bacteria in agrobiology: stress management. Springer, Berlin, pp 239–258

Aroca R, Ruiz-Lozano JM, Zamarreno A et al (2013) Arbuscular mycorrhizal symbiosis influences strigolactone production under salinity and alleviates salt stress in lettuce plants. J Plant Physiol 170:47–55

Arun B, Gopinath B, Sharma S (2012) Plant growth promoting potential of bacteria isolated on N free media from rhizosphere of *Cassia occidentalis*. World J Microbiol Biotechnol 28:2849–2857

Ashrafi E, Zahedi M, Razmjoo J (2014) Co-inoculations of arbuscular mycorrhizal fungi and rhizobia under salinity in alfalfa. Soil Sci Plant Nutr 60:619–629

Aung TT, Tittabutr P, Boonkerd N, Herridge D, Teaumroong N (2013) Co-inoculation effects of *Bradyrhizobium japonicum* and *Azospirillum* sp. on competitive nodulation and rhizosphere eubacterial community structures of soybean under rhizobia-established soil conditions. Afr J Biotechnol 12:2850–2862

Avis TJ, Gravel V, Antoun H, Tweddell RJ (2008) Multifaceted beneficial effects of rhizosphere microorganisms on plant health and productivity. Soil Biol Biochem 40:1733–1740

Barnawal D, Maji D, Bharti N, Chanotiya CS, Kalra A (2013) ACC deaminase-containing *Bacillus subtilis* reduces stress ethylene-induced damage and improves mycorrhizal colonization and rhizobial nodulation in *Trigonella foenum-graecum* under drought stress. J Plant Growth Regul 32:809–822

Berg G, Martinez JL (2015) Friends or foes: can we make a distinction between beneficial and harmful strains of the *Stenotrophomonas maltophilia* complex? Front Microbiol 6:241

Berg G, Egamberdieva D, Lugtenberg B, Hagemann M (2010) Symbiotic plant-microbe interactions: stress protection, plant growth promotion and biocontrol by *Stenotrophomonas*. In: Seckbach J, Grube M (eds) Symbioses and stress, cellular origin, life in extreme habitats and astrobiology, Springer Dordrecht, 17(4):445–460

Berg G, Zachow C, Müller H, Philipps J, Tilcher R (2013) Next-generation bio-products sowing the seeds of success for sustainable agriculture. Agronomy 3:648–656

Berg G, Grube M, Schloter M, Smalla K (2014) Unraveling the plant microbiome: looking back and future perspectives. Front Microbiol 5:148

Chaparro JM, Badri DV, Vivanco JM (2014) Rhizosphere microbiome assemblage is affected by plant development. The ISME J 8:790–803

Cho ST, Chang HH, Egamberdieva D et al (2015) Genome analysis of *Pseudomonas fluorescens* PCL1751: a rhizobacterium that controls root diseases and alleviates salt stress for its plant host. PLoS One 10(10):e0140231

Deaker R, Roughley RJ, Kennedy IR (2004) Legume seed inoculation technology – a review. Soil Biol Biochem 36:1275–1288

de Varennes A, Goss MJ (2007) The tripartite symbiosis between legumes, rhizobia and indigenous mycorrhizal fungi is more efficient in undisturbed soils. Soil Biol Biochem 39:2603–2607

Debez A, Chaibi W, Bouzid S (2001) Effect du NaCl et de regulatoeurs de croissance sur la germination d' *Atriplex halimus* L. Cahiers Agricultures 10:135–138

Egamberdieva D, Kamilova F, Validov SH et al (2008) High incidence of plant growth-stimulating bacteria associated with the rhizosphere of wheat grown on salinated soil in Uzbekistan. Environ Microbiol 10:1–9

Egamberdieva D, Renella G, Wirth S, Islam R (2010) Secondary salinity effects on soil microbial biomass. Biol Fertil Soils 46(5):445–449

Egamberdieva D, Berg G, Lindström K, Räsänen LA (2013) Alleviation of salt stress of symbiotic *Galega officinalis* L. (goat's rue) by co-inoculation of *Rhizobium* with root colonising *Pseudomonas*. Plant Soil 369(1):453–465

Egamberdieva D, Shurigin V, Gopalakrishnan S, Sharma R (2014) Growth and symbiotic performance of chickpea *(Cicer arietinum)* cultivars under saline soil conditions. J Biol Chem Res 31(1):333–341

Egamberdieva D, Jabborova D, Berg G (2015a) Synergistic interactions between *Bradyrhizobium japonicum* and the endophyte *Stenotrophomonas rhizophila* and their effects on growth, nodulation and nutrition of soybean under salt stress. Plant Soil 405(1):35–45

Egamberdieva D, Li L, Lindström K, Räsänen L (2015b) A synergistic interaction between salt tolerant *Pseudomonas* and *Mezorhizobium* strains improves growth and symbiotic performance of liquorice (*Glycyrrhiza uralensis* fish.) under salt stress. Appl Microbiol Biotechnol 100(6):2829–2841

Egamberdieva D, Wirth S, Behrendt U, Abd-Allah EF, Berg G (2016) Biochar treatment resulted in a combined effect on soybean growth promotion and a shift in plant growth promoting rhizobacteria. Front Microbiol 7:209

Egamberdiyeva D (2005) Characterisation of *Pseudomonas* species isolated from the rhizosphere of plants grown in serozem soil, semi arid region of Uzbekistan. The Sci World J 5:501–509

Egamberdiyeva D, Hoflich G (2003) Influence of growth promoting bacteria on the growth of wheat at different soils and temperatures. Soil Biol Biochem 35:973–978

Fatnassi IC, Chiboub M, Saadani O, Jebara M, Jebara SH (2015) Impact of dual inoculation with *Rhizobium* and PGPR on growth and antioxidant status of *Vicia faba* L. under copper stress. C R Biol 338:241–254
Fernandez I, Merlos M, López-Ráez JA et al (2014) Defense related phytohormones regulation in arbuscular mycorrhizal symbioses depends on the partner genotypes. J Chem Ecol 40:791–803
Ghosh UD, Saha C, Maiti M et al (2014) Root associated iron oxidizing bacteria increase phosphate nutrition and influence root to shoot partitioning of iron in tolerant plant *Typha angustifolia*. Plant Soil 381(1–2):279–295
Gill SS, Tuteja N (2010) Reactive oxygen species and antioxidant machinery in abiotic stress tolerance in crop plants. Plant Physiol Biochem 48:909–930
Goswami D, Vaghela H, Parmar S, Dhandhukia P, Thakker JN (2013) Plant growth promoting potentials of *Pseudomonas* spp. strain OG isolated from marine water. J Plant Interact 8(4):281–290
Gulash M, Ames P, Larosiliere RC, Bergman K (1984) Rhizobia are attracted to localized sites on legume roots. Appl Environ Microbiol 48:149–152
Guo JK, Ding YZ, Feng RW et al (2015) *Burkholderia metalliresistens* sp. nov., a multiple metal-resistant and phosphate-solubilising species isolated from heavy metal-polluted soil in Southeast China. Antonie Van Leeuwenhoek 107(6):1591–1598
Hameed A, Egamberdieva D, Abd-Allah EF et al (2014) Salinity stress and arbuscular mycorrhizal symbiosis in plants. In: Miransari M (ed) Use of microbes for the alleviation of soil stresses. Springer, New York, pp 139–159. doi:10.1007/978-1-4614-9466-9-7
Hashem A, Abd-Allah EF, Alqarawi AA et al (2014) Alleviation of adverse impact of salinity on faba bean (*Vicia faba* L.) by arbuscular mycorrhizal fungi. Pak J Bot 46:2003–2013
Hashem A, Abd-Allah EF, Alqarawi AA et al (2015) Induction of salt stress tolerance in cowpea (*Vigna unguiculata* [L.] Walp) by arbuscular mycorrhizal fungi. Legum Res 38(5):579–588. doi: 10.18805/lr.v38i5.5933
Hashem A, Abd-Allah EF, Alqarawi AA et al (2016) The interaction between arbuscular mycorrhizal fungi and endophytic bacteria enhances plant growth of Acacia gerrardii under salt stress. Front Microbiol 7:1089. doi:doi.org/10.3389/fmicb.2016.01089
Hassen AA, Hu J, Yang J (2007) Growth conditions of associative nitrogen fixing bacteria *Enterobacter cloace*, in rice plants. Agric J 2:672–675
Hou Q, Wang C, Guo H et al (2015) Draft genome sequence of *Delftia tsuruhatensis* MTQ3, a strain of plant growth-promoting rhizobacterium with antimicrobial activity. Genome Announc 3(4):e00822–e00815
Huang B, Lv C, Zhuang P, Zhang H, Fan L (2011) Endophytic colonisation of *Bacillus subtilis* in the roots of *Robinia pseudoacacia* L. Plant Biol 13(6):925–931
Kämpfer P, Ruppel S, Remus R (2005) *Enterobacter radicincitans* sp. nov., a plant growth promoting species of the family *Enterobacteriaceae*. Syst Appl Microbiol 28:213–221
Kerkeb L, Donaire JP, Venema K, Rodriguez-Rosales MP (2001) Tolerance to NaCl induces changes in plasma membrane lipid composition, fluidity and H+-ATPase activity of tomato calli. Physiol Plant 113:217–224
Khan AL, Halo BA, Elyassi A et al (2016) Indole acetic acid and ACC deaminase from endophytic bacteria improves the growth of *Solanum lycopersicum*. Electron J Biotechnol 21:58–64
Köberl M, Schmidt R, Ramadan EM, Bauer R, Berg G (2014) The microbiome of medicinal plants: diversity and importance for plant growth, quality and health. Front Microbiol 4:400
Kumar H, Dubey RC, Maheshwari DK (2011) Effect of plant growth promoting rhizobia on seed germination, growth promotion and suppression of *Fusarium* wilt of fenugreek (*Trigonella foenum-graecum* L.) Crop Prot 30:1396–1403
Laranjo M, Alexandre A, Oliveira S (2014) Legume growth-promoting rhizobia: an overview on the *Mesorhizobium* genus. Microbiol Res 169:2–17
Latef AAHA, Jan S, Abd-Allah EF et al (2016) Soybean under abiotic stress: proteomic approach. In: Azooz MM, Ahmad P (eds) Plant environment interaction: responses and approaches to mitigate stress. Wiley, Hoboken. doi:10.1002/9781119081005.ch2

Lin T, Zhao L, Yang Y, Guan Q, Gong M (2013) Potential of endophytic bacteria isolated from *Sophora alopecuroides* nodule in biological control against *Verticillium* wilt disease. Aust J Crop Sci 7:139–146

Lucas JA, Ramos Solano B, Montes F et al (2009) Use of two PGPR strains in the integrated management of blast disease in rice (*Oryza sativa*) in Southern Spain. Field Crop Res 114:404–410

Lugtenberg B, Kamilova F (2009) Plant growth-promoting rhizobacteria. Annu Rev Microbiol 63:541–556

Mahalakshmi M, Reetha D (2009) Assessment of plant growth promoting activities of bacterial isolates from rhizpsphere of tomato (*Lycopersicon esculantum* L.) Recent Res Sci Technol 1:26–29

Malfanova N, Kamilova F, Validov S et al (2011) Characterization of *Bacillus subtilis* HC8, a novel plant-beneficial endophytic strain from giant hogweed. Microb Biotechnol 4:523–532

Mateos PF, Baker DL, Petersen M et al (2001) Erosion of root epidermal cell walls by *Rhizobium* polysaccharide-degrading enzymes as related to primary host infection in the *Rhizobium*-legume symbiosis. Can J Microbiol 47(6):475–487

Mathesius U (2008) Auxin: at the root of nodule development? Funct Plant Biol 35(8):651–668

Mensah JK, Ihenyen J (2009) Effects of salinity on germination, seedling establishment and yield of three genotypes of mung bean (*Vigna mungo* L. Hepper) in Edo State. Nigeria Nigerian Ann Nat Sci 8(2):17–24

Messele B, Pant LM (2012) Effects of inoculation of *Sinorhizobium ciceri* and phosphate solubilizing bacteria on nodulation, yield and nitrogen and phosphorus uptake of chickpea (*Cicer arietinum* L.) in Shoa Robit area. J Biofert Biopest 3:129

Mhadhbi H, Jebara M, Limam F, Aouani ME (2004) Rhizobial strain involvement in plant growth, nodule protein composition and antioxidant enzyme activities of chickpea rhizobia symbioses: modulation by salt stress. Plant Physiol Biochem 42:717–722

Mishra RK, Prakash O, Alam M, Dikshit A (2010) Influence of plant growth promoting rhizobacteria (PGPR) on the productivity of *Pelargonium graveolens* L. herit. Recent Res Sci Technol 2(5):53–57

Mohamed HI, Gomaa EZ (2012) Effect of plant growth promoting *Bacillus subtilis* and *Pseudomonas fluorescens* on growth and pigment composition of radish plants (*Raphanus sativus*) under NaCl stress. Photosynthetica 50(2):263–272

Molla AH, Shamsuddin ZH, Halimi MS, Morziah M, Puteh AB (2001) Potential for enhancement of root growth and nodulation of soybean co-inoculated with *Azospirillum* and *Bradyrhizobium* in laboratory systems. Soil Biol Biochem 33:457–463

Nadeem SN, Ahmad M, Zahir ZA, Javaid A, Ashraf M (2014) The role of mycorrhizae and plant growth promoting rhizobacteria (PGPR) in improving crop productivity under stressful environments. Biotechnol Adv 32:429–448

Näsholm T, Kielland K, Ganeteg U (2009) Uptake of organic nitrogen by plants. New Phytol 182:31–48

Naz I, Bano A (2012) Assessment of phytohormones producing capacity of *Stenotrophomonas maltophilia* SSA and itsinteraction with *Zea mays* l. Pak J Bot 44(1):465–469

Ndoye F, Kane A, Diedhiou AG et al (2015) Effects of dual inoculation with arbuscular mycorrhizal fungi and rhizobia on *Acacia senegal* (L.) Willd seedling growth and soil enzyme activities in Senegal. Int J Biosci 6:36–48

Nosheen A, Bano A, Ullah A (2013) The role of plant growth promoting rhizobacteria on oil yield and biodiesel production of canola (*Brassica napus* L.) Energy Sources 35:1574–1581

Panjebashi M, Hadi MRHS, Darzi MT (2012) Effects of the *Rhizobium* and PGPRs bacterium on seed yield and yield components in chickpea (*Cicer arietinum*). Int J Agr Plant Prod 3:651–655

Pereira GVM, Karina TM, Emi RL, Thiago PS, Rosane FC (2012) A multiphasic approach for the identification of endophytic bacterial in strawberry fruit and their potential for plant growth promotion. Microb Ecol 63:405–417

Pereira SIA, Monteiro C, Vega AL, Castro PML (2016) Endophytic culturable bacteria colonizing *Lavandula dentata* L. plants: isolation, characterization and evaluation of their plant growth-promoting activities. Ecol Eng 87:91–97

Pliego C, Kamilova F, Lugtenberg B (2011) Plant growth-promoting bacteria: fundamentals and exploitation. In: Maheshwari DK (ed) Bacteria in agrobiology: crop ecosystems. Springer, Germany, pp 295–343

Porcel R, Aroca R, Ruiz-Lozano JM (2012) Salinity stress alleviation using arbuscular mycorrhizal fungi. Agron Sustain Dev 32:181–200

Porras-Soriano A, Soriano-Martin ML, Porras-Piedra A, Azcon R (2009) Arbuscular mycorrhizal fungi increased growth, nutrient uptake and tolerance to salinity in olive trees under nursery conditions. J Plant Physiol 166:1350–1359

Prakamhang J, Tittabutr P, Boonkerd N et al (2015) Proposed some interactions at molecular level of PGPR co-inoculated with *Bradyrhizobium diazoefficiens* USDA110 and *B. japonicum* THA6 on soybean symbiosis and its potential of field application. Appl Soil Ecol 85:38–49

Rabie GH, Aboul-Nasr MB, Al-Humiany A (2005) Increase salinity tolerance of cowpea plants by dual inoculation of AM *fungus Glomus clarum* and nitrogen- fixer *Azospirillum brasilense*. Mycobiology 33(1):51–61

Rasaei B, Ghobadi ME, Ghobadi M, Najafi A, Rasaei A (2012) The study of effects of some biological agents on chickpea (*Cicer arietinum* L.) under semi-dry conditions in Kermanshah. Eur J Exp Biol 2(4):1113–1118

Räsänen LA, Saijets S, Jokinen K, Lindström K (2003) Evaluation of the roles of two compatible solutes, glycine betaine and trehalose, for the *Acacia senegal–Sinorhizobium* symbiosis exposed to drought stress. Plant Soil 260:237–251

Rokhzadi A, Asgharzadeh A, Darvish F, Nour-Muhammadi G, Majidi E (2008) Influence of plant growth promoting rhizobacteria on dry matter accumulation and yield of chickpea (*Cicer arietinum* L.) under field conditions. Amer Eur J Agr Env Sci 3(2):253–257

Rosenzweig N, Bradeen JM, Tu ZJ, Kinkel LL (2013) Rhizosphere bacterial communities associated with long-lived perennial prairie plants vary in diversity, composition, and structure. Can J Microbiol 59(7):494–502

Ruiz-Lozano JM, Porcel R, Azcón C, Aroca R (2012) Regulation by arbuscular mycorrhizae of the integrated physiological response to salinity in plants: new challenges in physiological and molecular studies. J Exp Bot 63:4033–4044

Sadiki M, Rabih K (2001) Selection of chickpea (*Cicer arietinum*) for yield and symbiotic nitrogen fixation ability under salt stress. Agronomie 21:659–666

Saini R, Dudeja SS, Giri RR, Kumar V (2014) Isolation, characterization and evaluation of bacterial root and nodule endophytes from chickpea cultivated in northern India. J Basic Microbiol 55(1):74–81

Salwani S, Amir HG, Najimudin N (2012) Evidence of diazotrophic symbionts in the leguminous cover crop *Mucuna bracteata*. J Trop Agr Science 35:537

Santi C, Bogusz D, Franche C (2013) Biological nitrogen fixation in non-legume plants. Ann Bot 111(5):743–767

Sassi-Aydi S, Aydi S, Abdelly M (2012) Inoculation with the native *Rhizobium gallicum* 8a3 improves osmotic stress tolerance in common bean drought-sensitive cultivar. Acta Agriculturae Scandinavica Section B Soil Plant Science 62:179–187

Sessitsch A, Kuffner M, Kidd P et al (2013) The role of plant-associated bacteria in the mobilization and phytoextraction of trace elements in contaminated soils. Soil Biol Biochem 60:182–194

Shaharoona B, Arshad M, Zahir ZA (2006) Effect of plant growth promoting rhizobacteria containing ACC-deaminase on maize (*Zea mays* L.) growth under axenic conditions and on nodulation in mung bean (*Vigna radiataL.*) Lett Appl Microbiol 42(2):155–159

Shahzad SM, Khalid A, Arshad M, Tahir J, Mahmood T (2010) Improving nodulation, growth and yield of *Cicer arietinum* L. through bacterial ACC-deaminase induced changes in root architecture. Eur J Soil Biol 46:342–347

Sindhu SS, Dadarwal KR (2001) Chitinolytic and cellulolytic *Pseudomonas* sp. antagonistic to fungal pathogens enhances nodulation by *Mesorhizobium cicer* in chickpea. Microbiol Res 156:353–358

Singh SC, Sinha RP, Hader DP (2002) Role of lipids and fatty acids in stress tolerance in cyanobacteria. Acta Protozool 41:297–308

Singh A, Jain A, Sarma BK, Upadhyay RS, Singh HB (2014) Beneficial compatible microbes enhance antioxidants in chickpea edible parts through synergistic interactions. LWT Food Sci Technol 56:390–397

Sofo A, Scopa A, Hashem A, Abd-Allah EF (2016) Lipid metabolism and oxidation in plants subjected to abiotic stresses. In: Azooz MM, Ahmad P (eds) Plant-environment interaction: responses and approaches to mitigate stress. Wiley, Hoboken. doi:10.1002/9781119081005.ch11

Susilowati A, Wahyudi AT, Lestari Y, Suwanto A, Wiyono S (2011) Potential of *Pseudomonas* isolated from soybean rhizosphere as biocontrol against soilborne phytopathogenic fungi. HAYATI J Biosciences 18:51–56

Talaat NB, Shawky BT (2013) 24-Epibrassinolide alleviates salt-induced inhibition of productivity by increasing nutrients and compatible solutes accumulation and enhancing antioxidant system in wheat (*Triticum aestivum* L.) Acta Physiol Plant 35:729–740

Tamilarasi S, Nanthakumar K, Karthikeyan K, Lakshmanaperumalsamy P (2008) Diversity of root associated microorganisms of selected medicinal plants and influence of rhizo-microorganisms on the antimicrobial property of *Coriandrum sativum*. J Environ Biol 29:127–134

Tejera NA, Soussi M, Lluch C (2006) Physiological and nutritional indicators of tolerance to salinity in chickpea plants growing under symbiotic conditions. Envir Exper Botany 58:17–24

Tilak KVB, Ranganayaki N, Manoharachari C (2006) Synergistic effects of plant-growth promoting rhizobacteria and *Rhizobium* on nodulation and nitrogen fixation by pigeon pea (*Cajanus cajan*). Eur J Soil Sci 57:67–71

Tomar NS, Agarwal RM (2013) Influence of treatment of *Jatropha curcas* L. leachates and potassium on growth and phytochemical constituents of wheat (*Triticum aestivum* L.) Amer J Plant Sci 4:1134–1150

Upadhyay SK, Singh JS, Singh DP (2011) Exopolysaccharide-producing plant growth promoting rhizobacteria under salinity condition. Pedosphere 2:214–222

Vasebi Y, Safaie N, Alizadeh A (2013) Biological control of soybean charcoal root rot disease using bacterial and fungal antagonists in vitro and greenhouse condition. J Crop Prot 2:139–150

Vikram A, Hamzehzarghani H (2008) Effect of phosphate solubilizing bacteria on nodulation and growth parameters of greengram (*Vigna radiata* L. Wilczek). Res J Microbiol 3:62–72

Waqas M, Khan AL, Kamran M et al (2012) Endophytic fungi produce gibberellins and indoleacetic acid and promotes host-plant growth during stress. Molecules 17:10754–10773

Yadav J, Verma JP (2014) Effect of seed inoculation with indigenous *Rhizobium* and plant growth promoting rhizobacteria on nutrients uptake and yields of chickpea (*Cicer arietinum* L.) Eur J Soil Sci 63:70–77

Yadegari M, Rahmani A (2010) Evaluation of bean (*Phaseolus vulgaris*) seeds inoculation with *Rhizobium phaseoli* and plant growth promoting *Rhizobacteria* (PGPR) on yield and yield components. Afr J Agric Res 5:792–799

Yinsuo J, Vincent MG, Colin JS (2004) The influence of *Rhizobium* and arbuscular mycorrhizal fungi on nitrogen and phosphorus accumulation by *Vicia faba*. Ann Bot 94:251–258

Yooyongwech S, Samphumphuang T, Tisarum R, Theerawitaya C, Cha-Um S (2016) Arbuscular mycorrhizal fungi (AMF) improved water deficit tolerance in two different sweet potato genotypes involves osmotic adjustments via soluble sugar and free proline. Sci Hortic 198:107–117

Yuan ZL, Zhang CL, Lin FC (2010) Role of diverse non-systemic fungal endophytes in plant performance and response to stress: progress and approaches. J Plant Growth Regul 29:116–126

Zaheer A, Mirza BC, Mclean JE et al (2016) Association of plant growth-promoting *Serratia* spp. with the root nodules of chickpea. Res Microbiol. doi:10.1016/j.resmic.2016.04.001

Zahran HH (1999) *Rhizobium*-legume symbiosis and nitrogen fixation under severe conditions and in an arid climate. Microbiol Mol Biol Rev 63:968–989

Zakria M, Ohsaco A, Saeki Y, Yamamoto A, Akao S (2008) Colonization and growth promotion characteristics of *Enterobacter* sp. and *Herbaspirillum* sp. on *Brassica oleracea*. Soil Sci Plant Nutr 54:507–516

Zarea MJ, Karimi N, Goltapeh EM, Ghalavand A (2011) Effect of cropping systems and arbuscular mycorrhizal fungi on soil microbial activity and root nodule nitrogenase. J Saudi Soc Agric Sci 10:109–120

Zhu XC, Song FB, Xu HW (2010) Influence of arbuscular mycorrhiza on lipid peroxidation and antioxidant enzyme activity of maize plants under temperature stress. Mycorrhiza 20:325–332

Zhu B, Liu H, Tian WX et al (2012) Genome sequence of *Stenotrophomonas maltophilia* RR-10, isolated as an endophyte from rice root. J Bacteriol 194(5):1280–1281

Frankia and Actinorhizal Plants: Symbiotic Nitrogen Fixation

12

Thanh Van Nguyen and Katharina Pawlowski

Abstract

Two types of symbioses are known where nitrogen-fixing soil bacteria induce the formation of special organs, i.e. nodules, on the roots of their dicotyledonous host plants; legume-rhizobia symbioses and actinorhizal symbioses. The later are the symbioses between actinobacteria of the genus *Frankia* and a group of mostly woody plant species from eight families and three different orders (Fagales, Rosales, Cucurbitales). While so far, research has mostly focused on legume-rhizobia symbioses, actinorhizal symbioses with their wider phylogenetic range are more likely to hold the key to understanding the common principles underlying the evolution of an intracellular plant-bacterial symbiosis. In contrast with the unique stem-like structure of legume nodules, actinorhizal nodules are composed of modified lateral roots with infected cells in the expanded cortex. In contrast with rhizobia, *Frankia* strains can protect the oxygen-sensitive nitrogenase enzyme complex, and thus nitrogen fixation, from oxygen. Therefore, oxygen protection systems established in actinorhizal nodules from different host plants involve contributions of both symbiotic partners. In this chapter, structural and developmental features of actinorhizal symbioses are described.

12.1 Introduction

Nitrogen is the element that most often limits plant growth. Biosphere nitrogen is continuously lost to the atmosphere by denitrification and can only be repleted by nitrogen fixation. Only some prokaryotes can form the enzyme complex

T. Van Nguyen • K. Pawlowski (✉)
Department of Ecology, Environment and Plant Sciences, Stockholm University, Stockholm, Sweden
e-mail: katharina.pawlowski@su.se

S. Mehnaz (ed.), *Rhizotrophs: Plant Growth Promotion to Bioremediation*, Microorganisms for Sustainability 2, DOI 10.1007/978-981-10-4862-3_12

nitrogenase that catalyzes the reduction of air dinitrogen to ammonia and thus moves nitrogen from the atmosphere to the biosphere. Several plant species can enter symbioses with nitrogen-fixing soil bacteria. The most efficient of those are root nodule symbioses. Rhizobial symbioses are entered between a polyphyletic group of Gram-negative soil proteobacteria, rhizobia, and species from the legume family as well as from one non-legume genus, *Parasponia* (Cannabaceae). Actinorhizal symbioses are between Gram-positive soil actinobacteria of the genus *Frankia* and 25 genera of dicotyledonous plants, from eight families belonging to three different orders, collectively called actinorhizal plants. All plants able to form root nodule symbioses go back to a common ancestor which is supposed to have acquired a unique predisposition based on which a root nodule could evolve (Soltis et al. 1995). How often root nodule symbioses evolved, and how often the symbiotic capacity was lost, is controversial (Werner et al. 2014).

In these symbioses, the host plants form special organs, the root nodules, upon signal exchange with the microsymbionts. In root nodules, the microsymbionts fix nitrogen while stably internally accommodated within nodule cells and export the products of nitrogen fixation to the host plant, thereby rendering it independent of soil nitrogen sources. With one exception, *Datisca glomerata*, actinorhizal plants are woody shrubs or trees, and actinorhizal nodules are perennial organs. These nodules consist of multiple lobes, each of which represents a modified lateral root without root cap and with infected cells in the expanded cortex.

Actinorhizal nodules were first described by Meyen (1829), but in 1895 it could be shown that these nodules contributed to the plants' nitrogen nutrition (Hiltner 1895). Due to their symbiosis, actinorhizal plants mostly represent pioneer plants and are often used in reforestation or soil reclamation (Diem and Dommergues 1990).

Legume symbioses, in particular those involving the two model species *Medicago truncatula* and *Lotus japonicus*, are the best examined root nodule symbioses (Oldroyd 2013). However, legume root nodules have features that distinguish them from all other root nodule symbioses including that of *Parasponia* species; legume nodules represent stem-like organs with peripheral vascular systems and infected cells in the expanded cortex (Mylona et al. 1995). For identifying the common principles of nitrogen-fixing symbioses, these principles have to be distinguished from plant family-specific characteristics. In this chapter, legume symbioses will be mentioned for comparative purposes.

12.2 The Microsymbionts: *Frankia* Strains

In 1886, J. Brunchorst used the name *Frankia* for the microsymbiont of alder trees to honor his mentor, Swiss biologist A.B. Frank (Quispel 1990). But at that time, both of them thought the microorganism was a fungus. The genus was later reclassified as member of a new family *Frankiaceae*, of the order *Actinomycetales*. Based on their host plants, ten species, i.e., *Frankia alni*, *F. elaeagni*, *F. brunchorstii*, *F. discariae*, *F. casuarinae*, *F. ceanothi*, *F. coriariae*, *F. dryadis*, *F. purshiae*, and *F. cercocarpi*, were assigned (Becking 1970).

Members of the genus *Frankia* have been found on all continents except Antarctica (Dawson 2008). These strains infect host plants in a wide range of climates from glacial bays (Lawrence et al. 1967) to volcanic soils (Burleigh and Dawson 1994). Many studies have shown that *Frankia* can be found not only under host plants (Huss-Danell 1997), *Frankia* strains also occur in the soils with no recent presence (Burleigh and Dawson 1994; Huss-Danell and Frej 1986; Zitzer and Dawson 1992) or devoid of actinorhizal plants (Smolander and Sundman 1987; Maunuksela et al. 1999; Gauthier et al. 2000; Jeong 2001).

Phylogenetically, symbiotic *Frankia* strains can be divided into three main clusters (Normand et al. 1996; Clawson et al. 2004). Strains from *Frankia* cluster I nodulate members of the actinorhizal plant families Betulaceae, Casuarinaceae (with the exception of the genus *Gymnostoma*), and Myricaceae (Normand et al. 1996). *Frankia* cluster III strains nodulate plants from two families of order Rosales, i.e., Rhamnaceae (with the exception of the genus *Ceanothus*) and Elaeagnaceae, and two genera from the order Fagales (*Gymnostoma* and *Morella*) (Huguet et al. 2004). Strains from the cluster II nodulate the broadest range of host plants which belong to four families from two different orders, including Rosaceae and the rhamnaceous genus *Ceanothus* from the Rosales, and Datiscaceae and Coriariaceae from the order Curcubitales (Normand et al. 1996; Vanden Heuvel et al. 2004). This cluster also forms the basal group of the symbiotic *Frankia* clusters (Normand et al. 1996). With one exception, no member of the cluster II could be cultured; the only cultured strain, *Frankia* sp. BMG5.1, in contrast with all other known *Frankia* strains, is alkaliphilic (Gtari et al. 2015).

Cluster IV *Frankia* strains were isolated from nodules but cannot induce nodules or fix nitrogen on their own (Fix^-/Nod., Ramírez-Saad et al. 1998). Presumably, these Fix^-/Nod^- strains colonize the nodule periderm and occasionally escape surface sterilization. Such strains have to be distinguished from Fix^-/Nod^+ strains which can induce ineffective, i.e., non-nitrogen-fixing nodules on certain host plants but cannot fix nitrogen (Baker et al. 1980; Wolters et al. 1997).

Frankia can form three cell types: hyphae, sporangia, and vesicles (Torrey and Callaham 1982). The width of septate hyphae of free-living *Frankia* cells ranges from 0.5 to 1.5 μm. In culture, the hyphae form multiple branches and produce multilocular sporangia (Schwintzer 1990). Under aerobic conditions and nitrogen limitation, vesicles are produced at the tips of growing vegetative hyphae or short side hyphae (Tjepkema et al. 1980; Fontaine et al. 1984). In these vesicles, the oxygen-sensitive nitrogenase enzyme complex is formed and nitrogen fixation takes place (Lechevalier 1994). Under microaerobic conditions and nitrogen limitation, *Frankia* expresses nitrogenase in hyphae (Murry et al. 1985).

All isolated strains of *Frankia* can produce sporangia in culture. Sporangia can be terminal or intercalary. Depending on the strain, the number of spores per sporangium can range from a few to several hundreds. Some strains can form sporangia within nodules; none of those strains could be cultured to date (VandenBosch and Torrey 1985). It was reported that inoculant from nodules that contain spores was much more infective than inoculant from nodules in which *Frankia* does not form spores (Burleight and Torrey 1990).

Table 12.1 List of *Frankia* genomes sequenced

Cluster	Strain	Host*	Size (Mbp)	Accession number
I	ACN14a	*Alnus crispa*	7.49	CT573213
	ACN1[ag]	*Alnus viridis*	7.52	LJPA00000000
	AvcI.1	*Alnus viridis*	7.74	LJFZ00000000
	BMG5.23	*Casuarina glauca*	5.27	JDWE00000000
	CcI3	*Casuarina cunninghamiana*	5.43	CP000249
	CcI6	*Casuarina cunninghamiana*	5.57	AYTZ00000000
	CeD	*Casuarina equisetifolia*	5.00	JPGU00000000.1
	CpI1-S	*Comptonia peregrina*	7.62	JYFN00000000
	CpI1-P	*Comptonia peregrina*	7.61	LJJX00000000
	QA3	*Alnus nitida*	7.59	AJWA00000000
	Thr	*Casuarina cunninghamiana*	5.30	JENI00000000
II	Dg1	*Datisca glomerata*	5.32	CP002801
	BMG5.1	*Coriaria myrtifolia*	5.79	JWIO00000000
III	EAN1pec	*Elaeagnus angustifolia*	8.98	CP000820.1
	BCU110501	*Discaria trinevis*	7.89	ARDT00000000
	BMG5.12	*Elaeagnus* sp.	7.58	ARFH00000000
	R43	Elaeagnaceae	10.45	LFCW00000000
IV	CN3	Nonsymbiotic	9.97	AGJN00000000
	DC12	Nonsymbiotic	6.88	LANG00000000

**Frankia* strains were isolated from nodules of these plants

Analysis of *Frankia* strains was always impeded by the fact that so far, these strains cannot be transformed. Only when genome sequences started to become available in 2007 (Normand et al. 2007), *Frankia's* full biochemical capacities could be assessed. Some features and the references of the currently available genomes of *Frankia* strains are summarized in Table 12.1. The sizes of *Frankia* genomes show an unusual variation; these range from 5.0 Mbp for a *Casuarina*-infective strain (CeD) to 10.45 Mbp for a strain that infects Elaeagnaceae (R43). Phylogenetic analysis has shown that cluster II is basal in the genus (Fig. 12.1; Sen et al. 2014; Gtari et al. 2015; Persson et al. 2015). These strains have genome sizes between 5 and 6 Mbp. Strains belonging to cluster IV, which neither can enter a root nodule symbiosis nor fix nitrogen, have genome sizes between 6.9 and 10 Mbp. Strains in the most derived clusters I and III show different genome sizes: cluster III genomes range between 9 and 10.45 Mbp, while in cluster I, one subgroup (Ic), the *Casuarina*-infective strains, shows strong genome reduction with 5–6 MB while the other subgroup, strains infecting *Alnus* sp. and the Myricaceae (Ia), shows less genome reduction with 7–8 Mbp.

Keeping in mind that the only cultured cluster II strain is alkaliphilic (Gtari et al. 2015), a feature that is unlikely to have evolved after the strain became a root symbiont, this raises the question of whether the precursors of the *Frankia* genus were extremophiles with genomes in the range of 5–6 Mbp, and genome size was

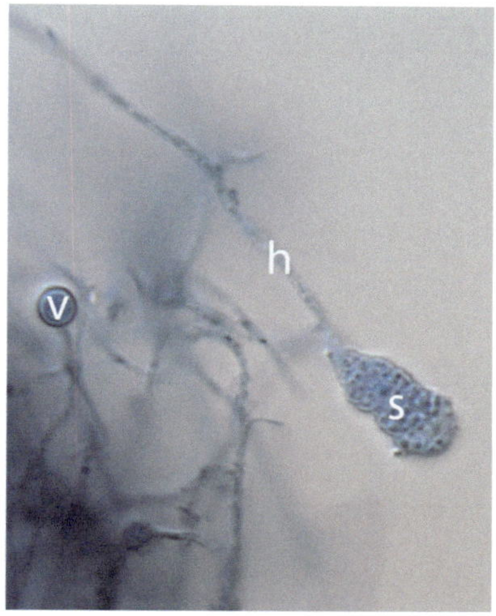

Fig. 12.1 The three cell types of *Frankia*. The photograph shows *Frankia alni* ACN14a grown under normal oxygen tension and nitrogen-limiting conditions, stained with trypan blue. *h* hypha, *v* vesicle, *s* sporangium (The photograph was kindly provided by Anke Sirrenberg (University of Göttingen, Göttingen, Germany))

extended when a subgroup of them adapted to moderate environments (cluster IV) or whether the ancestors of *Frankia* had genomes in the 10 Mbp range and cluster II strains underwent genome reduction. Given that genome reduction in symbiosis is also observed in cluster I, the second hypothesis seems more likely.

12.3 Actinorhizal Plants

More than 200 species of dicotyledonous plants, distributed in 24 genera belonging to eight families from three different orders, can enter a root nodule symbiosis with *Frankia* (Fig. 12.2). Generally, all species of an actinorhizal genus can form root nodules. The only exception is *Dryas*, the ancestral genus in the tribe Dryadoideae, the actinorhizal subgroup of the Rosaceae family, where the species *Dryas octopetala* has never been found nodulated (Uemura 1971; Bond 1976).

Thanks to their symbiosis, actinorhizal plants can grow on marginal soils and have been used in soil reclamation, erosion control, agroforestry, and dune stabilization (Diem and Dommergues 1990). *Hippophae rhamnoides* is currently being domesticated since its fruits are very nutritious, rich in vitamin C and carotenes, and the seed oil is highly unsaturated and has properties that make it a promising ingredient in cosmetics and phytopharmaceuticals (Suryakumar and Gupta 2011).

12.4 Nodule Structure

Root nodules of actinorhizal plants and legumes show some similarities but differ significantly in many respects (Fig. 12.3). Legume nodules represent stem-like organs with peripheral vascular system and infected cells in the central tissue. Legume nodules can be indeterminate or determinate. Indeterminate nodules have a

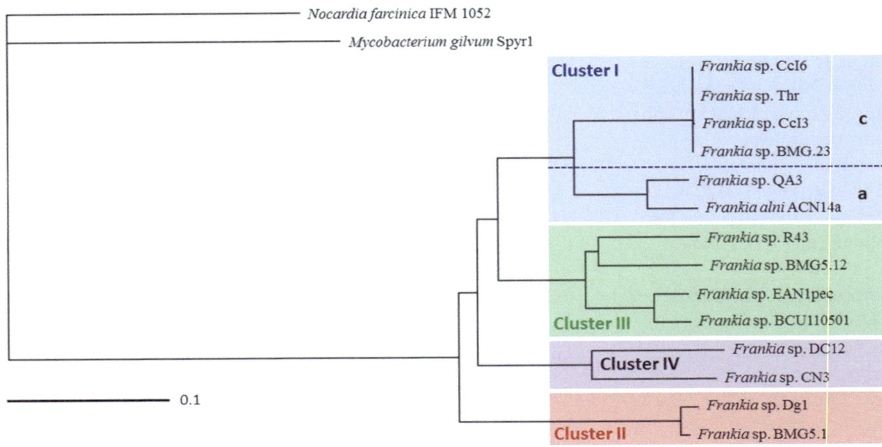

Order	Family	Genus
Fagales	Betulaceae	***Alnus***
	Myricaceae	***Comptonia***
		Morella
		Myrica
	Casuarinaceae	***Allocasuarina***
		Casuarina
		Ceuthostoma
		Gymnostoma
Rosales	Rhamnaceae	***Ceanothus***
		Colletia
		Discaria
		Kentrothamnus
		Retanilla
		Talguenea
		Trevoa
	Elaeagnaceae	***Elaeagnus***
		Hippophae
		Shepherdia
	Rosaceae	***Cercocarpus***
		Chamaebatia
		Cowania
		Dryas
		Purshia
Cucurbitales	Coriariaceae	***Coriaria***
	Datiscaceae	***Datisca***

persistent apical meristem, so the infected cells are arranged in a spatial developmental gradient. In determinate legume nodules, no persistent meristem exists, and the spatial developmental gradient is replaced by a temporal one. In contrast, all nodules of non-legumes – actinorhizal nodules and nodules of *Parasponia* sp. – are coralloid organs consisting of multiple lobes, each lobe representing a modified lateral root with central vascular system and infected cells in the expanded cortex (Pawlowski and Bisseling 1996). Due to their apical meristem of each lobe, the infected cortical cells are arranged in a developmental gradient. Right below the meristem is the infection zone where the cells are becoming filled with branching *Frankia* hyphae in infection thread-like structures. The next zone is the nitrogen fixation zone where in most host plants, *Frankia* vesicles have differentiated and nitrogen is fixed. In mature nodules, a senescence zone is present which contains infected cells with inactive *Frankia* bacteria which are degraded by the plant (Pawlowski and Demchenko 2012).

Actinorhizal nodules show a remarkable anatomic diversity. The distribution of infected cells in the cortex depends on the host plant. In Fagales and Rosales, the infected cortical cells are interspersed with uninfected cortical cells. In nodules of actinorhizal Cucurbitales, however, the infected cells make up an uninterrupted region, kidney shaped in cross section, on one side of the acentric vascular bundle (Newcomb and Pankhurst 1982; Berg et al. 1999).

Frankia morphology in nodules varies significantly, depending on the host. With the exception of nodules of *Casuarina* and *Allocasuarina*, *Frankia* strains fix nitrogen in vesicles within infected cells. The shape, septation, and subcellular position of the vesicles depend on the host plant species, i.e., the same strain can form different types of vesicles in different host plants (Huss-Danell 1997). The alder vesicle style – a septate sphere with a stalk – appears in *Alnus*, the family Elaeagnaceae and some members of the family Rhamnaceae (Berg 1994). This vesicle type has the closest similarity to vesicles formed in the free-living state. In alder nodules, vesicles are located at the periphery of the infected cortical cells (Lalonde and Knowles 1975). Vesicles in *Ceanothus* sp. are nonseptate, pear shaped, and have no stalk

Fig. 12.2 Phylogeny and host specificity of *Frankia*. Comparison of core genomes of 14 sequenced *Frankia* strains (Normand et al. 2007; Persson et al. 2011; Ghodhbane-Gtari et al. 2013; Nouioui et al. 2013; Sen et al. 2013; Wall et al. 2013; Ghodbhane-Gtari et al. 2014; Hurst IV et al. 2014; Mansour et al. 2014; Gtari et al. 2015; Pujic et al. 2015; Tisa et al. 2015) using EDGAR (Blom et al. 2009). Outgroups were two actinobacterial genomes: *Nocardia farcinica* (Ishikawa et al. 2004) and *Mycobacterium gilvum* Spyr1 (Kallimanis et al. 2011). The phylogenetic tree was deduced from concatenated core gene alignments using PHYLIP (Felsenstein 2005). The bar below the phylogenetic tree represents the scale of sequence divergence. The phylogenetic tree was kindly provided by Daniel Wibberg (University of Bielefeld, Germany) and Jochen Blom (Justus Liebig University, Gießen, Germany). Host specificity is indicated in the table. Genera the members of which can enter symbioses with cluster I strains are depicted in *blue*. Hosts of cluster II strains are depicted in *red*, and hosts of cluster III strains are given in *green*. Strains of cluster IV are not able to induce root nodules

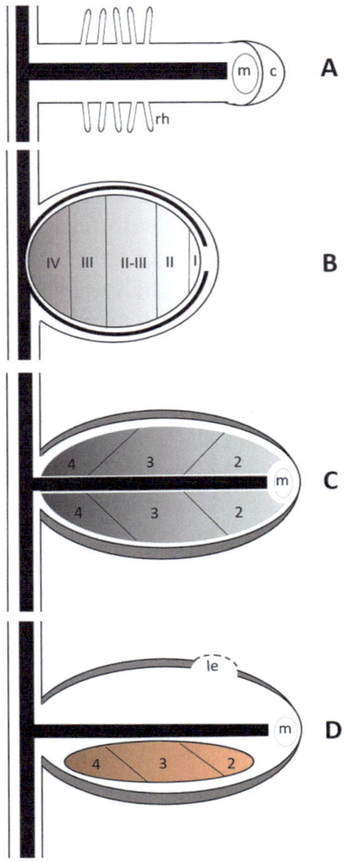

Fig. 12.3 Comparison of longitudinal sections of different lateral root organs. Panel (**a**) shows a standard lateral root with root hairs (*rh*), apical meristem (*m*), and calyptra (*c*). The vascular system is depicted in *black*. Panel (**b**) shows an indeterminate legume nodule with peripheral vascular system and infected cells in the inner tissue. Due to the activity of the apical meristem (*I*), the cells of the inner tissue (*shaded in gray*) are arranged in a spatial developmental gradient (Vasse et al. 1990): The prefixation zone (*II*) is followed by the interzone (*II–III*) where nitrogen fixation commences. Zone III is the nitrogen fixation zone and Zone IV is the zone of senescence. Panel (**c**) shows a lobe of an actinorhizal nodule formed by a member of the Fagales or Rosales. Like a lateral root, the lobe has a central vascular bundle, but it is surrounded by a superficial periderm given in *dark gray*. The infected cells are located in the expanded cortex (*shaded in gray*), interspersed with uninfected cells. Due to the activity of the apical meristem (*m*), they are arranged in a developmental gradient: (*2*) zone of infection, (*3*) zone of nitrogen fixation, (*4*) zone of senescence. Panel (**d**) shows a nodule formed by a member of the Cucurbitales. The infected cells (*shaded in orange*) form a continuous section in the cortex, kidney shaped in cross section, on one side of the acentric stele, and are not interspersed with uninfected cells. Again, the activity of the apical meristem (*m*) leads to a spatial developmental gradient of infected cells: (*2*) zone of infection, (*3*) zone of nitrogen fixation, and (*4*) zone of senescence. The periderm can be interrupted by lenticels (*le*) which are always located opposite to the side of the infected cells

(Berg 1994). Like the alder-type, these vesicles are also formed at the periphery of the host cell. In *Morella cerifera*, the vesicles are septate, elongated, and club shaped (Berg 1994). In nodules of actinorhizal Cucurbitales, the vesicles are rod shaped, arranged in radial orientation, and form a sphere around the central vacuole of the infected cell (Newcomb and Pankhurst 1982; Berg et al. 1999). *Casuarina* sp. and *Allocasuarina* sp. are unique among host plant genera in that *Frankia* does not form vesicles in root nodules even though the corresponding *Frankia* strains are capable of doing so when grown in culture (Berg and McDowell 1987).

12.5 Nodule Physiology

On the whole-plant level, nodules represent carbon sinks and nitrogen sources. Nodules need assimilated carbon for growth and maintenance, for supporting bacterial N_2 fixation, for ammonium assimilation and transport of nitrogenous solutes, and for starch biosynthesis. Uninfected cortical cells of actinorhizal nodules tend to contain large starch grains, but the function of nodule starch is not known. Nodules are supplied with sucrose via the phloem.

Legume nodule metabolism, and the exchange of metabolites between host plants and microsymbionts, has been analyzed in detail (reviewed by Udvardi and Poole 2013). In legume nodules, sucrose from the phloem is cleaved by sucrose synthases and metabolized further via the glycolytic pathway. The role of sucrose synthase could be confirmed for *Alnus glutinosa* (Van Ghelue et al. 1996), *Casuarina glauca* (Schubert et al. 2013), and *Datisca glomerata* (Schubert et al. 2011). Rhizobial bacteroids are supplied with carbon sources in the form of dicarboxylates, especially malate. This also seems to be the case for *Frankia* in actinorhizal Fagales, since a dicarboxylate transporter has been identified in the perisymbiont membrane of nodules of *A. glutinosa* (Jeong et al. 2004).

In legume nodules, rhizobial bacteroids fix nitrogen but avoid assimilation of the resulting ammonia. Instead, ammonia is exported and assimilated in the cytosol of the infected cells, mostly by the glutamine synthetase (GS)/glutamate synthase (GOGAT) pathway and also by aspartate amino transferase (Udvardi and Poole 2013). The cytosolic assimilation of ammonia in infected cells could be confirmed for actinorhizal Fagales, i.e., *A. glutinosa* (Guan et al. 1996). However, the situation in actinorhizal Cucurbitales – i.e., *D. glomerata* – is different in that the ammonia resulting from nitrogen fixation is assimilated by *Frankia*, and an assimilated form of nitrogen – most likely arginine – is exported to the plant cytosol (Berry et al. 2004, 2011).

In all root nodule symbioses, assimilated ammonium is exported from the nodule via the xylem. In actinorhizal symbioses, the transport forms are amino acids (glutamate, glutamine, aspartate, asparagine) or ureides (citrulline, arginine) (Schubert 1986; Guan et al. 1996; Persson et al. 2016).

12.6 Oxygen Protection Mechanisms

The process of nitrogen fixation is oxygen sensitive in that nitrogenase is rapidly irreversibly denatured by oxygen (Gallon 1981). However, the process of nitrogen fixation requires high amounts of energy in the form of ATP, preferably provided by aerobic respiration. This causes the so-called oxygen dilemma of nitrogen fixation. The microsymbionts of legume-rhizobia symbioses rely on their hosts to solve this problem, which is achieved by providing a microaerobic environment in the infected cells while providing an oxygen-binding protein, leghemoglobin, to achieve efficient supply of oxygen to the respiratory chain (Minchin 1997). On the other hand, *Frankia* can fix nitrogen under microaerobic conditions in the free-living state by expressing nitrogenase in vesicles (Parsons et al. 1987). Vesicles have been confirmed as the site of nitrogen fixation by (i) showing that acetylene reduction activity occurs within the vesicles developed in culture and effective nodules (Tjepkema et al. 1980) with the nitrogenase activity of vesicle fraction was found to be 100-fold higher than that of the hyphae (Noridge and Benson 1986) and (ii) immunolabeling of bacterial cryosections proved that the localization of nitrogenase is restricted to the vesicles (Meesters et al. 1987). Vesicles are surrounded by envelopes consisting of multiple layers of hopanoids, bacterial steroid lipids (Berry et al. 1993). Since the number of the layers of the vesicle envelope increases with the external oxygen tension, it was concluded that the vesicle envelop forms an oxygen diffusion barrier (Parsons et al. 1987).

In a nitrogen-free culture grown under aerobic conditions, the hyphal termini can swell and form structures called provesicles (Fontaine et al. 1984). Provesicles are spherical and do not show nitrogenase activity. Subsequently, provesicles become mature vesicles, where nitrogenase is formed and nitrogen fixation occurs. Nitrogenase performance peaks after 5–6 days. Then the vesicles lose their function and become ghosts with deformed structure and empty appearance (Fontaine et al. 1984).

In nodules of *(Allo-)Casuarina* sp., the plant provides an oxygen protection system for the bacterial nitrogenase complex similar to the situation in legume nodules. The infected cells in nodules of *Casuarina glauca* have walls impregnated with a special lignin which provides an oxygen diffusion barrier, leading to microaerobic conditions in the infected cells (Berg and McDowell 1988; Schubert et al. 2013). Infected cells contain large amounts of a class II hemoglobin just like legume nodules (Jacobsen-Lyon et al. 1995). The problem is that in legume nodules, vascular system which needs energy for loading and unloading process is peripheral, so it is possible to place an oxygen diffusion barrier between the vascular system and the central tissue containing the infected cells. Since actinorhizal nodules have a central vascular system oxygen access to which should not be impaired, each infected cell needs its own oxygen diffusion barrier. The lignin in the primary walls of infected cells *C. glauca* nodules causes the apoplastic isolation of infected cells and also affects plasmodesmata, thereby gradually interfering with symplastic transport (Schubert et al. 2013).

In nodules of *Alnus* sp., it could be shown that the number of vesicle layers increases with the external oxygen tension and that the hopanoid composition is dependent on the depth of soil where *Frankia* is growing (Kleemann et al. 1994; Nalin et al. 2000). Hence, it seems that here, the bacterial oxygen protection system of nitrogenase is used. In nodules of *Datisca glomerata*, vesicle envelopes are thin but the positioning of the vesicles in the infected cells insures minimal oxygen access (Berg et al. 1999). Furthermore, the nodule periderm forms an oxygen diffusion barrier, and lenticels are always present at the side of the acentric vascular bundle, not at the side of the infected cells (Fig. 12.3). The presence of a bacterial hemoglobin could also contribute to shuttle oxygen to the sites of respiration (Pawlowski et al. 2007). The thickness of the suberized periderm that surrounds *Coriaria* nodules increases at elevated O_2 concentrations (Silvester and Harris 1989). A single, large lenticel on the uninfected side of the nodule lobe limits the gas diffusion pathway to the infected cells to the narrow gap between the inner periderm and the steel (Silvester and Harris 1989).

In summary, in actinorhizal systems, both symbiotic partners can contribute to oxygen protection of nitrogenase. In *Alnus* sp., the bacterial contribution dominates, while in *(Allo-)Casuarina* sp., the plant contribution dominates. Mixed contributions are used in actinorhizal Cucurbitales.

Gas (oxygen/nitrogen) access to nodules can become limiting when the host plants are growing in wetlands. In well-drained soils, nodules of *Alnus* sp. are well aerated since their periderm is interrupted by lenticels and their outer cortex contains large intercellular spaces (Wheeler et al. 1979). In waterlogged soil, gas is transported thermo-osmotically from the aerial parts to the roots (Schröder 1989). Other actinorhizal plants growing in wet or waterlogged soils have developed a special mechanism for gas transport to their nodules: species of *Casuarina*, *Gymnostoma*, *Myrica*, and *Comptonia* provide oxygen to nodules via air spaces in the so-called nodule roots (Silvester et al. 1990). Nodule roots are formed at the tips of nodule lobes (the nodule lobe meristem turns into a nodule root meristem) and grow upward; their length is negatively correlated with the aeration of the growth substrate (Tjepkema 1978). Nodules of *Datisca cannabina* form nodule roots (Silvester et al. 1990), while nodules of *D. glomerata* form nodule roots in hydroculture or waterlogged soil and lenticels in well-drained soil (Pawlowski and Demchenko 2012).

12.7 Nodule Induction

In actinorhizal as in legume symbioses, the infection pathway is determined by the host plant (Miller and Baker 1985; Racette and Torrey 1989). For actinorhizal symbioses, two ways have been described for *Frankia* to enter the plant roots, intracellular via root hairs in Fagales or intercellular by penetration between epidermal cells in Rosales (Table 12.2). The infection pathway of Cucurbitales has not been analyzed yet.

Table 12.2 Mechanisms of root nodule induction by *Frankia*

	Fabales	Fagales	Rosales	Cucurbitales
Infection mechanism	Intracellular (most common) Crack entry (rare)	Intracelluar	Intercellular	Unknown
Root hair deformation	Yes (with intracellular infection), host specific	Yes, not host specific	No	No
Bacteria enter the root	Infection threads (most common) Crack entry exploiting gaps in the epidermis (rare)	Infection thread-like structures	Dissolution of the middle lamella between adjacent epidermal cells	Unknown
Induction of cortical cell divisions	Yes	Yes	No[a]	No
Bacteria colonize the root/nodule primodium/ nodule via	Transcellular infection thread growth (most common) Intercellularly (rare)	Transcellular infection thread growth	Intercellularly: *Frankia* colonizes the apoplast and infects new cells from the apoplast	Transcellular infection thread growth but mechanism differs from that in Fagales and Fabales
Stable internal accommodation of bacteria in plant cells	Symbiosomes (most common) Branching fixation threads (rare)	Branching fixation threads	Branching fixation threads	Branching fixation threads

The mechanisms employed by rhizobia in legumes (Fabales) have been included for comparison
[a]Induction of cortical cell divisions has been described for *Ceanothus* sp., but in contrast with prenodule cells of Fagales, these cells were not infected by *Frankia* (Berry and Sunell 1990)

12.7.1 Intracellular Infection via Root Hairs in the Fagales

This process is quite similar to the root hair infection process described for the model legumes *M. truncatula* and *L. japonicus* (Oldroyd 2013). The first response of the plant to the presence of the microsymbiont is the deformation and branching of growing root hairs (Torrey 1976; Callaham and Torrey 1977; Callaham et al. 1979; Berry et al. 1986). Only in a few root hairs, a *Frankia* hypha is entrapped in a root hair curl, and an infection thread-like structure is formed by dissolution of the cell wall and invagination of the root hair plasma membrane. Within this infection thread-like structure, the hypha is embedded in a plant-derived cell wall-like pectin-rich matrix, the so-called encapsulation (Lalonde and Knowles 1975; Callaham et al. 1979; Berry and Torrey 1983; Berry et al. 1986; Berg 1990). These actinorhizal infection thread-like structures have a smaller diameter than infection threads in legume nodules since they contain only one hypha.

The transcellular growth of infection threads in actinorhizal Fagales resembles infection thread growth in legume nodules: in both cases, before an infection thread crosses a cortical cell, a so-called preinfection thread (PIT) is formed in that cell (Berg 1999a). During this process, the nucleus moves to the center of the cell, and microtubules and cytoplasm rearrange to form a phragmoplast-like structure (van Brussel et al. 1992). These structures are polarized; most of the cytoplasm as well as the endomembranes are located at the outer side. This polarization of the cytoplasm is required for tip growth; root hairs, pollen tubes, and infection threads are the only plant structures showing tip growth (Van Brussel et al. 1992).

Concomitantly with the formation of an infection thread-like structure in a root hair, the formation of the so-called prenodule is initiated by cell divisions in the root cortex close to the infected root hair. The infection thread-like structures grow toward the prenodule by cell-to-cell passage and infect some, but not all, prenodule cells by extensive branching within these cells, filling them from the center outward (Schwintzer et al. 1982). This process – the branching of infection threads – does not involve PITs. Infected prenodule cells become hypertrophic, while uninfected prenodule cells accumulate starch (Callaham and Torrey 1977). *Frankia* can fix nitrogen in infected prenodule cells (Angulo Carmona 1974; Laplaze et al. 2000). Studies on *C. glauca* prenodules have shown that these structures represent primitive symbiotic organs consisting of three cell types with unique differentiation features equivalent to their counterparts in mature nodule lobes. These cell types are (1) infected cells harboring *Frankia*, (2) uninfected cells showing the same features as uninfected cortical cells in the mature nodule lobe while differing from root cortical cells (Laplaze et al. 2000), and (3) polyphenol-containing cells with gene expression pattern of which resembles those of polyphenol-containing cortical cells of mature nodule lobes while differing from that of polyphenol-containing root cortical cells (Smouni et al. 2002).

Nevertheless, the prenodule is only an intermediate stage in Fagales nodule development. While the prenodule developing, the formation of the nodule lobe primordium is initiated in the root pericycle near the infection site, opposite to a protoxylem pole, and *Frankia* hyphae in infection thread-like structures grow from the prenodule to the nodule primordium, again by cell-to-cell passage (transcellularly) and infect primordium cells.

Infection does not always lead to an effective symbiosis. Some strains can induce nitrogen-fixing nodules on one plant species but only ineffective (i.e., non-nitrogen fixing) nodules on another: these strains are incompatible with the second species (VandenBosch and Torrey 1983).

12.7.2 Intercellular Infection in the Rosales

During intercellular infection, *Frankia* hyphae penetrate the middle lamella between adjacent cells of the root epidermis and progressively colonize the intercellular spaces of the root cortex (Miller and Baker 1985; Racette and Torrey 1989; Berry and Sunell 1990; Liu and Berry 1991a, 1991b; Valverde and Wall 1999). Epidermal and cortical cells secrete pectin-rich material into apoplast; this material is likely to represent the equivalent of the cell wall-like material encapsulating *Frankia* hyphae in infection thread-like structures formed during intracellular infection (Liu and Berry 1991b). Concomitantly, the formation of a nodule lobe primordium is initiated in the root pericycle, and *Frankia* hyphae infect primordium cells from the apoplast. During this process, the plasma membrane of the infected cells invaginates, and the hyphae are embedded in cell wall-like material as infection thread-like structures. In the Rosales, infection threads do not show transcellular growth, and no PIT formation is observed.

A comparison between the infection mechanisms in Fagales and Rosales shows that there are two types of infection thread-like structures: those that show transcellular growth connected with PIT formation and those that do not show transcellular growth and whose growth does not involve PIT formation. Berg (1999a, 1991b) coined the terms "invasive hyphae" for infection thread-like structures showing transcellular growth and "vegetative hyphae" for the others. Intracellular infection involves both types of infection thread-like structures, while intercellular infection involves only the "vegetative hyphae."

12.7.3 Unknown Infection Mechanism in the Cucurbitales

As mentioned above, the infection mechanism of actinorhizal Cucurbitales has not yet been analyzed. However, detailed cytological studies of mature nodules have been performed (Newcomb and Pankhurst 1982; Hafeez et al. 1984; Mirza et al. 1994; Berg et al. 1999). The absence of prenodules would lead to the assumption that Cucurbitales are infected intercellularly. Yet, infection threads show transcellular growth (Berg et al. 1999); this transcellular growth, however, did not involve the formation of PITs (Berg et al. 1999). Furthermore, in actinorhizal Cucurbitales, infected cells are filled with branching infection thread-like structures from the periphery inward and a large central vacuole is retained, while in actinorhizal Fagales as well as Rosales, the central vacuole is fragmented during infection (Berg et al. 1999; Pawlowski and Demchenko 2012). Altogether, the infection thread growth mechanism in actinorhizal Cucurbitales is clearly different from that in actinorhizal Fagales, in legumes, and from that in actinorhizal Rosales.

12.8 Signal Exchange Between Microsymbiont and Host Plant

The signal exchange between microsymbionts and host plant has been studied extensively in legume-rhizobia symbioses. Flavonoids from the host plant root exudate bind the rhizobial NodD protein, a transcriptional activator. In consequence, NodD activates the transcription of a number of nodulation (*nod, nol, noe*) genes that are required for the synthesis of the bacterial signal molecules, lipochitooligosaccharide (LCO) Nod factors, which when perceived by plant receptors cause changes in the roots. The basic structure of Nod factors consists of a backbone of β-1,4-linked N-acetyl glucosamines carrying a fatty acid on the nonreducing end (Mylona et al. 1995). This basic structure is synthesized by the canonical Nod proteins NodA, NodB, and NodC. NodC is a chitin synthase and NodB an oligosaccharide deacetylase; both represent subfamilies of bacterial chitin synthases and deacetylases, respectively. NodA represents an acyl transferase that attaches a fatty acid to the deacetylated sugar residue and was considered unique to rhizobia (Atkinson et al. 1994). Nod factors differ with regard to the polymerization degree of the chito-oligosaccharide, the fatty acid, and the type of substitutions at both ends of the chitin oligomer; many different Nod (Nol, Noe) proteins are responsible for these individual modifications (Mergaert et al. 1997).

It has long been known that Nod factor signal transduction had recruited modules from arbuscular mycorrhizal (AM) fungi signaling systems (Markmann and Parniske 2009). Signaling of both rhizobia and AM fungi occurs via the common symbiotic signaling pathway (CSSP; Fig. 12.4). Nod factors bind to heterodimeric LysM

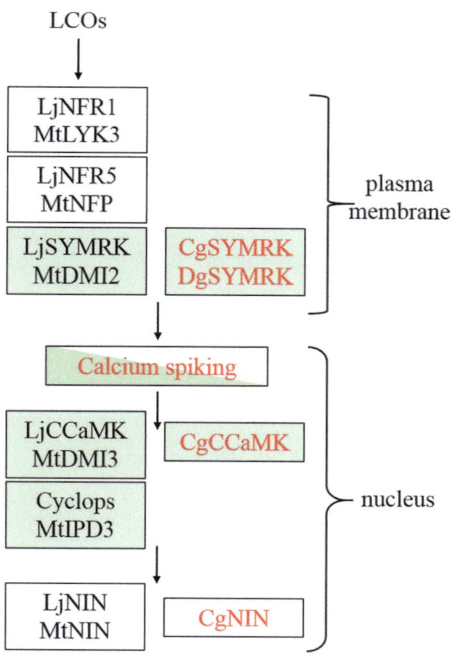

Fig. 12.4 The common symbiotic signal transduction pathway. The *left* column shows the components known for legume-rhizobia symbioses, using the nomenclature for *L. japonicus* (*top*) and *M. truncatula* (*bottom*). All components which also participate in the Myc factor signal transduction pathway are highlighted in *green*. All components the function of which has been demonstrated for an actinorhizal symbiosis are given in *red*. Using *L. japonica* nomenclature, rhizobial LCO Nod factors bind to the LysM receptor kinases NFR1/NFR5 in the plasma membrane, which signal to SymRK. Signal transduction leads to nuclear Ca^{2+} spiking. The same happens during LCO Myc factor signaling, but the Ca^{2+} signature is different. In the nucleus, the Ca^{2+} signature is read by CCaMK. When the signature was caused by Nod factors, CCaMK phosphorylates Cyclops, and the CCaMK/Cyclops complex activates the transcription of *NIN*

receptor kinases in the plant root plasma membrane, Nod factor receptor 1 (NFR1) and Nod factor receptor 5 (NFR5) in *Lotus japonicus* and LysM receptor kinase 3 (LYK3) and Nod factor perception (NFP) in *Medicago truncatula* (Limpens et al. 2003; Madsen et al. 2003). These receptors evolved from chitin receptors (Zhang et al. 2009). The recognition of the Nod factor also involves a plasma membrane receptor kinase (LjSYMRK/MtDMI2/MsNORK; Endre et al. 2002; Stracke et al. 2002) which is also required for the signal exchange between AM fungi and plant roots. The signal transduction pathway leads to nuclear calcium spiking read by a complex of calcium/calmodulin-dependent kinase (LjCCaMK/MtDMI3) (Lévy et al. 2004) and the transcription factor LjCyclops/MtIPD3 (Yano et al. 2008; Horváth et al. 2011). This complex activates the transcription of *nodule inception* (*LjNIN/MtNIN*; Marsh et al. 2007; Singh et al. 2014) which encodes a transcription factor which will activate the expression of further transcription factors (Oldroyd 2013).

All components between, and including, SymRK and Cyclops are shared between arbuscular mycorrhizal and rhizobial signaling: these components form the CSSP. The signal factors of AM fungi are LCOs (Maillet et al. 2011) or chito-oligosaccharides (COs; Sun et al. 2015) which, like rhizobial Nod factors, are also perceived by LysM receptor kinases. However, with the exception of *Parasponia andersonii*, no overlap was found between Nod factor-binding and Myc factor-binding LysM receptor kinases (Op den Camp et al. 2011).

Studies using RNAi in chimeric plants with transgenic root systems have shown for one actinorhizal member of the Cucurbitales (*D. glomerata*) and one of the Fagales (*C. glauca*) that this pathway is also involved in the communication between *Frankia* and their host plants (Gherbi et al. 2008; Markmann et al. 2008). RNAi studies have confirmed this further by showing that a CCaMK/DMI3 homolog is part of symbiotic signal transduction in *C. glauca* (Svistoonoff et al. 2013), and Granqvist et al. (2015) showed that culture supernatants of a homologous *Frankia* strain led to the induction of Ca^{2+} spiking in *Alnus glutinosa* root hairs. Furthermore, *NIN* has a similar function in *C. glauca* as in legumes (Clavijo et al. 2015). Hence, unsurprisingly, also *Frankia* signals via the CSSP.

12.8.1 *Frankia* Signal Factors

In the context of symbiotic signaling via the CSSP, one would expect *Frankia* signal molecules to represent LCOs since LCOs are used by rhizobia as well as by AM fungi. However, genomes of cluster I and cluster III *Frankia* strains do not contain the canonical *nod* genes *nodABC*.

The commonly used bioassay for rhizobial Nod factors is based on the induction of root hair deformation. During legume nodulation, a rhizobium attached to a growing root hair and producing Nod factors will cause the reorientation of root hair growth leading to the formation of a so-called shepherd's crook which entraps the rhizobium. This will lead to the formation of an infection thread (Esseling et al. 2003). Purified Nod factors first block root hair extension and then re-induce it in a random position of the root hair, thereby leading to the formation of deformed root hairs (Heidstra et al. 1994). However, this bioassay could not be applied to *Frankia* signal factors. Knowlton et al. (1980) showed that also non-*Frankia* bacteria could induce root hair deformation on actinorhizal plants. Similarly, Van Ghelue et al. (1997) showed that *Frankia* strains that could induce root hair deformation on an actinorhizal plant could not necessarily nodulate this plant species. Cérémonie et al. (1999) tried to use a root hair deformation assay to isolate the signal factors of *Frankia* sp. ACoN24d Nod factors and could demonstrate that the compound(s) inducing *A. glutinosa* root hair deformation did not share the solubility features of rhizobial Nod factors. Recent data suggest that *Frankia* sp. CcI3 can produce hydrophilic and chitinase-resistant molecules that trigger Ca^{2+} spiking and activate the *NIN* promoter in its host plant *C. glauca* (Chabaud et al. 2016).

LysM receptor kinases not only recognize chitin and LCOs but also peptidoglycan (Willmann et al. 2011) and exopolysaccharides (Kawaharada et al. 2015). Hence, it seems plausible that *Frankia* cluster I and cluster III strains use signal molecules other than LCOs and that these signal molecules are recognized by LysM receptor kinases; however, the nature of these signal molecules remains to be examined.

There is evidence that the signals of cluster I strains can be imitated by other organisms. The fungus *Penicillium nodositatum* can induce so-called myconodules on the alder roots (Sequerra et al. 1994, 1995). Structurally, myconodules resemble ineffective, i.e., non-nitrogen-fixing actinorhizal nodules which typically remain mostly single lobes and contain large amounts of polyphenols. As in ineffective nodules, the infection of the host plant by the fungus does not elicit a strong resistance response (Sequerra et al. 1994, 1995). The nodule induction process used by the fungus – intracellular infection via root hairs – resembles that employed by *Frankia* (Sequerra et al. 1994), indicating that *P. nodositatum* produces compounds that activate symbiotic signaling.

Interestingly, the first genome of a *Frankia* cluster II strain to be sequenced, *Candidatus* Frankia datiscae Dg1, contained the canonical *nod* genes *nodABC* which were expressed in symbiosis (Persson et al. 2011, 2015). A series of rather diverse *nodA* homologs was identified in the phylum Actinobacteria, while *nodA* genes are otherwise only present in rhizobia (alpha- and beta-Proteobacteria) where *nodA* displayed much less sequence diversity than in Actinobacteria. These new data suggest that *nodA* evolved in Actinobacteria and was laterally transferred to rhizobia (Persson et al. 2015). Yet, the question whether *Frankia* cluster II strains actually use LCO Nod factors to nodulate their host plants is still open.

12.9 Concluding Remarks

Over the last decade, evidence has emerged about the similarities between legume and actinorhizal symbioses in that both symbioses involve bacterial signaling via the CSSP which leads to the formation of the first dedicated transcription factor in root nodule organogenesis, NIN. Suzuki et al. (2013) have postulated that the duplication of transcription factors involved in the response to nitrate (NIN-like proteins, NLPs), which yielded NIN, and NIN's subsequent loss of nitrate sensitivity was one of the events necessary for the evolution of symbiotic nitrogen fixation in legumes. Soyano et al. (2015) could confirm that in *L. japonicus*, NLPs and NIN indeed act antagonistically. Since NIN is also involved in the formation of actinorhizal nodules (*C. glauca*, Casuarinaceae, Fagales; Clavijo et al. 2015), its evolution must have preceded the separation of Fabales and Fagales and might represent the common predisposition acquired by the progenitor of the symbiotic clade (Soltis et al. 1995). Yet, NIN is not the only transcription factor specific to root nodule organogenesis, and much has to be learned about the network of transcription factors in both legumes and actinorhizal plants before the different steps in the evolution of both symbioses are understood.

Apart from the dissimilarity of the microsymbionts, there are two striking differences between actinorhizal and legume-rhizobia symbioses. The former have a much wider phylogenetic range, while the latter involve far more plant species – legumes are a very diverse family with ca. 20,000 species (Doyle and Luckow 2003), thanks to a burst of speciation ca. 60–50 million years ago (Lavin et al. 2005), i.e., probably after evolving a root nodule symbiosis. Why did no such burst of speciation occur in any actinorhizal symbiosis? One explanation could be that with one exception (*Datisca glomerata*), actinorhizal plants are woody, and net diversification rates are higher for herbaceous annuals than for woody perennials (Soltis et al. 2013). However, plant growth form can change in the course of evolution (Beaulieu et al. 2013). So the question remains why actinorhizal plants did not evolve to leave their ecological niches although this often seems to have happened in legume evolution.

References

Angulo Carmona AF (1974) La formation des nodules fixateurs d'azote chez *Alnus glutinosa* (L.) Acta Bot Neerl 23:257–303. doi:10.1111/j.1438-8677.1974.tb00944.x

Atkinson EM, Palcic MM, Hindsgaul O et al (1994) Biosynthesis of *Rhizobium meliloti* lipo oligosaccharide Nod factors: NodA is required for an N-acyltransferase activity. Proc Natl Acad Sci U S A 91:8418–8422

Baker D, Newcomb W, Torrey JG (1980) Characterization of an ineffective actinorhizal microsymbiont, *Frankia* sp. EuI1 (Actinomycetales). Can J Microbiol 26:1072–1089. doi:10.1139/m80-180

Beaulieu JM, O'Meara BC, Donoghue MJ (2013) Identifying hidden rate changes in the evolution of a binary morphological character: the evolution of plant habit in campanulid angiosperms. Syst Biol 62:725–737. doi:10.1093/sysbio/syt034

Becking JH (1970) *Frankiaceae* fam. nov. (Actinomycetales) with one new combination and six new species of the genus *Frankia* Brunchorst 1886, 174. Int J Syst Bacteriol 20:201–220. doi:10.1099/00207713-20-2-201

Berg RH (1990) Cellulose and xylans in the interface capsule in symbiotic cells of actinorhizae. Protoplasma 159:35–43. doi:10.1007/BF01326633

Berg RH (1994) Symbiotic vesicle ultrastructure in high pressure-frozen, freeze-substituted actinorhizae. Protoplasma 183:37–48. doi:10.1007/BF01276811

Berg RH (1999a) Cytoplasmic bridge formation in the nodule apex of actinorhizal root nodules. Can J Bot 77:1351–1357. doi:10.1139/b99-078

Berg RH (1999b) *Frankia* forms infection threads. Can J Bot 77:1327–1333

Berg RH, McDowell L (1987) Endophyte differentiation in *Casuarina* Actinorhizae. Protoplasma 136:104–117. doi:10.1007/BF01276359

Berg RH, McDowell L (1988) Cytochemistry of the wall of infected cells in *Casuarina actinorhizae*. Can J Bot 66:2038–2047. doi:10.1139/b88-279

Berg RH, Langenstein B, Silvester WB (1999) Development in the *Datisca-Coriaria* nodule type. Can J Bot 77:1334–1350. doi:10.1139/b99-076

Berry AM, Sunell LA (1990) The infection process and nodule development. In: Schwintzer CR, Tjepkema JD (eds) The biology of *Frankia* and actinorhizal plants. Academic, San Diego, pp 61–81

Berry AM, Torrey JG (1983) Root hair deformation in the infection process of *Alnus rubra*. Can J Bot 61:2863–2876. doi:10.1139/b83-319

Berry AM, McIntyre L, McCully ME (1986) Fine structure of root hair infection leading to nodulation in the *Frankia-Alnus* symbiosis. Can J Bot 64:292–305. doi:10.1139/b86-043

Berry AM, Harriott OT, Moreau RA et al (1993) Hopanoid lipids compose the *Frankia vesicle* envelop, presumptive barrier of oxygen diffusion to nitrogenase. Proc Natl Acad Sci USA 90:6091–6094. doi:10.1073/pnas.90.13.6091

Berry AM, Murphy TM, Okubara PA et al (2004) Novel expression pattern of cytosolic Gln synthetase in nitrogen-fixing root nodules of the actinorhizal host, *Datisca glomerata*. Plant Physiol 135:1849–1862. doi:10.1104/pp.103.031534

Berry AM, Mendoza-Herrera A, Guo Y-Y et al (2011) New perspectives on nodule nitrogen assimilation in actinorhizal symbioses. Funct Plant Biol 38:645–652. doi:10.1071/FP11095

Blom J, Albaum SP, Doppmeier D et al (2009) EDGAR: A software framework for the comparative analysis of prokaryotic genomes. BMC Bioinforma 10:154. doi:10.1186/1471-2105-10-154

Bond G (1976) The results of the IBP survey of root nodule formation in non-leguminous angiosperms. In: Nutman PS (ed) Symbiotic nitrogen fixation in plants. Cambridge University Press, London, pp 443–474

Burleigh SH, Dawson JO (1994) Occurrence of *Myrica*-nodulating *Frankia* in Hawaiian volcanic soils. Plant Soil 164:283–289. doi:10.1007/BF00010080

Burleight S, Torrey JG (1990) Effectiveness of different *Frankia* cell types as inocula for the actinorhizal plant *Casuarina*. Appl Environ Microbiol 8:2565–2567

Callaham D, Torrey JG (1977) Prenodule formation and primary nodule development in roots of *Comptonia* (Myricaceae). Can J Bot 51:2306–2318

Callaham D, Newcomb W, Torrey JG et al (1979) Root hair infection in actinomycete-induced root nodule initiation in *Casuarina, Myrica,* and *Comptonia*. Bot Gaz (Suppl) 140:S1–S9

Cérémonie H, Debelle F, Fernandez MP (1999) Structural and functional comparison of *Frankia* root hair deforming factor and rhizobia Nod factor. Can J Bot 77:1293–1301. doi:10.1139/b99-060

Chabaud M, Gherbi H, Pirolles E et al (2016) Chitinase-resistant hydrophilic symbiotic factors secreted by *Frankia* activate both Ca^{2+} spiking and *NIN* gene expression in the actinorhizal plant *Casuarina glauca*. New Phytol 209:86–93. doi:10.1111/nph.13732

Clavijo F, Diedhiou I, Vaissayre V et al (2015) The *Casuarina NIN* gene is transcriptionally activated throughout *Frankia* root infection as well as in response to bacterial diffusible signals. New Phytol 208:887–903. doi:10.1111/nph.13506

Clawson ML, Bourret A, Benson DR (2004) Assessing the phylogeny of *Frankia* – actinorhizal plant nitrogen-fixing root nodule symbioses with *Frankia* 16S rRNA and glutamine synthetase gene sequences. Mol Phylogenet Evol 31:131–138. doi:10.1016/j.ympev.2003.08.001

Dawson JO (2008) Ecology of actinorhizal plants. In: Pawlowski K, Newton WE (eds) Nitrogen-fixing actinorhizal symbioses. Springer, New York, pp 199–234

Diem HG, Dommergues YR (1990) Current and potential uses and management of Casuarinaceae in tropics and subtropics. In: Schwintzer CR, Tjepkema JD (eds) The biology of *Frankia* and actinorhizal plants. Academic Press, San Diego, pp 317–342

Doyle LL, Luckow MA (2003) The rest of the iceberg. Legume diversity and evolution in a phylogenetic context. Plant Physiol 131:900–910. doi:10.1104/pp.102.018150

Endre G, Kereszt A, Kevei Z et al (2002) A receptor kinase gene regulating symbiotic nodule development. Nature 417:962–966. doi:10.1038/nature00842

Esseling JJ, Lhuissier FG, Emons AM (2003) Nod factor-induced root hair curling: continuous polar growth towards the point of Nod factor application. Plant Physiol 132:1982–1988. doi:10.1104/pp.103.021634

Felsenstein J (2005) PHYLIP (Phylogeny Inference Package). Available from: http://evolution.genetics.washington.edu/phylip.html

Fontaine MS, Lancelle SA, Torrey JG (1984) Initiation and ontogeny of vesicles in cultured *Frankia* sp. strain HFPArI3. J Bacteriol 160:921–927

Gallon JR (1981) The oxygen sensitivity of nitrogenase: a problem for biochemists and microorganisms. Trends Biol Sci 6:19–23. doi:10.1016/0968-0004(81)90008-6

Gauthier D, Jaffre T, Prin Y (2000) Abundance of *Frankia* from *Gymnostoma* spp. in the rhizosphere of *Alphitonia neocaledonica*, a non-nodulated Rhamnaceae endemic to New Caledonia. Eur J Soil Biol 36:169–175. doi:10.1016/S1164-5563(00)01061-X

Gherbi H, Markmann K, Svistoonoff S et al (2008) SymRK defines a common genetic basis for plant root endosymbioses with arbuscular mycorrhiza fungi, rhizobia, and *Frankia bacteria*. Proc Natl Acad Sci USA 105:4928–4932. doi:10.1073/pnas.0710618105

Ghodhbane-Gtari F, Beauchemin N, Bruce D et al (2013) Draft genome sequence of *Frankia* sp. strain CN3, an atypical, noninfective (Nod$^-$) ineffective (Fix$^-$) isolate from *Coriaria nepalensis*. Genome Announc 1:e00085–e00013. doi:10.1128/genomeA.00085-13

Ghodhbane-Gtari F, Hurst SG IV, Oshone R et al (2014) Draft genome sequence of *Frankia* sp. strain BMG5.23, a salt tolerant nitrogen-fixing actinobacterium isolated from the root nodules of *Casuarina glauca* grown in Tunisia. Genome Announc 2:e00520-14. doi:10.1128/genomeA.00520-14

Granqvist E, Sun J, Op den Camp R et al (2015) Bacterial-induced calcium oscillations are common to nitrogen-fixing associations of nodulating legumes and nonlegumes. New Phytol 207:551–558. doi:10.1111/nph.13464

Gtari M, Ghodhbane-Gtari F, Nouioui I et al (2015) Cultivating the uncultured: growing the recalcitrant cluster-2 *Frankia* strains. Sci Rep 5:13112. doi:10.1038/srep13112

Guan C, Ribeiro A, Akkermans AD et al (1996) Nitrogen metabolism in actinorhizal nodules of *Alnus glutinosa:* expression of glutamine synthetase and acetylornithine transaminase. Plant Mol Biol 32:1177–1184. doi:10.1007/BF00041403

Hafeez F, Akkermans ADL, Chaudhary AH (1984) Observations on the ultrastructure of *Frankia* sp. in root nodules of *Datisca cannabina* L. Plant Soil 79:383–402. doi:10.1007/BF02184330

Heidstra R, Geurts R, Franssen H et al (1994) Root hair deformation activity of nodulation factors and their fate on *Vicia sativa*. Plant Physiol 105:787–797

Hiltner L (1895) Über die Bedeutung der Wurzelknöllchen von *Alnus glutinosa* für die Stickstoffernährung dieser Pflanze. Landwirtschaftliche Verständnisstudien 46:153–161

Horváth B, Yeun LH, Domonkos A et al (2011) *Medicago truncatula* IPD3 is a member of the common symbiotic signaling pathway required for rhizobial and mycorrhizal symbioses. Mol Plant-Microbe Interact 24:1345–1358. doi:10.1094/MPMI-01-11-0015

Huguet V, Mergeay M, Cervantes E et al (2004) Diversity of *Frankia* strains associated to *Myrica gale* in Western Europe: impact of host plant (*Myrica vs. Alnus*) and of edaphic factors. Environ Microbiol 6:1032–1041. doi:10.1111/j.1462-2920.2004.00625.x

Hurst SG IV, Oshone R, Ghodhbane-Gtari F et al (2014) Draft genome sequence of *Frankia* sp. strain Thr, a nitrogen-fixing actinobacterium isolated from the root nodules of *Casuarina cunninghamiana* grown in Egypt. Genome Announc 2:e00493–14. doi:10.1128/genomeA.00493-14

Huss-Danell K (1997) Actinorhizal symbioses and their N_2 fixation. Tansley review no. 93. New Phytol 136:375–405. doi:10.1046/j.1469-8137.1997.00755.x

Huss-Danell K, Frej AK (1986) Distribution of *Frankia* strains in forest and afforestation sites in Northern Sweden. Plant Soil 90:407–418. doi:10.1007/BF02277412

Ishikawa J, Yamashita A, Mikami Y et al (2004) The complete genomic sequence of *Nocardia farcinica* IFM 10152. Proc Natl Acad Sci USA 101:14925–14930. doi:10.1073/pnas.0406410101

Jacobsen-Lyon K, Jensen EO, Jørgensen JE et al (1995) Symbiotic and nonsymbiotic hemoglobin genes of *Casuarina glauca*. Plant Cell 7:213–223. doi:10.1105/tpc.7.2.213

Jeong SC (2001) Population size and diversity of *Frankia* in soils of *Ceanothus velutinus* and Douglas-fir stands. Soil Biol Biochem 33:931–941. doi:10.1016/S0038-0717(00)00241-8

Jeong J, Suh S, Guan C et al (2004) A nodule-specific dicarboxylate transporter from alder is a member of the peptide transporter family. Plant Physiol 134:969–978. doi:10.1104/pp.103.032102

Kallimanis A, Karabika E, Mavromatis K et al (2011) Complete genome sequence of *Mycobacterium* sp. strain (Spyr1) and reclassification to *Mycobacterium gilvum* Spyr1. Stand Genomic Sci 5:144–153. doi:10.4056/sigs.2265047

Kawaharada Y, Kelly S, Nielsen MW et al (2015) Receptor-mediated exopolysaccharide perception controls bacterial infection. Nature 523:308–312. doi:10.1038/nature14611

Kleemann G, Kellner R, Poralla K (1994) Purification and properties of the squalene-hopene cyclase from *Rhodopseudomonas palustris*, a purple non-sulfure bacterium producing hopanoids and tetrahymanol. Biochim Biophys Acta 1210:317–320. doi:10.1016/0005-2760(94)90235-6

Knowlton S, Berry A, Torrey JG (1980) Evidence that associated soil bacteria may influence root hair infection of actinorhizal plants by *Frankia*. Can J Microbiol 26:971–977. doi:10.1139/m80-228

Lalonde M, Knowles R (1975) Ultrastructure, composition and biogenesis of the encapsulation material surrounding the endophyte in *Alnus crispa* var. *mollis* root nodules. Can J Bot 53:1951–1971. doi:10.1139/b75-219

Laplaze L, Duhoux E, Franche C et al (2000) *Casuarina glauca* prenodule cells display the same differentiation as the corresponding nodule cells. Mol Plant-Microbe Interact 13:107–112. doi:10.1094/MPMI.2000.13.1.107

Lavin M, Herendeen P, Wojciechowski MF (2005) Evolutionary rates analysis of Leguminosae implicates a rapid diversification of lineages during the tertiary. Syst Biol 54:530–549. doi:10.1080/10635150590947131

Lawrence DB, Schoenike RE, Quispel A et al (1967) The role of *Dryas drummondii* in vegetation development following ice recession at Glacier Bay, Alaska, with special reference to its nitrogen fixation by root nodules. J Ecol 55:793–813. doi:10.2307/2258426

Lechevalier MP (1994) Taxonomy of the genus *Frankia* (Actinomycetales). Int J Syst Bacteriol 44:1–8. doi:10.1099/00207713-44-1-1

Lévy J, Bres C, Geurts R et al (2004) A putative Ca^{2+} and calmodulin-dependent protein kinase required for bacterial and fungal symbioses. Science 303:1361–1364. doi:10.1126/science.1093038

Limpens E, Franken C, Smit P et al (2003) LysM domain receptor kinases regulating rhizobial Nod factor-induced infection. Science 302:630–633. doi:10.1126/science.1090074

Liu Q, Berry AM (1991a) The infection process and nodule initiation in the *Frankia-Ceanothus* root nodule symbiosis: a structural and histochemical study. Protoplasma 163:82–92. doi:10.1007/BF01323332

Liu Q, Berry AM (1991b) Localization and characterization of pectic polysaccharides in roots and root nodules of *Ceanothus* spp. during intercellular infection by *Frankia*. Protoplasma 164:93–101. doi:10.1007/BF01323333

Madsen EB, Madsen LH, Radutoiu S et al (2003) A receptor kinase gene of the LysM type is involved in legume perception of rhizobial signals. Nature 425:637–640. doi:10.1038/nature02045

Maillet F, Poinsot V, André O et al (2011) Fungal lipochitooligosaccharide symbiotic signals in arbuscular mycorrhiza. Nature 469:58–63. doi:10.1038/nature09622

Mansour SR, Oshone R, Hurst SG et al (2014) Draft genome sequence of *Frankia* sp. strain CcI6, a salt-tolerant nitrogen-fixing actinobacterium isolated from the root nodule of *Casuarina cunninghamiana*. Genome Announc 2:e01205–e01213. doi:10.1128/genomeA.01205-13

Markmann K, Parniske M (2009) Evolution of root endosymbiosis with bacteria: how novel are nodules? Trends Plant Sci 14:77–86. doi:10.1016/j.tplants.2008.11.009

Markmann K, Giczey G, Parniske M (2008) Functional adaptation of a plant receptor-kinase paved the way for the evolution of intracellular root symbioses with bacteria. PLoS Biol 6:e68. doi:10.1371/journal.pbio.0060068

Marsh JF, Rakocevic A, Mitra RM et al (2007) *Medicago truncatula* NIN is essential for rhizobial-independent nodule organogenesis induced by autoactive calcium/calmodulin-dependent protein kinase. Plant Physiol 144:324–335. doi:10.1104/pp.106.093021

Maunuksela L, Zepp K, Koivula T et al (1999) Analysis of *Frankia* populations in three soils devoid of actinorhizal plants. FEMS Microbiol Ecol 28:11–21. doi:10.1111/j.1574-6941.1999.tb00556.x

Meesters TM, Van Vliet WM, Akkermans ADL (1987) Nitrogenase is restricted to the vesicles in *Frankia* strain EANlpec. Physiol Plant 70:267–271

Mergaert P, Van Montagu M, Holsters M (1997) Molecular mechanisms of Nod factor diversity. Mol Microbiol 25:811–817. doi:10.1111/j.1365-2958.1997.mmi526.x

Meyen J (1829) Über das Hervorwachsen parasitischer Gebilde aus den Wurzeln anderer Pflanzen. Flora (Jena) 12:49–64

Miller IM, Baker DD (1985) The initiation development and structure of root nodules in *Elaeagnus angustifolia* (Elaeagnaceae). Protoplasma 128:107–119. doi:10.1007/BF01276333

Minchin FR (1997) Regulation of oxygen diffusion in legume nodules. Soil Biol Biochem 29:881–888. doi:10.1016/S0038-0717(96)00204-0

Mirza S, Pawlowski K, Hafeez FY et al (1994) Ultrastructure of the endophyte and localization of *nifH* transcripts in root nodules of *Coriaria nepalensis* Wall. by *in situ hybridization*. New Phytol 126:131–136. doi:10.1111/j.1469-8137.1994.tb07538.x

Murry MA, Zhang Z, Torrey JG (1985) Effect of O_2 on vesicle formation, acetylene reduction, and O_2-uptake kinetics in *Frankia* sp. HFPCcI3 isolated from *Casuarina cunninghamiana*. Can J Microbiol 31:804–809. doi:10.1139/m85-151

Mylona P, Pawlowski K, Bisseling T (1995) Symbiotic nitrogen fixation. Plant Cell 7:869–885. doi:10.1105/tpc.7.7.869

Nalin R, Putra SR, Domenach A-M, Rohmer M, Gourbiere F, Berry AM (2000) High hopanoid/total lipids ratio in *Frankia* mycelia is not related to the nitrogen status. Microbiology 146:3013–3019. doi:10.1099/00221287-146-11-3013

Newcomb W, Pankhurst CE (1982) Fine structure of actinorhizal nodules of *Coriaria arborea* (Coriariaceae). New Zeal J Bot 20:93–103. doi:10.1080/0028825X.1982.10426409

Noridge NA, Benson DR (1986) Isolation and nitrogen-fixing activity of *Frankia* sp. strain CpI1 vesicles. J Bacteriol 166:301–305. doi:10.1155/2014/568549

Normand P, Orso S, Cournoyer B et al (1996) Molecular phylogeny of the genus *Frankia* and related genera and emendation of the family Frankiaceae. Int J Syst Bacteriol 46:1–9

Normand P, Lapierre P, Tisa LS et al (2007) Genome characteristics of facultatively symbiotic *Frankia* sp. strains reflect host range and host plant biogeography. Genome Res 17:7–15. doi:10.1101/gr.5798407

Nouioui I, Beauchemin N, Cantor MN et al (2013) Draft genome sequence of *Frankia* sp. strain BMG5.12, a nitrogen-fixing actinobacterium isolated from Tunisian soils. Genome Announc 1:e00468–13. doi:10.1128/genomeA.00468-13

Oldroyd GE (2013) Speak, friend, and enter: signalling systems that promote beneficial symbiotic associations in plants. Nat Rev Microbiol 11:252–263. doi:10.1038/nrmicro2990

Op den Camp R, Streng A, De Mita S et al (2011) LysM-type mycorrhizal receptor recruited for rhizobium symbiosis in nonlegume *Parasponia*. Science 331:909–912. doi:10.1126/science.1198181

Parsons R, Silvester WB, Harris S et al (1987) *Frankia* vesicles provide inducible and absolute oxygen protection for nitrogenase. Plant Physiol 83:728–731. doi:10.1104/pp.83.4.728

Pawlowski K, Bisseling T (1996) Rhizobial and actinorhizal symbioses: what are the shared features? Plant Cell 8:1899–1913. doi:10.1105/tpc.8.10.1899

Pawlowski K, Demchenko KN (2012) The diversity of actinorhizal symbiosis. Protoplasma 249:967–979. doi:10.1007/s00709-012-0388-4

Pawlowski K, Jacobsen KR, Alloisio N et al (2007) Truncated hemoglobins in actinorhizal nodules of *Datisca glomerata*. Plant Biol 9:776–785. doi:10.1055/s-2007-965258
Persson T, Benson DR, Normand P et al (2011) Genome sequence of "*Candidatus* Frankia datiscae" Dg1, the uncultured microsymbiont from nitrogen-fixing root nodules of the dicot *Datisca glomerata*. J Bacteriol 193:7017–7018. doi:10.1128/JB.06208-11
Persson T, Battenberg K, Demina IV et al (2015) *Candidatus* Frankia datiscae Dg1, the actinobacterial microsymbiont of *Datisca glomerata,* expresses the canonical *nod* genes *nodABC* in symbiosis with its host plant. PLoS One 10:e0127630. doi:10.1371/journal.pone.0127630
Persson T, Nguyen TV, Alloisio N et al (2016) The N-metabolites of roots and actinorhizal nodules from *Alnus glutinosa* and *Datisca glomerata*: can *D. glomerata* change N-transport forms when nodulated? Symbiosis. doi:10.1007/s13199-016-0407-x
Pujic P, Bolotin A, Fournier P et al (2015) Genome sequence of the atypical symbiotic *Frankia* R43 strain, a nitrogen-fixing and hydrogen-producing actinobacterium. Genome Announc 3:e01387–e01315. doi:10.1128/genomeA.01387-15
Quispel A (1990) Discoveries, discussions and trends in research on actinorhizal root nodule symbioses before 1978. In: Schwintzer CR, Tjepkema JD (eds) The biology of *Frankia* and actinorhizal plants. Academic, San Diego, pp 15–33
Racette S, Torrey JG (1989) Root nodule initiation in *Gymnostoma* (Casuarinaceae) and *Shephardia* (Elaeagnaceae) induced by *Frankia* strain HFPGpI1. Can J Bot 67:2873–2879. doi:10.1139/b89-368
Ramírez-Saad H, Janse JD, Akkermans ADL (1998) Root nodules of *Ceanothus caeruleus* contain both the N_2-fixing *Frankia* endophyte and a phylogenetically related Nod^-/Fix^- actinomycete. Can J Microbiol 44:140–148. doi:10.1139/cjm-44-2-140
Schröder P (1989) Aeration of the root system in *Alnus glutinosa* L. Gaertn. Ann Sci Forest 46:310–314. doi:10.1051/forest:19890571
Schubert KR (1986) Products of biological nitrogen fixation in higher plants: synthesis, transport and metabolism. Annu Rev Plant Physiol 37:539–574. doi:10.1146/annurev.pp.37.060186.002543
Schubert M, Koteeva NK, Wabnitz PW et al (2011) Plasmodesmata distribution and sugar partitioning in nitrogen-fixing root nodules of *Datisca glomerata*. Planta 233:139–152. doi: 0.1007/s00425-010-1285-8
Schubert M, Koteyeva NK, Zdyb A et al (2013) Lignification of cell walls of infected cells in *Casuarina glauca* nodules that depend on symplastic sugar supply is accompanied by reduction of plasmodesmata number and narrowing of plasmodesmata. Physiol Plant 147:524–540. doi:10.1111/j.1399-3054.2012.01685.x
Schwintzer CR (1990) Spore-positive and spore-negative nodules. In: Schwintzer CR, Tjepkema JD (eds) The biology of *Frankia* and actinorhizal plants. Academic Press, San Diego, pp 177–193
Schwintzer CR, Berry AM, Disney LD (1982) Seasonal patterns of root nodule growth, endophyte morphology, nitrogenase activity and shoot development in *Myrica gale*. Can J Bot 60:746–757
Sen R, Lahudkar S, Durairaj G, Bahumik SR (2013) Functional analysis of Bre1p, an E3 ligase for histone H2B ubiquitylation, in regulation of RNA polymerase II association with active genes and transcription in vivo. J Biol Chem 288(14):9619–9633
Sen A, Daubin V, Abrouk D et al (2014) Phylogeny of the class Actinobacteria revisited in the light of complete genomes. The orders 'Frankiales' and Micrococcales should be split into coherent entities: proposal of Frankiales ord. nov., Geodermatophilales ord. nov., Acidothermales ord. nov. and Nakamurellales ord. nov. Int J Syst Evol Microbiol 64:3821–3832. doi:10.1099/ijs.0.063966-0
Sequerra J, Capellano A, Faure-Raynaud M et al (1994) Root hair infection process and myconodule formation on *Alnus incana* by *Penicillium nodositatum* Valla. Can J Bot 72:955–975. doi:10.1139/b94-121
Sequerra J, Capellano A, Gianinazzi-Pearson V (1995) Ultrastructure of cortical root cells of *Alnus incana* infected by *Penicillium nodositatum*. New Phytol 130:545–555. doi:10.1111/j.1469-8137.1995.tb04331.x

Silvester WB, Harris SL (1989) Nodule structure and nitrogenase activity of *Coriaria arborea* in response to varying pO_2. Plant Soil 118:97–109. doi:10.1007/BF02232794

Silvester WB, Harris SL, Tjepkema JD (1990) Oxygen regulation and hemoglobin. In: Schwintzer CR, Tjepkema JD (eds) The biology of *Frankia* and actinorhizal plants. Academic Press, San Diego, pp 157–176

Singh S, Katzer K, Lambert J et al (2014) CYCLOPS, a DNA-binding transcriptional activator, orchestrates symbiotic root nodule development. Cell Host Microbe 15:139–152. doi:10.1016/j.chom.2014.01.011

Smolander A, Sundman V (1987) *Frankia* in acid soils of forests devoid of actinorhizal plants. Physiol Plant 70:297–303. doi:10.1111/j.1399-3054.1987.tb06147.x

Smouni A, Laplaze L, Bogusz D et al (2002) The 35S promoter is not constitutively expressed in the transgenic tropical actinorhizal tree *Casuarina glauca*. Funct Plant Biol 29:649–656. doi:10.1071/PP01121

Soltis DE, Soltis PS, Morgan DR et al (1995) Chloroplast gene sequence data suggest a single origin of the predisposition for symbiotic nitrogen fixation in angiosperms. Proc Natl Acad Sci USA 92:2647–2651. doi:10.1073/pnas.92.7.2647

Soltis DE, Mort ME, Latvis M et al (2013) Phylogenetic relationships and character evolution analysis of Saxifragales using a supermatrix approach. Am J Bot 100:916–929. doi:10.3732/ajb.1300044

Soyano T, Shimoda Y, Hayashi M (2015) Nodule inception antagonistically regulates gene expression with nitrate in *Lotus japonicus*. Plant Cell Physiol 56:368–376. doi:10.1093/pcp/pcu168

Stracke S, Kistner C, Yoshida S et al (2002) A plant receptor-like kinase required for both bacterial and fungal symbiosis. Nature 417:959–962. doi:10.1038/nature00841

Sun J, Miller JB, Granqvist E, Wiley-Kalil A et al (2015) Activation of symbiosis signaling by arbuscular mycorrhizal fungi in legumes and rice. Plant Cell 27:823–838. doi:10.1105/tpc.114.131326

Suryakumar G, Gupta A (2011) Medicinal and therapeutic potential of sea buckthorn (*Hippophae rhamnoides* L.) J Ethnopharmacol 138:269–278. doi:10.1016/j.jep.2011.09.024

Suzuki W, Konishi M, Yanagisawa S (2013) The evolutionary events necessary for the emergence of symbiotic nitrogen fixation in legumes may involve a loss of nitrate responsiveness of the NIN transcription factor. Plant Signal Behav 8:e25975. doi:10.4161/psb.25975

Svistoonoff S, Benabdoun FM, Nambiar-Veetil M et al (2013) The independent acquisition of plant root nitrogen-fixing symbiosis in Fabids recruited the same genetic pathway for nodule organogenesis. PLoS One 8:e64515. doi:10.1371/journal.pone.0064515

Tisa LS, Beauchemin N, Cantor MN, Furnholm T et al (2015) Draft genome sequence of *Frankia* sp. strain DC12, an atypical, noninfective, ineffective isolate from *Datisca cannabina*. Genome Announc 3:e00889–e00815. doi:10.1128/genomeA.00889-15

Tjepkema JD (1978) The role of diffusion from the shoots and nodule roots in nitrogen fixation by root nodules of *Myrica gale* L. Am J Bot 70:59–63. doi:10.1139/b78-156

Tjepkema JD, Ormerod W, Torrey JG (1980) On vesicle formation and *in vitro* acetylene reduction by *Frankia*. Nature 287:633–635

Torrey JG (1976) Initiation and development of root nodules of *Casuarina* (Casuarinaceae). Am J Bot 63:335–344

Torrey JG, Callaham D (1982) Structural features of the vesicle of *Frankia* sp. cell in culture. Can J Microbiol 28:749–757. doi:10.1139/m82-114

Udvardi M, Poole PS (2013) Transport and metabolism in legume-rhizobia symbioses. Annu Rev Plant Biol 64:781–805. doi:10.1146/annurev-arplant-050312-120235

Uemura S (1971) Non-leguminous root nodules in Japan. Plant Soil 29:349–350. doi:10.1007/BF02661863

Valverde C, Wall LG (1999) Time course of nodule development in the *Discaria trinervis* (Rhamnaceae)–*Frankia* symbiosis. New Phytol 141:345–354. doi:10.1046/j.1469-8137.1999.00345.x

Van Brussel AAN, Bakhuizen R, van Spronsen PC et al (1992) Induction of pre-infection thread structures in the leguminous host plant by mitogenic lipooligosaccharides of *Rhizobium*. Science 257:70–72. doi:10.1126/science.257.5066.70

Van Ghelue M, Ribeiro A, Solheim B et al (1996) Sucrose synthase and enolase expression in actinorhizal nodules of *Alnus glutinosa*: comparison with legume nodules. Mol Gen Genet 250:437–446. doi:10.1007/s004380050096

Van Ghelue M, Løvaas E, Ringø E et al (1997) Early interactions between *Alnus glutinosa* and *Frankia* strain ArI3. Production and specificity of root hair deformation factor(s). Physiol Plant 99:579–587. doi:10.1111/j.1399-3054.1997.tb05360.x

Vanden Heuvel BD, Benson DR, Bortiri E et al (2004) Low genetic diversity among *Frankia* spp. strains nodulating sympatric populations of actinorhizal species of Rosaceae, Ceanothus (Rhamnaceae) and *Datisca glomerata* (Datiscaceae) west of the Sierra Nevada (California). Can J Microbiol 50:989–1000. doi:10.1139/W04-079

VandenBosch KA, Torrey JB (1983) Host-endophyte interactions in effective and ineffective nodules induced by the endophyte of *Myrica gale*. Can J Bot 61:2898–2909. doi:10.1139/b83-323

VandenBosch KA, Torrey JG (1985) Development of endophytic *Frankia* sporangia in field- and laboratory-grown nodules of *Comptonia peregrina* and *Myrica gale*. Am J Bot 72:99–108

Vasse J, de Billy F, Camut S et al (1990) Correlation between ultrastructural differentiation of bacteroids and nitrogen fixation in alfalfa nodules. J Bacteriol 172:4295–4306

Wall LG, Beauchemin N, Cantor MN et al (2013) Draft genome sequence of *Frankia* sp. strain BCU110501, a nitrogen-fixing actinobacterium isolated from nodules of *Discaria trinervis*. Genome Announc 1:e00503–e00513. doi:10.1128/genomeA.00503-13

Werner GD, Cornwell WK, Sprent JI et al (2014) A single evolutionary innovation drives the deep evolution of symbiotic N_2-fixation in angiosperms. Nat Commun 5:4087. doi:10.1038/ncomms5087

Wheeler CT, Gordon JC, Ching TM (1979) Oxygen relations of the root nodules of *Alnus rubra* bong. New Phytol 82:449–457. doi:10.1111/j.1469-8137.1979.tb02671.x

Willmann R, Lajunen H, Erbs G et al (2011) *Arabidopsis* lysin-motif proteins LYM1/LYM3/CERK1 mediate bacterial peptidoglycan sensing and immunity to bacterial infection. Proc Natl Acad Sci USA 108:19824–19829. doi:10.1073/pnas.1112862108

Wolters DJ, Van Dijk C, Zoetendal EG, Akkermans AD (1997) Phylogenetic characterization of ineffective *Frankia* in *Alnus glutinosa* (L.) Gaertn. nodules from wetland soil inoculants. Mol Ecol 6:971–981. doi:10.1046/j.1365-294X.1997.00265.x

Yano K, Yoshida S, Müller J et al (2008) CYCLOPS, a mediator of symbiotic intracellular accommodation. Proc Natl Acad Sci USA 105:20540–20545. doi:10.1073/pnas.0806858105

Zhang XC, Cannon SB, Stacey G (2009) Evolutionary genomics of *LysM* genes in land plants. BMC Evol Biol 9:183. doi:10.1186/1471-2148-9-183

Zitzer SF, Dawson JO (1992) Soil properties and actinorhizal vegetation influence nodulation of *Alnus glutinosa* and *Elaeagnus angustifolia* by *Frankia*. Plant Soil 140:197–204. doi:10.1007/BF00010597

Application of Protists to Improve Plant Growth in Sustainable Agriculture

13

Alexandre Jousset

Abstract

Plant health and growth are largely dependent on root-associated microbiota. Several bacteria and fungi can provide important services to plants, such as nutrient mineralization or protection against diseases. To date, most of our knowledge is centered on bacterial and fungal taxa. This chapter presents protists as an essential yet often overlooked component of the rhizosphere microbiome, where they play a crucial role in structuring microbial populations. Protists are a keystone group, functioning as predators of bacteria and fungi. They exert a strong pressure on plant-associated microbial communities and shape their functional and phylogenetic composition. They further enhance nutrient turnover and activate bacterial genes needed for pathogen suppression. Protists offer thus new venues to manage plant-associated microbial communities to enhance their functionality and ability to support a high plant growth in agricultural context. This chapter presents the main functional groups of soil protists and explains their distribution and importance for soil fertility. Finally, their applications in biotechnological settings aiming at reducing pesticide and fertilizer input in sustainable agriculture, are discussed.

13.1 Introduction

The growing human population calls for new strategies to improve agricultural yields. Engineering the rhizosphere microbiome to enhance plant yield and health, forms one of the cornerstones of current agricultural research. Plants live in

A. Jousset (✉)
Department of Biology, University of Utrecht,
Padualaan 8, 3584 CH Utrecht, The Netherlands
e-mail: a.l.c.jousset@uu.nl

S. Mehnaz (ed.), *Rhizotrophs: Plant Growth Promotion to Bioremediation*, Microorganisms for Sustainability 2, DOI 10.1007/978-981-10-4862-3_13

association with diverse microbial communities that colonize plant roots and provide essential services to the plant. Plant-beneficial microbial functions encompass, for instance, nutrient mineralization, fine-tuning of the hormonal balance, or pathogen suppression. Together, these functions can contribute to higher plant growth and enhanced nutrition, resistance to abiotic stressors, or reduction of disease symptoms. They may thus offer a great potential to improve agricultural yield while reducing the input of fertilizers or pesticides.

Most current approaches to improve microbiome functionality rely on the introduction of beneficial bacteria or fungi, carrying one or more desired functions. These approaches have led to impressive results, as illustrated in the other chapters of this book. However, they are limited by the survival of the introduced microbiota, which have to compete with the indigenous flora to be able to develop their beneficial activity. In this chapter, I suggest that protists may be used as a keystone group to enhance selected microbes and boost the functionality of the rhizosphere microbiome. I will give an overview of the phylogenetic affiliation of protists, functional groups present in the soil, their potential impact on microbial communities, and applications of protists in the agriculture.

13.2 Phylogeny of Protists

The term "protist" encompasses all eukaryotes with the exception of green plants, fungi, and animals. This paraphyletic concept originates from the scientific tradition of the nineteenth century and has persisted today. Protists were the first organisms discovered by Leeuwenhoek, who named them animalcules. Later, different organisms were investigated by different scientific communities. Algae were investigated by botanists and mobile protists by zoologists, who coined the term protozoa ("primitive animals"). For a long time, all protists were grouped into one separate kingdom (kingdom Protista). This vision radically changed with the development of molecular methods, which brought a complete shift in eukaryote classification. According to the latest phylogenies (Adl et al. 2012; Pawlowski 2013), protists now encompass most eukaryotes, with plants, animal, and fungi appearing as three tiny branches in the eukaryotic tree of life. Eukaryotes contain dozen more of phyla encompassing several lineages with diverse lifestyle and ecological relevance. The whole field of protistology is still influenced by its separate historical background and has developed concepts distinct from the remaining microbiology. These historical legacies still persist in the scientific vocabulary. Protists are for instance often been referred as "microfauna," small animals, in soil studies, a remnant of ancient scientific traditions. Yet protists are not miniature animals, they encompass phylogenetic groups often unrelated to animals, and are thousands of times more diverse. Similarly, bacterivorous protists have been (and are still being) referred to as grazers, a poetic analogies to cows feeding on a meadow, that is unfortunately in discrepancy with the standard ecological nomenclature. Grazing refers to the partial consumption of the prey, which remains alive and can regrow. Bacterivore prey consumption is best described as predation, which results in prey death, which had very distinct consequences on population dynamics and evolutionary processes.

13.3 Functional Groups of Soil Protists

Soil is a hotspot of microbial diversity, including protists (Ekelund and Ronn 1994). Besides a few parasitic taxa such as apicomplexan or oomycetes, as well as phototrophic algae, most protists groups are heterotrophs, feeding on bacteria and fungi (Foissner 1987). Protists are often classified on the base of morphological characteristics. Even if these characteristics are not always useful from a taxonomic perspective, with for instance, morphologically almost identical amoebae found in extremely distinct lineages (Pawlowski 2013), they nonetheless offer interesting insights in functionality of protists and their importance for soil fertility (Ekelund and Ronn 1994).

13.3.1 Ciliates

Ciliated protozoa, the only morphotype matching with one defined phylogenetic group (within the phylum alveolates), can consume huge amounts of prey. They are among the biggest protists in soil, ranging from 5 to 500 μm. They are covered with cilia, allowing them to swim rapidly, and possess a large mouth-like opening enabling the rapid consumption of several preys. Ciliates contain, for instance, the model species *Tetrahymena pyriformis* and are one of the most commonly used organisms in food web models (Fussmann et al. 2014; Altermatt et al. 2015). Thanks to their high rate of consumption, it is possible to monitor accurately their interactions with preys under various conditions. Ciliate are excellent bio-indicators of soil conditions, allowing for instance discriminating polluted from pristine soils (Jousset et al. 2009a). Thanks to the large amount of data available, species found in one sample can be readily classified as r or K strategist or biofilm dwelling or planktonic swimmer. Ciliates are an extremely diverse group (Foissner et al. 2004) but depend on the presence of water films for their activity. As a result, ciliate density is low in most soils (Darbyshire et al. 1989), with exception of paddy soils, in which ciliates seem to play an important role as regulator of bacterial communities. Most described ciliates are bacterivorous; however, some species have evolved stylet-like structures allowing them to suck the content of fungal hyphae.

13.3.2 Flagellates

This denomination encompasses several groups of different morphology and phylogenetic affiliation. Flagellates are typically 2–10 μm big and share the property of having one or more flagella, which can serve for locomotion or catching preys. In soils, some of the most common groups of flagellates can be found in the phylum

Rhizaria, with the genera *Cercomonas*, *Heteromita*, in the phylum Euglenozoa, with for instance the kinetoplastida genus *Bodo*, or the excavates, comprising species such as *Jakoba*. Some protists can switch between a flagellate and amoeboid stage during their different life stages. One of the best described example is the amoeba of the genus *Naegleria* (Fritz-Laylin and Fulton 2016).

13.3.3 Naked Amoeba

Naked amoeba shows an irregular shape, with bursting pseudopods. Amoebae are one of the most abundant soil protist functional groups. It also covers several phyla, including Rhizaria, Amoebozoa, and Excavata. Note that the term "Amoebozoa" only refers to one specific eukaryotic phylum, covering only a fraction of all the amoeboid organisms. Isolates of the Amoebozoa, genus *Acanthamoeba*, have been used as model organism in several studies, as they are comparatively easy to isolate and cultivate axenically (Bonkowski 2004; Rosenberg et al. 2009; Jousset and Bonkowski 2010; Neidig et al. 2010).

13.3.4 Testate Amoebae

These amoebae are protected by a shell made of debris, minerals, or secretions from the amoeba. Testate amoebae are particularly abundant in acidic soils such as bogs but can be found in most soils. The empty shells remain identifiable for a long time after the death of the organism, providing useful paleo-records, helping redraw the history of a given site. They are further very sensitive to environmental changes such as fertilization (Krashevska et al. 2014). The exact importance of testate amoeba for soil fertility is not elucidated yet, and experiments are complicated because of the difficulty to cultivate many species. No doubt, however, that new insights will come soon.

13.4 Functionality of Protists and Impact on Microbial Communities

Protists have a keystone function in the soil ecosystem. They are primary consumers in the food web, making the link between primary producers and higher soil food web levels such as collembolans and mites. Recent studies have in addition extended this view and protists are now recognized to also feed on animals such as nematodes (Neidig et al. 2010; Geisen et al. 2015; Geisen 2016), showing that the importance of protists may be spread across several trophic levels. This section will focus on the interaction between protists and bacteria, as most of the literature is centered on these taxa.

Protists are, together with nematodes, the main consumers of bacteria in soil (Bonkowski 2004). Predation is so intense that many scientists consider rhizosphere communities to be top-down regulated: The main constraint on bacterial fitness is not to find nutrients but to escape predation (Moore et al. 2003). Bacterivorous protists appear thus as keystone element that can control microbiome structure and function. Protist populations have long been put in relation with soil fertility and may be used as inoculant in the agriculture to boost soil microbiome functionality. By providing a selective advantage to indigenous microbes that may otherwise be rare and inactive, protists may unlock different functions beneficial for plant growth. Protists can have for instance following effects on the soil microbiome:

13.4.1 Nutrient Turnover

Protists consume bacteria and fungi and typically release the excess of nutrients such as nitrogen, phosphorus, or micronutrients. These nutrients are often limiting in soil and would be without predation kept locked in dormant microbial cells, preventing their use to active ones. One of the most striking effects of protists is their stimulation of nitrogen mineralization and nitrification (Kuikman et al. 1991; Alphei et al. 1996). Since nitrate is the favored nitrogen source by many plants, protists may stimulate nitrogen uptake in plant, helping in use of resources more efficiently (Alphei et al. 1996; Kreuzer et al. 2006).

13.4.2 Selection of Specific Bacteria

Due to their activity, protists select for r strategists (Swallow et al. 2013). In addition, protists carefully select their prey and discriminate between edible and non-edible bacteria on the base of several characteristics including surface properties or size (Montagnes et al. 2008). Closely related species can have distinct effects on microbial communities. Protist predators play further a key role in promoting toxic microorganisms in the rhizosphere. Secondary metabolites linked to pathogen suppression and may further add protection against predation by protists, which avoid these toxic bacteria and confer them a growth advantage. The productions of broad-spectrum bioactive secondary metabolites such as alkaloids (Klapper et al. 2016), lipopeptides (Andersen and Winding 2004; Jousset et al. 2006; Mazzola et al. 2009) or polyketides (Jousset et al. 2006), and addition of amoebae multiplied the fitness of an introduced biocontrol *Pseudomonas* in rice rhizosphere by a factor of three, by preferentially feeding on nontoxic competitors of the introduced bacteria (Jousset et al. 2008). Similarly, production of gluconic acid, compound-helping bacteria mineralizing phosphorus, can protect bacteria against predation by protozoa from very distinct taxonomic affiliation (Gomez et al. 2010). Thanks to this overlap, protists may thus function as a booster of introduced microbes, ensuring their survival.

13.4.3 Shifts in Microbiome Composition

In addition of the bacterivorous amoeba, *Acanthamoeba castellanii*, to rhizosphere microbial communities results in shift in phylogenetic composition, promoting, for example, actinobacteria or *Herbaspirillum*, two bacterial groups linked to plant growth promotion (Rosenberg et al. 2009).

13.4.4 Enhancement of Plant-Microbe Symbioses

Arbuscular mycorrhizae are an important symbiont of almost all terrestrial plants. They can greatly enhance plant yield and stress tolerance by extending the contact surface with the soil, allowing foraging of regions unreachable for roots alone. Koller and colleagues (2013) showed that mycorrhiza function is largely dependent on protists. Amoebae increased nitrogen turnover around hyphae and stimulated its transfer to the plant. Likely, mycorrhiza fungi themselves were not able to produce the required enzymes required for mineralizing the soil organic material. Instead they secreted plant-derived carbon in their surroundings, fueling associated microbial communities. Without predation, these communities would get blocked by nutrient limitation, a problem solved by adding amoebae to the system. Protists may thus be important to shift plant-mycorrhiza interaction from parasitism (the fungus taking up plant-derived carbon without delivering nutrients) to mutualism in which the fungus provides the plant with the required nutrients such as nitrogen or phosphorus. Protists can further maintain microbiome functioning over evolutionary scales by preventing the emergence of bacterial cheats that consume plant-derived resources but do not contribute to plant health. Since such cheats also become more vulnerable to predation, plant-bacteria cooperation can be maintained by predation by bacterivorous amoebae (Jousset et al. 2009b).

13.4.5 Manipulation of Plant Hormone Balance

Predation by protists favors bacteria producing auxin, a hormone-stimulating root development (Bonkowski and Brandt 2002). Plants co-inoculated with amoebae show for instance a much more ramified root system than control plants (Kreuzer et al. 2006). Such a ramified root system helps plant take up soil nutrients and reducing the use for fertilizer. Although no explanatory mechanism for this selection is available to date, this process appears as a "hormonal" microbial loop (Bonkowski 2004), in which predation by protists such as amoebae cause an increase in nitrate and auxin, which forces the plant to invest more in the root system, feeding more bacteria and ultimately amoebae, completing the cycle.

13.4.6 Stimulation of Beneficial Trait Expression

Some bacterial traits linked to plant growth promotion, including for instance the production of siderophores or toxic secondary metabolites, are strongly affected by the presence of bacterivorous protists. For instance, siderophore production in *Pseudomonas fluorescens* can be stimulated in the presence of amoebae (Levrat et al. 1992). Cyclic lipopeptides (Mazzola et al. 2009) or 2,4-DAPG (Jousset and Bonkowski 2010; Jousset et al. 2010) production increased after confronting bacteria with amoebae or their supernatant. Although the exact nature of the signals involved in this interaction is not known yet, small molecules (<3 kDa) secreted by amoebae are required for the recognition of protists by bacteria (Jousset et al. 2010). Many bacteria can react to chemical cues from predatory amoebae and flagellates (Corno and Jurgens 2006) and such chemical communication may play an important role in structuring rhizosphere communities. Further studies are needed to assess how specific bacteria can recognize and respond to predators. Together, these findings suggest that protozoa may be used to promote the activity of soil microbes. This is illustrated by studies showing that adding protists increased the antagonism of a biocontrol *Pseudomonas* against the plant-pathogenic fungus *Fusarium oxysporum* (Levrat et al. 1991).

13.5 Application of Protists as Microbiome Enhancers

As mentioned above, protists can improve several functions of the soil microbiome relevant to plant health, including making nutrients available to the plant, stimulating plant growth and suppressing diseases. Thanks to the variety of shape and function, they provide a formidable biotechnological pool to improve various soil processes linked to fertility and sustainable crop production (Chen et al. 2007).

Protists cover most of eukaryotes lineages and are thus a huge and untapped source of genetic diversity. The effect of different groups of protists on microbial communities remains to be elucidated, but the sensitivity of protists to secondary metabolites seems to correlate with high level taxonomy (Pedersen et al. 2011). As a result, screening protozoa across the tree of life may help discovering that which group supports which type of antibiotics. In contrast, both closely related *Cercomonas* species and more distantly related protists from different phyla may have very distinct effects on microbial community composition (Rønn et al. 2002; Glucksman et al. 2010). A rapid coevolution process may thus be occurring in soil, potentially on the base of prey recognition receptors and bacterial antigens (Wildschutte et al. 2004). These contrasting results indicate that a huge pool of protists may be used to favor selected functions in the soil microbiome.

Protozoa can be applied in several ways. They can function as an enhancer for introduced plant-beneficial bacteria or fungi: The survival of introduced microbes is often a limiting factor for their impact on plant growth and health. Protists can improve the survival of introduced biocontrol *Pseudomonas* spp. by a 200 % (Jousset et al. 2006), by consuming indigenous species. This effect may be best obtained with biocontrol agents producing toxic secondary metabolites, yet other traits such as hard cells, biofilm, or filament formation may also provide a competitive advantage (Matz and Kjelleberg 2005; Jousset 2012). For instance, bacteria of the genus *Arthrobacter* seem to be fostered under protists predation, suggesting that several taxa can be enhanced by adding the right protist (Rønn et al. 2002). Different species may be more or less sensitive to specific bacterial metabolites, so that custom pairs of protists and bacteria may best work together. New screenings are needed to uncover the appropriate combinations (Pedersen et al. 2011). Protists may further serve as general enhancer of microbiome function. Protists stimulate nutrient turnover and accelerate the mineralization of organic fertilizer. They are already included in first commercial products, where they speed up nutrient release from organic fertilizer at low temperature, which may be particularly relevant for spring conditions in cold climates.

Finally, protists may be used to directly consume pathogens. Fungivorous amoebae have long been suspected for instance of being able to induce suppressiveness against *Fusarium* (Levrat et al. 1991), although more experimental proofs are needed.

13.6 Protist Preparation

Protists can be grown either on undefined bacteria co-isolated with the species, mono-axenically on one reference bacteria, or axenically in a sterile culture medium (Weekers et al. 1993). Axenic growth is the best option, as it allows high yields, but can be tedious to obtain as protists are typically associated with various bacteria and may not be cultivable in absence of a prey.

Most – if not all – soil protists build cysts, a resistance stage allowing survival in extreme conditions. This property is most useful for biotechnology purposes as it makes dry formulation possible. Once introduced to soil, the cyst hatches and the trophozoites start multiplying. Some cysts can carry bacteria in them. This property has long been known as an issue for potential pathogens (Molmeret et al. 2005) but may as well serve as vector for otherwise vulnerable beneficial soil microbes.

13.7 Precautions

Most protists are free-living organisms. The few obligate parasites species, such as *Trypanosoma* or *Plasmodium*, are not relevant for biotechnological applications aiming at improving soil fertility. However, some bacterivorous species are known to be opportunistic pathogens. As with other opportunistic pathogen, they are ubiquitous in the environment and do not pose objective hazards to healthy individuals. Nonetheless, avoiding them would avoid unnecessary danger for immune-compromised patients and prevent bureaucratic hassles during the registration process.

Some amoebae of the genus *Acanthamoeba* can cause keratitis, a rare but hard to cure eye disease typically associated with poor contact lenses hygiene. Even if only a very few genotypes can cause disease, regulation agencies may not be easy to convince (Siddiqui and Khan 2012). Further, *Naegleria* spp. living in warm waters can cause deadly brain diseases in immune-compromised patient, calling for caution when cultivating them.

13.8 Conclusion

Protists offer new venues to manipulate the soil microbiome and enhance plant health. Several studies on taxonomy and function could be linked together to provide robust biotechnological applications.

References

Adl SM, Simpson AG, Lane CE et al (2012) The revised classification of eukaryotes. J Eukaryot Microbiol 59(5):429–493

Alphei J, Bonkowski M, Scheu S (1996) Protozoa, nematoda and lumbricidae in the rhizosphere of *Hordelymus europaeus* (Poaceae): faunal interactions, response of microorganisms and effects on plant growth. Oecologia 106:111–126

Altermatt F, Fronhofer EA, Garnier A et al (2015) Big answers from small worlds: a user's guide for protist microcosms as a model system in ecology and evolution. Methods Ecol Evol 6:218–231

Andersen KS, Winding A (2004) Non-target effect of bacterial biological control agents on soil protozoa. Biol Fertil Soils 40:230–236

Bonkowski M (2004) Protozoa and plant growth: the microbial loop in soil revisited. New Phytol 162:617–631

Bonkowski M, Brandt F (2002) Do soil protozoa enhance plant growth by hormonal effects? Soil Biol Biochem 34:1709–1715

Chen X, Liu M, Hu F, Mao X, Li H (2007) Contributions of soil micro-fauna (protozoa and nematodes) to rhizosphere ecological functions. Acta Ecol Sin 27:3132–3143

Corno G, Jurgens K (2006) Direct and indirect effects of protist predation on population size structure of a bacterial strain with high phenotypic plasticity. Appl Environ Microbiol 72:78–86

Darbyshire JF, Griffiths BS, Davidson MS, Mchardy WJ (1989) Ciliate distribution amongst soil aggregates. Eur J Soil Biol 26:47–56

Ekelund F, Ronn R (1994) Notes on protozoa in agricultural soil with emphasis on heterotrophic flagellates and naked amoebae and their ecology. FEMS Microbiol Rev 15:321–353

Foissner W (1987) Soil protozoa: fundamental problems, ecological significance, adaptations in ciliates and testaceans, bioindicators, and guide to the literature. Prog Protistol 2:69–212

Foissner W, Berger H, Xu K, Zechmeister-Boltenstern S (2004) A huge, undescribed soil ciliate (Protozoa: Ciliophora) diversity in natural forest stands of central Europe. Biodivers Conserv 14:617–701

Fritz-Laylin LK, Fulton C (2016) Naegleria: a classic model for de novo basal body assembly. Cilia 5:10

Fussmann KE, Schwarzmuller F, Brose U, Jousset A, Rall BC (2014) Ecological stability in response to warming. Nat Clim Chang 4:206–210

Geisen S (2016) The bacterial-fungal energy channel concept challenged by enormous functional versatility of soil protists. Soil Biol Biochem (in press)

Geisen S, Rosengarten J, Koller R et al (2015) Pack hunting by a common soil amoeba on nematodes. Environ Microbiol 17:4538–4546
Glucksman E, Bell T, Griffiths RI, Bass D (2010) Closely related protist strains have different grazing impacts on natural bacterial communities. Environ Microbiol 12:3105–3113
Gomez W, Buela L, Castro LT et al (2010) Evidence for gluconic acid production by *Enterobacter intermedium* as an efficient strategy to avoid protozoan grazing. Soil Biol Biochem 42:822–830
Jousset A (2012) Ecological and evolutive implications of bacterial defences against predators. Environ Microbiol 14:1830–1843
Jousset A, Bonkowski M (2010) The model predator *Acanthamoeba castellanii* induces the production of 2,4- DAPG by the biocontrol strain *Pseudomonas fluorescens* Q2-87. Soil Biol Biochem 42:1647–1649
Jousset A, Lara E, Wall LG, Valverde C (2006) Secondary metabolites help biocontrol strain *Pseudomonas fluorescens* CHA0 to escape protozoan grazing. Appl Environ Microbiol 72:7083–7090
Jousset A, Scheu S, Bonkowski M (2008) Secondary metabolite production facilitates establishment of rhizobacteria by reducing both protozoan predation and the competitive effects of indigenous bacteria. Funct Ecol 22:714–719
Jousset A, Lara E, Nikolausz M, Harms H, Chatzinotas A (2009a) Application of the denaturing gradient gel electrophoresis (DGGE) technique as an efficient diagnostic tool for ciliate communities in soil. Sci Total Environ 408:1221–1225
Jousset A, Rochat L, Keel C, Pechy-Tarr M, Scheu S, Bonkowski M (2009b) Predators promote toxicity of rhizosphere bacterial communities by selective feeding on non-toxic cheaters. ISME J 3:666–674
Jousset A, Rochat L, Scheu S, Bonkowski M, Keel C (2010) Predator-prey chemical warfare determines the expression of antifungal genes by rhizosphere pseudomonads. Appl Environ Microbiol 76:5263–5268
Klapper M, Götze S, Barnett R, Willing K, Stallforth P (2016) Bacterial alkaloids prevent amoebal predation. Angew Chem 128(31):9090–9093
Koller R, Rodriguez A, Robin C, Scheu S, Bonkowski M (2013) Protozoa enhance foraging efficiency of arbuscular mycorrhizal fungi for mineral nitrogen from organic matter in soil to the benefit of host plants. New Phytol 199(1):203–211
Krashevska V, Sandmann D, Maraun M, Scheu S (2014) Moderate changes in nutrient input alter tropical microbial and protist communities and belowground linkages. ISME J 8:1126–1134
Kreuzer K, Adamczyk J, Iijima M et al (2006) Grazing of a common species of soil protozoa (*Acanthamoeba castellanii*) affects rhizosphere bacterial community composition and root architecture of rice (*Oryza sativa* L.) Soil Biol Biochem 38:1665–1672
Kuikman PJ, Jansen AG, Vanveen JA (1991) ^{15}N-nitrogen mineralization from bacteria by protozoan grazing at different soil-moisture regimes. Soil Biol Biochem 23:193–200
Levrat P, Pussard M, Steinberg C, Alabouvette C (1991) Regulation of *Fusarium oxysporum* populations introduced into soil – the amoebal predation hypothesis. FEMS Microbiol Ecol 86:123–129
Levrat P, Pussard M, Alabouvette C (1992) Enhanced bacterial metabolism of a *Pseudomonas* strain in response to the addition of culture filtrate of a bacteriophagous amoeba. Eur J Protistol 28:79–84
Matz C, Kjelleberg S (2005) Off the hook – how bacteria survive protozoan grazing. Trends Microbiol 13:302–307
Mazzola M, de Bruijn I, Cohen MF, Raaijmakers JM (2009) Protozoan-induced regulation of cyclic lipopeptide biosynthesis is an effective predation defense mechanism for *Pseudomonas fluorescens*. Appl Environ Microbiol 75:6804–6811
Molmeret M, Horn M, Wagner M, Santic M, Abu Kwaik Y (2005) Amoebae as training grounds for intracellular bacterial pathogens. Appl Environ Microbiol 71:20–28
Montagnes DJS, Barbosa AB, Boenigk J et al (2008) Selective feeding behaviour of key free-living protists: avenues for continued study. Aquat Microb Ecol 53:83–98

Moore JC, McCann K, Setälä H, De Ruiter PC (2003) Top-down is bottom-up: does predation in the rhizosphere regulate aboveground dynamics? Ecology 84:846–857

Neidig N, Jousset A, Nunes F et al (2010) Interference between bacterial feeding nematodes and amoebae relies on innate and inducible mutual toxicity. Funct Ecol 24:1133–1138

Pawlowski J (2013) The new micro-kingdoms of eukaryotes. BMC Biol 11:1–3

Pedersen AL, Winding A, Altenburger A, Ekelund F (2011) Protozoan growth rates on secondary-metabolite-producing *Pseudomonas* spp. correlate with high-level protozoan taxonomy. FEMS Microbiol Letters 316:16–22

Rønn R, McCaig A, Griffiths B, Prosser J (2002) Impact of protozoan grazing on bacterial community structure in soil microcosms. Appl Environ Microbiol 68:6094–6105

Rosenberg K, Bertaux J, Scheu S, Bonkowski M (2009) Soil amoeba rapidly change bacterial community composition in *Arabidopsis thaliana* rhizosphere. ISME J 3:675–684

Siddiqui R, Khan NA (2012) Biology and pathogenesis of *Acanthamoeba*. Parasit Vectors 5:6

Swallow MJB, Quideau SA, Norris CE (2013) Ciliate dependent production of microbial anthranilic acid occurring within aspen litter. Soil Biol Biochem 60:113–121

Weekers PHH, Bodelier PLE, Wijen JPH, Vogels GD (1993) Effects of grazing by the free-living soil amoebae *Acanthamoeba castellanii, Acanthamoeba polyphaga*, and *Hartmannella vermiformis* on various bacteria. Appl Environ Microbiol 59:2317–2319

Wildschutte H, Wolfe DM, Tamewitz A, Lawrence JG (2004) Protozoan predation, diversifying selection, and the evolution of antigenic diversity in *Salmonella*. Proc Natl Acad Sci U S A 101:10644–10649

Bioremediation of Heavy Metals for Sustainable Agriculture

14

Zakia Latif and Aatif Amin

Abstract

Due to industrial revolution, pollution of agricultural lands by toxic pollutants has become a great concern, worldwide. Naturally, heavy metals are present in the earth's crust. There are several/many toxic pollutants that can be changed into various oxidation states easily and cause many deleterious effects in several physiological processes in plants. Plants growing in these polluted soils show a reduction in growth, performance, and yield. Therefore, there is an urgent need to realize the heavy metal-induced toxicity in plants and animals and the harmful effects caused by the consumption of contaminated foods in humans. Bioremediation is an effective, suitable, cost-effective, and non-disturbing method of soil remediation; it is useful for the treatment of heavy metal-polluted soils. Microorganisms and plants employ different mechanisms for the bioremediation of polluted soils. Several microorganisms have been successfully used to reduce the toxicity of heavy metals. These microbes encode several detoxification processes to modify toxic metallic ions to nontoxic elemental state. Using plants for the treatment of polluted soils is a more common approach in this regard. Combining microorganisms and plants, for bioremediation, ensures a more efficient cleanup of heavy metal-polluted soils. This chapter presents the review of a comprehensive study of literature about heavy metal-induced toxicity in plants and its detoxification processes to provoke for advance research in the field of sustainable agriculture.

Z. Latif (✉) • A. Amin
Department of Microbiology and Molecular Genetics, University of the Punjab, Lahore 54590, Pakistan
e-mail: zakia.mmg@pu.edu.pk

S. Mehnaz (ed.), *Rhizotrophs: Plant Growth Promotion to Bioremediation*, Microorganisms for Sustainability 2, DOI 10.1007/978-981-10-4862-3_14

14.1 Introduction

14.1.1 Heavy Metals and Their Sources

Highly electronegative metals with a density greater than 5 g/cm^3 are termed "heavy metals." Arsenic is an exception. Although it is lighter and nonmetallic, it resembles with heavy metals and exerts chemico-ecological effects (Duffus 1981). All heavy metals have relatively smaller cationic site because of their heavier nucleus and increasingly compressed electrons. The "electron thirst" ability makes many of them to bond with those atoms having loosely held outer shell electron like sulfur, thus forming sulfides as in lead, tin, copper, and mercury, or prevents them from yielding electron-hungry elements like oxygen (oxidation) and preserving their "nobleness" as gold, platinum, silver, etc. It is the fundamental chaleophile character of the heavy metals that threatens the sulfide bridge between giant protein molecules in the living system. For instance, mercaptan or mercury poisoning, known as Minamata disease, is due to the entry of mercury into organisms (Kim and Choi 1995; Harada et al. 1999). Lead, zinc, copper, cadmium, arsenic, mercury, and nickel, the ecologically more significant heavy metals, play similar disruptive role when they enter the body system in amounts higher than what is required. Heavy metals include the elements with atomic numbers greater than 20, excluding alkali metal, alkaline earths, lanthanoids, and actionoids. Metallic elements are intrinsic components of the environment. Although the natural background or metal fraction in air, soil, and plants is highly variable, there are some anomalous areas or high levels caused by anthropogenic pollution due to mining of metal-rich ores by metal-smelting industries. The anthropogenic sources of eight common heavy metals are shown in Table 14.1.

The natural sources of heavy metal contamination include the dried water droplets from oceans, dust particles from volcanoes, erosion from soil, weathering of rocks, and forest fires. Biodegradation of dead animals and plants also contribute

Table 14.1 Anthropogenic sources of eight common heavy metals (Dixit et al. 2015)

Source	Hg	As	Cd	Cr	Cu	Pb	Ni	Zn
Mining and ore processing	+	+	+	–	+	–	–	+
Metallurgy	+	+	+	+	+	+	+	+
Chemical industry	+	+	+	+	+	+	–	+
Alloy industry	–	–	–	–	–	+	–	–
Paint industry	–	–	+	+	–	+	–	+
Glass industry	+	+	–	–	–	+	–	–
Paper and pulp mills	+	–	–	+	+	+	+	–
Leather tanning	+	+	–	+	–	–	–	+
Textile dyeing and printing	+	+	+	–	+	+	+	+
Chemical fertilizer industry	+	+	+	+	+	+	+	+
Chlor-alkali industry	+	+	+	+	–	+	–	+
Petroleum refining	+	+	+	+	+	+	–	+
Coal burning	+	+	+	+	+	+	–	–

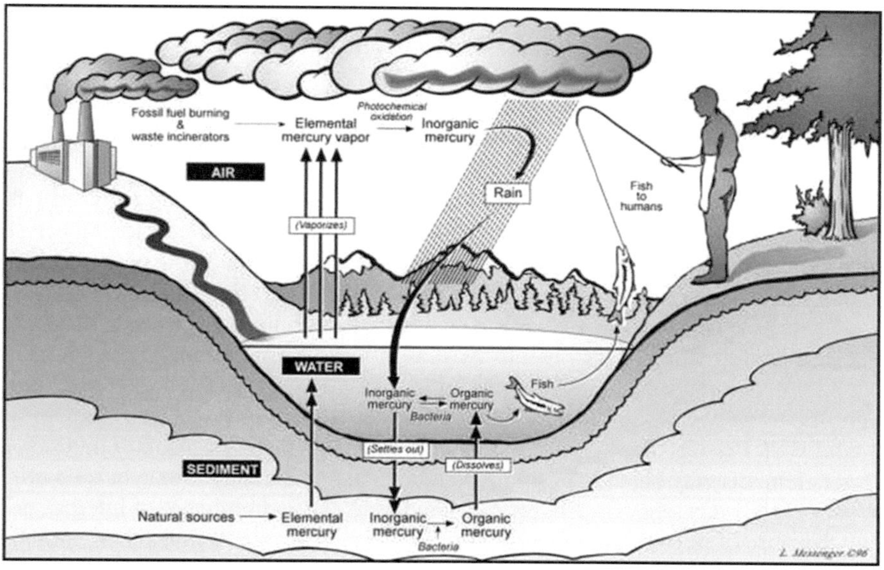

Fig. 14.1 The global biogeochemical cycling of mercury (https://people.uwec.edu 2014)

significantly to background levels of metals in waters (Callender 2003; Järup 2003; Wilson and Pyatt 2007). Many studies have reported different sources and spreading of heavy metals, especially mercury in the environment as shown in Fig. 14.1.

14.1.2 Heavy Metals to Agricultural Soil

Agricultural soils have become a big reservoir of heavy metals due to the extensive usage of different agrochemicals like fungicides, herbicides, and phosphate fertilizers, organic manure, and decaying plant and animal residues (Uwah et al. 2009). The use of sewage sludge and industrial waste water for irrigation further increases the concentration of heavy metals in agricultural soils (Sharma et al. 2007). Agricultural runoff together with soil erosion is the potential source of heavy metals in aquatic bodies. The ecological equilibrium is mainly disrupted by heavy metal-polluted rhizospheric soils. In nature, soils constitute a large variety of metallic elements with different concentrations and as variable chemical species. Some metallic elements have no biological importance, while some elements, known as essential trace elements, have very important role in biological ecosystem. These essential metallic elements become toxic when crossing a certain concentration level. The relevance of concentrations and bioavailabilities of metallic elements in nature is indicated by their reaction with negatively charged soil particles (Benedetti et al. 1995). In order to maintain essential metals available and at a certain concentration level, microorganisms living in rhizosphere constantly regulate their activities (Khan et al. 2000). The soil microorganisms adapt physiological pathways and proceed their evolution process under selection pressure imposed by heavy metals by taking into account the space and time variability of soils (Orcutt 2000).

14.1.3 Phytotoxic Effects of Heavy Metals

Toxic levels of metals in soil may be caused by natural soil properties or by agricultural, manufacturing, mining, and waste disposal procedures. Metal toxicity is an important growth-limiting factor for plants in many acid soils below pH 5.0. Metal toxicity in certain crops is aggravated by high temperature (Foy et al. 1978).

Any heavy metal can be toxic at source level of solubility and has been observed to cause phytotoxicity. Heavy metals exist as inorganic compounds or are bound to organic matter clays or hydrous oxides in soils. Due to this precipitation and sorption, the toxicity of many metals such as Zn, Cu, and Ni has occurred frequently. Toxicity of Pb, Co, Be, As, and Cd occurs only under very unusual conditions. Other elements may be toxic in solution cultures but are not phytotoxic in soils even at very high levels (e.g., Cr, Ag, Sn, Ga, Ge). Lead and cadmium are of interest, not only because of phytotoxicity but because their uptake by plants move them into the food chain. Thus most research on toxicity of heavy metals has involved Zn, Cu, Ni, Cd, Pb, and Hg (Ashraf et al. 2016; Clemens and Ma 2016; Gao et al. 2016; Versieren et al. 2016).

Due to the presence of heavy metals in parent materials, toxic metals also occur naturally in soils. Geochemical studies such as the science of biogeochemistry and the art of biogeochemical prospecting have confirmed the presence of metal residues and the enrichment of metal in plants and soils over or near an ore deposit.

The introduction of Hg in plant systems has principle importance due to its application in fertilizers, herbicides, and seed disinfectants (Cavallini et al. 1999). Few mercury species are being used on tree foliage as fungicides, and they can be transferred, relocated, and redistributed in plants. At the cellular and subcellular level, the processes by which metals may prove lethal include obstruction of biologically significant molecules (e.g., enzymes, polynucleotides), transportation of micronutrients, displacement or substitution of metal ions from biomolecules (e.g., Mg from chlorophyll), deformation and inactivation of enzymatic proteins, and compromise of cell membrane integrity. The possible processes causing Hg-induced phytotoxicity are modifications in the porosity of the cell membrane; high affinity for sulfhydryl (–SH) groups, phosphate groups, and reactive groups of adenosine diphosphate (ADP) or adenosine triphosphate (ATP); and displacement of essential ions and its capability in the disruption of several functions involving critical proteins (Patra and Sharma 2000; Patra et al. 2004). Toxic Hg^{+2} also disrupts the antioxidant defense mechanism by altering the modulation of intracellular nonprotein thiols (NPSH); reduced glutathione (GSH), which is a nonenzymatic antioxidant; ascorbate peroxidase (APX) and glutathione reductase (GR); and superoxide dismutase (SOD), an antioxidant enzyme (Ortega-Villasante et al. 2005; Sparks 2005; Israr et al. 2006; Calgaroto et al. 2010).

The evidence of mercury phytotoxicity has been studied in various grain crops like *Oryza sativa* and *Triticum aestivum*. The primary effects of Hg compounds are on the embryo and secondary on endosperm. Hg compounds cause the breakdown of –SH-system by interfering in biological systems resulting in the production of –S-Hg-S-bridge which may influence germination and embryo development (rich in SH ligands). In *Oryza sativa* and *Zea mays*, $HgCl_2$ is involved in the obstruction of primary root elongation as compared to shoots (Patra and Sharma 2000; Patra et al. 2004).

Hg influences both, light and dark reactions, of photosynthesis by substituting the central atom of chlorophyll (Mg) by Hg, in vivo, which is an important damaging mechanism. It also reduces the transpiration rate, water uptake, and chlorophyll synthesis. Toxic mercuric cations are involved in the loss of magnesium, potassium, manganese, and deposition of iron which lead to the modifications in cell membrane porosity (Boening 2000). The cellular and molecular mechanisms that are involved in Hg-induced toxicity in plants are practically unknown due to scarce studies considering Hg genotoxicity. However, it has been shown that mercury can insert harmful genetic effects to different plant species (De Flora et al. 1994).

In earlier experiments, multinucleated cells in the root tips of corn seedlings, exposed to solution of Ceresan (ethyl mercuric phosphate, a fungicide), resulted in the formation of polyploidy, aneuploidy, and c-tumors through c-mitosis (Kostoff 1939, 1940). C-mitosis (colchicine treated), sister chromatid exchanges, chromosomal aberrations, and spindle alterations can be stimulated by several compounds at similar dosage, but butyl mercury bromide is most notable in this respect (FiskesjÖ 1969). It has been reported that inorganic mercury poisoning in *Allium cepa* and *Allium sativum* resulted in the reduction of mitotic index in the root tip cells and an increment in chromosomal aberrations that depend on concentration and time of exposure. $HgCl_2$ was concluded as more cytotoxic as compared to mercurous chloride, and lowest effective concentration tested (LECT) was measured as 10 ppm. The greater tolerance of *A. sativum* than *A. cepa* was attributed to the presence of high levels of heterochromatin in the former and low amount of sulfur in the later (Patra and Sharma 2000; Patra et al. 2004).

For other metals which are beneficial to plants, concentration in small amount of these metals in the soil could actually improve plant growth and development. However, at higher concentrations of these metals, reduction in plant growth had been recorded. Uptake of low amount of heavy metals increased in plant growth, nutrient content, biochemical content, and antioxidant enzyme activities for plant. Improvements in growth and physiology of cluster beans have also been reported by Manivasagaperumal et al. (2011) at medium Zn concentration of the soil solution. On the other hand, excess concentration of Zn has adverse effects on plant growth. It is also reported that the combination of Pb and Cu at both high concentration and low concentration resulted in a rapid and complete death of the leaves and stem of *Lythrum salicaria* (Brennan and Shelley 1999). Some plants are able to tolerate these metals through three mechanisms: (i) exclusion of heavy metal in the shoot over a wide range of soil concentrations, (ii) inclusion of heavy metal in the shoot reflecting those in the soil solution through a linear relationship, and (iii) bioaccumulation of metals in the shoot and roots of plants at both low and high soil concentrations.

14.1.4 Remediation of Heavy Metals by Microorganisms

Soil microorganisms are involved in interaction with soil constituents and roots of plant (Attitalla et al. 2004; Khan et al. 2009). The regions on the surface of roots and those around them are nutrient rich, and because of the availability of nutrients, the activities of microorganisms are higher in the rhizosphere, as compared to other areas of plants (Dessaux et al. 2009). Plant growth-promoting bacteria present in soil are group of different bacteria involved in improving plant growth while directly and indirectly bioremediating heavy metals like Hg, Cd and Co, etc. (Hayat et al. 2010; Yu et al. 2014). The direct effect depends on the production of hormones, nutrient availability, and increase in plant defense processes against pathogens (Choudhary 2011). Indole acetic acid improves plant root growth and supply of phosphorus to plants (Marschner et al. 2011). *Bacillus* and *Paenibacillus* sp. have the ability to produce spores, and as spores are resistant, so these bacteria can be more persistent in soil environment (Nicholson et al. 2000; Lal and Tabacchioni 2009). Soil microbes convert the insoluble form of phosphate into its soluble form and make it available for the plants to promote their growth (Rodríguez et al. 2006). Bacteria that are involved in phosphate solubilization and nitrogen fixation can be used in biofertilizers (Cakmakci et al. 2007). Some bacteria such as *Bacillus* and *Rhodococcus* are reported to be involved in the siderophore production (Tian et al. 2009). Microorganisms that are present in rhizospheric environment improve the plant growth and directly or indirectly involved in yield increase of the plant (Dimkpa et al. 2009).

Bacillus is involved in growth promotion of plants by producing auxin and siderophore (Kumar et al. 2012). Beneficial microorganisms can be used as biofertilizers, hence minimizing the use of chemical fertilizers. The usage of microorganisms as a biofertilizer is cost-effective, reduces pollution caused by chemical fertilizers, and helps to preserve the natural environment (Stefan et al. 2008).

Release of heavy metals from natural sources and anthropogenic sources poses a major menace to the soil environment (de Oliveira et al. 2001). Generally, heavy metals cannot be degraded by biological mechanisms and exist in the environment to an indefinite extent. After their accumulation in the soils, the lethal heavy metals adversely influence the soil microflora, including plant growth-promoting rhizobacteria (PGPR) in the rhizosphere, and their physiological processes. Furthermore, the elevated concentrations of heavy metals and their uptake by plants also pose adverse effects on plant growth (Han et al. 2006), symbiotic relationships, and ultimately crop yields by disrupting cell organelles and disintegrating the membranes, serving as genotoxic substance that disrupts photosynthetic and respiration processes (Piehler et al. 1999; Perez-Sanz et al. 2012). Therefore, the bioremediation of heavy metal-polluted sites has become an urgent need, as these lands have covered large areas which have been interpreted inapplicable for sustainable agriculture.

Amin and Latif (2015) have provided a comprehensive study of literature about Hg-induced toxicity in plants and its detoxification processes to provoke the advance research in this field. Two extensively studied bioremediation systems based on clustered genes on *Mer* operon and also *Met* gene allow microorganisms to detoxify

Hg^{+2} into volatile Hg^0 and to precipitate it into nontoxic HgS by encoding mercuric reductase and also sulfhydrylase (SHLase) enzymes, respectively (Ray et al. 1989; Ono et al. 1991, 1996). Detoxification mechanisms that employ different microbes to take off environmental contaminants have obtained a profound interest in the recent years (Gupta and Ali 2004).

Many bacterial and yeast genera are being commonly used in the bioremediation of heavy metals (Patra and Sharma 2000; Patra et al. 2004). Rafique et al. (2015) have reported species of *Cronobacter*, *Pseudomonas*, and *Bacillus* which are capable to bioremediate mercury up to 95% in mercuric chloride supplemented in YEM medium. The ability of mercury-resistant nitrogen-fixing bacteria (NFB) to remediate it from the synthetic medium, containing 20 μg/ml $HgCl_2$, was determined. Figure 14.2 indicates that *Cronobacter* species (ZM12 and ZM36) are more efficient to remove mercury from the medium as compared to *Pseudomonas* (ZM24, ZM45, and ZM50) and *Bacillus* sp. (ZM2, ZM40, and ZM57). It is also clear from the observations that H_2S producing NFB with minimum zone of inhibition on Hg amended agar plates are more resistant to mercury and remediate up to 95% of total mercury supplemented in synthetic YEM medium.

Tariq et al. (2015) have also reported *Pseudomonas* spp. on the basis of biochemical characterization and single-sequence repeat (SSR) phylogenetic analysis that possess dual characteristics such as detoxification of mercury pollutants and fixation of atmospheric nitrogen (N_2). The phylogenetic tree was constructed in order to check the percentage homology of different *Pseudomonas* species which

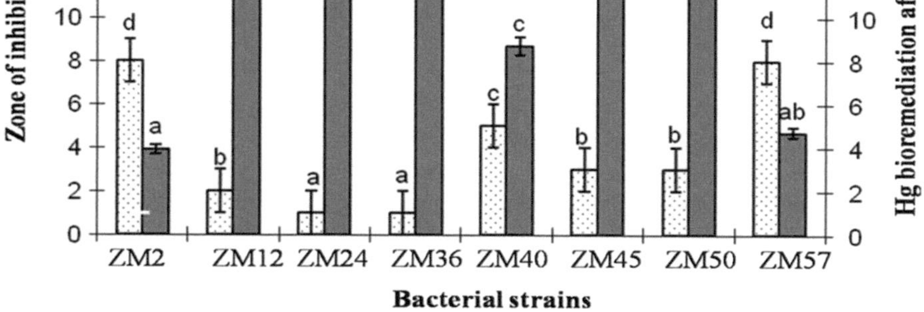

Fig. 14.2 Correlation between zone of inhibition (mm) by well-plate method (*dotted bars*) and removal of mercury (μg/ml) in culture medium after 36 h incubation at 37 °C quantified by dithizone method (*black bars*). The $p < 0.05$ was calculated by ANOVA, and different letters indicate significant difference between means of each treatments calculated by DMRT at probability level 0.05

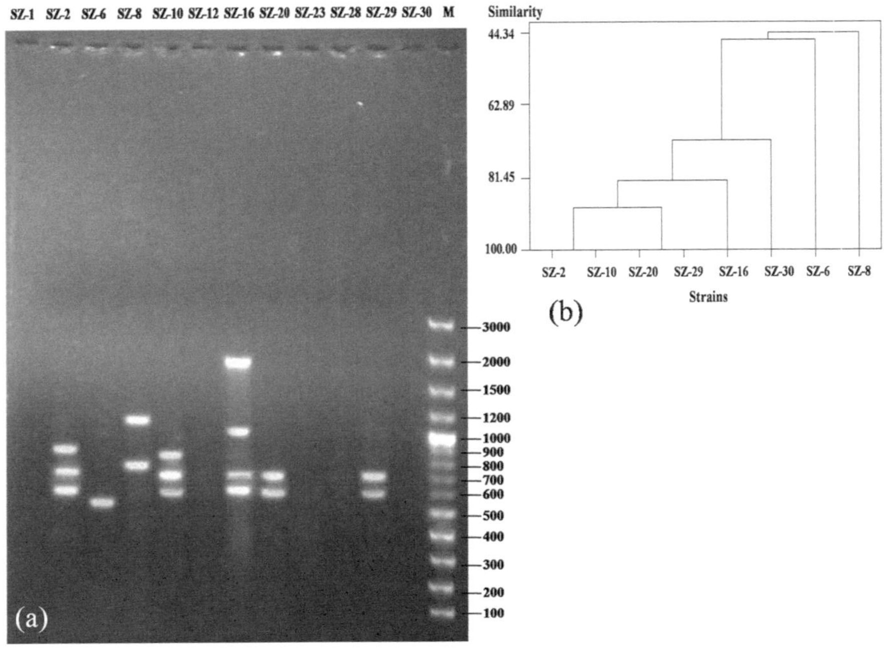

Fig. 14.3 Genetic diversity in mercury-resistant *Pseudomonas* species using SSR $(GACA)_4$: (**a**) Gel electrophoresis (**b**) Dendrogram constructed by using SSR banding pattern

showed that strains SZ-2, SZ-10, SZ-20, and SZ-29 (cluster 1) showed highest homology (100%) with others on the basis of banding pattern while strain SZ-16 showed 81.45% similarity with strains of cluster 1. Similarly, strains SZ-30, SZ-6, and SZ-8 showed 71%, 46%, and 44% homology with strains of cluster 1 and SZ-16, respectively (Fig. 14.3).

Agronomic strategy for using these microorganisms is helpful for obtaining sustainable agriculture. No doubt, a continued work in this area of research is needed to explore the potential of PGPRs and their ecological, genetic, and biochemical relationships in habitat.

Among yeasts, *Candida xylopsoci* and *Pichia kudriavzevii* have the potential to detoxify mercury by 95% and 94.5%, respectively, from enriched medium containing mercury (Amin and Latif 2011).

The study suggests that both strains may have significant biotechnological role in the treatment of contaminants, containing mercury, before they discharge into the soil environment to make it friendly for living organisms. In another study, Amin and Latif (2013) have reported that immobilization of yeast cells responsible for the detoxification of mercury has numerous advantages over free suspended culture (Fig. 14.4). The immobilization of yeast cells has advantages over the free cells such as the reuses of entrapped yeast strains to remediate mercury remain constant after using multiple times (Fig. 14.5).

Fig. 14.4 Na-alginate (synthetic) beads of hydrogen sulfide producing yeast strains

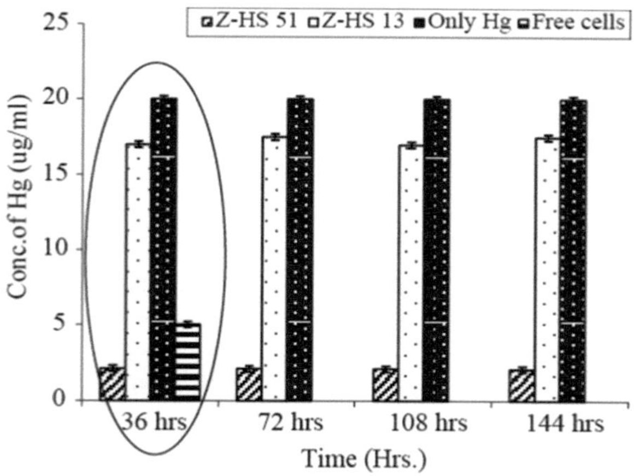

Fig. 14.5 Repeated use of immobilized *Candida xylopsoci* (Z-HS51) and *C. rugosa* (Z-HS 13) to check the potential of mercury remediation (four constitutive cycles). Reduction in mercury concentration (μg/ml) is shown by *bars*, and *encircled* one shows the comparison of immobilized beads with free cells of *C. xylopsoci* for the remediation of mercury from the culture medium supplemented with 20 μg/ml $HgCl_2$

The most important finding of this study is that no residues remain in the medium because they redissolved at the end. The same strategy can be applied in any polluted reservoir because the immobilized cells never lose their ability to reduce the pollutants from the environment and also there would not be any need to dispose entrapped microorganism from the bioreactor because they redissolve within the system. The immobilization does not affect the shelf life of microbes but provides favorable microenvironmental conditions for the organisms, protects against harsh environment, improves genetic stability, and can be transferred easily and safely at any time and place.

Thus, by applying these microorganisms as a biofertilizer to heavy metal-contaminated soils, the toxicity of heavy metal can be reduced resulting in the enhancement of soil fertility and crop productivity which aids in sustainable agriculture.

14.1.5 Remediation of Heavy Metals by Plants

Phytoremediation uses different types of green plants to clean up hazardous waste from contaminated soil polluted by heavy metals. It is an important form of bioremediation and is suitable for pollutants that cover a large area and are within the root zone of the plant (Padmavathiamma and Li 2007). There are different remediating mechanisms of heavy metals by plants.

14.1.5.1 Phytoextraction

Phytoextraction is primarily being used for the remediation of heavy metal-polluted rhizospheric soils. In this treatment, specific plant species, also known as higher accumulator, absorb and precipitate the higher concentrations of heavy metals from polluted soils and accumulate them into their aerial parts. Padmavathiamma and Li (2007) found that some plants have a great potential to extract the concentrated heavy metals into their roots and translocate them into their aerial parts which results in the production of increased plant biomass. Plants used for phytoextraction usually have the following characteristics: rapid growth rate, high biomass, extensive root system, and ability to take high amounts of heavy metals. Generally, there are different criteria being used for hyperaccumulators:

1. The concentration of metal in the shoot must be higher than 0.1% for Al, As, Co, Cr, Cu, Ni, and Se, higher than 0.01% for Cd, and higher than 1.0% for Zn.
2. The ratio of shoot to root concentration must be consistently higher than 1; this indicates the capability to transport metals from roots to shoots, the existence of hypertolerance ability, and the degree of plant metal uptake.

In most cases, plants absorb metals that are readily available in the soil solution. Some metals are present in soil in soluble forms for plant uptake, whereas others occur as insoluble precipitates and are thus unavailable for plant uptake.

14.1.5.2 Phytostabilization

Phytostabilization is also being used for the treatment of heavy metal-polluted soils, sediments, and sludges. By this method, heavy metals such as arsenic (As), cadmium (Cd), chromium (Cr), copper (Cu), and zinc (Zn) are being remediated by plant roots which limit the heavy metals in the rhizosphere via mobility and bioavailability mechanisms (Sharma and Sharma 1993). The plants prohibit the root epidermis via soil matrix and act as barrier which results in the decrease of water percolation and also prevent direct contact with the polluted rhizosphere. It may also prevent the soil from reducing their bioavailability through erosion, leaching, and distribution of the toxic heavy metal to other areas. It is helpful in the treatment of contaminated land areas affected by mining activities. Plants help stabilize the soil through their root systems that prevent leaching, and hence erosion, via reduction of water percolation through the soil. Plants used for phytostabilization should have the following characteristics: dense rooting system, ability to tolerate soil conditions, ease of establishment and maintenance under field conditions, rapid growth to provide adequate ground coverage, longevity, and ability to self-propagate. Soil and organic amendments are used for contaminated soil to reduce the toxicity of heavy metals, increase availability of nutrients for plant growth, and improve the physical properties of soil. Several studies have suggested that phytostabilization may detoxify metal toxicity by converting soluble oxidation state to an insoluble oxidation state of metal, e.g., plants have converted available toxic Cr (VI) to unavailable and less toxic Cr (III) (Salt et al. 1995).

14.1.5.3 Phytovolatilization

Phytovolatilization is a method being used for the treatment of heavy metal-polluted soils. In this remediation process, plants take up heavy metals from rhizosphere and transform their metallic forms into volatile forms which are then released into the atmosphere by transformation. The growing trees and other plants may use the xylem vessels for passing heavy metal contaminants from rhizosphere toward the leaves where transformation from toxic to nontoxic forms may occur and then may finally volatilize into the atmosphere. It is basically used for As, Hg, and Se which exist as gaseous species in the contaminated soil. In the recent years, scientists have found natural and genetically modified plants which have capability to absorb toxic forms of these metals and then biologically converting them into gaseous states for releasing them into the atmosphere. Phytovolatilization is a controversial technology in the field of phytoremediation because Hg and Se are toxic so there is uncertainty about the biosafety of these elements into the atmosphere (Suszcynsky and Shann 1995; Sakakibara et al. 2010). In Se phytovolatilization, the gaseous Se is produced from inorganic or organic Se compounds (McGrath et al. 2002). Moreover, Se pollution is a worldwide problem, so its volatilization into the atmosphere is an attractive phytoremediation technology. Furthermore, many researchers have made considerable efforts to inert mercuric ion (Hg^{+2}) reductase into plants for Hg-volatilization (Rugh et al. 1998; Bizily et al. 1999).

14.2 Remediation of Heavy Metals by Combination of Plants and Microbes

The combined use of both microorganisms and plants for the remediation of polluted soils results in a faster and more efficient cleanup of the polluted area. Mycorrhizal fungi have been used in several remediation techniques of heavy metal-polluted soils. Increased mycorrhizal efficiency have resulted into decreased metal accumulation and increased the growth of white clover growing in heavy metal (Zn)-polluted soil.

Phytoextraction is the best method for the accumulation of heavy metals in plants, and other methods improve phytostabilization through metal immobilization and reduction of metal concentration in plants (Abhilash et al. 2012).

In general, the benefits derived from mycorrhizal associations, which range from increased nutrient and water acquisition to the provision of a stable soil, for plant growth and increase in plant resistance to diseases are believed to aid the survival of plants growing in polluted soils and thus help in the vegetation and revegetation of remediated soils. In addition of certain species of mycorrhizal fungi, arbuscular mycorrhizal fungi can be more sensitive to pollutants compared to plants. Other microorganisms apart from mycorrhizal fungi have also been used in conjunction with plants for the remediation of heavy metal-polluted soils. Most of these microbes are the PGPR that are usually found in the rhizosphere. Several microbes stimulate plant growth by some mechanisms such as production of phytohormones, siderophores, and other chelating agents specific for enzyme activity, supplying nutrients, N fixation, and reduction of ethylene production to encourage root growth (Divya and Kumar 2011).

Enhanced accumulation of heavy metals such as Cd and Ni by hyperaccumulators (*Brassica juncea* and *Brassica napus*) has been observed when the plants were inoculated with *Bacillus* spp. (Khalid and Tinsley 1980). On the other hand, increased plant growth due to reduction in the accumulation of Cd and Ni in the shoot and root tissues of tomato plant was observed when it was inoculated with *Methylobacterium oryzae*.

14.3 Conclusion

This chapter reveals that heavy metals are hazardous contaminants associated with serious problems in plants and animals because they can be easily spread through many ecosystems. Unfortunately, very less knowledge is available about phytotoxicity caused by heavy metals, processes by which heavy metals are absorbed by plant cells and detoxification mechanisms by which they are modified from toxic to nontoxic form in soil through microorganisms. Although plants attribute a significant role as the base of several trophic levels in food chain, particularly of humankind subsistence and thriftiness, it is an urgent necessity to upgrade the knowledge about the mechanisms of heavy metal uptake by plants, its phytotoxicity, and bioremediation mechanisms of these pollutants. Combining both plants and microorganisms in bioremediation increases the efficiency of remediation. The literature presented here provides a worthy rootage for other scientists engaged in research on heavy metal-induced phytotoxicity and its modification or bioremediation processes to stimulate foster research in this field.

References

Abhilash P, Powell JR, Singh HB, Singh BK (2012) Plant-microbe interactions: novel applications for exploitation in multipurpose remediation technologies. Trends Biotechnol 30(8):416–420

Amin A, Latif Z (2011) Isolation and characterization of H_2S producing yeast to detoxify mercury containing compounds. Intl Res J Microbiol 2(12):517–525

Amin A, Latif Z (2013) Detoxification of mercury pollutant by immobilized yeast strain *Candida xylopsoci*. Pak J Bot 45(4):1437–1442

Amin A, Latif Z (2015) Phytotoxicity of Hg and its detoxification through microorganisms in soil. Adv Life Sci 2(2):98–105

Ashraf MY, Roohi M, Iqbal Z et al (2016) Cadmium (Cd) and lead (Pb) induced changes in growth, some biochemical attributes and mineral accumulation in two cultivars of mung bean [*Vigna radiata* (L.) Wilczek]. Commun Soil Sci Plant Anal 47(4):405–413

Attitalla IH, Fatehi J, Brishammar S (2004) Biology and partial sequencing of an endophytic *Fusarium oxysporum* and plant defense complex. Dissertation, University of Upsala

Benedetti MF, Milne CJ, Kinniburgh DG et al (1995) Metal ion binding to humic substances: application of the non-ideal competitive adsorption model. Environ Sci Technol 29(2):446–457

Bizily SP, Rugh CL, Summers AO, Meagher RB (1999) Phytoremediation of methylmercury pollution: *mer*B expression in *Arabidopsis thaliana* confers resistance to organomercurials. Proc Nat Acad Sci USA 96(12):6808–6813

Boening DW (2000) Ecological effects, transport, and fate of mercury: a general review. Chemosphere 40(12):1335–1351

Brennan MA, Shelley ML (1999) A model of the uptake, translocation, and accumulation of lead (Pb) by maize for the purpose of phytoextraction. Ecol Eng 12(3):271–297

Cakmakci R, Dönmez MF, Erdoğan Ü (2007) The effect of plant growth promoting rhizobacteria on barley seedling growth, nutrient uptake, some soil properties, and bacterial counts. Turk J Agri Forest 31(3):189–199

Calgaroto NS, Castro GY, Cargnelutti D et al (2010) Antioxidant system activation by mercury in *Pfaffia glomerata* plantlets. Biometals 23(2):295–305

Callender E (2003) Heavy metals in the environment-historical trends A2- Turekian. In: Heinrich D, Holland KK (eds) Treatise on geochemistry. Pergamon, Oxford, pp 67–105

Cavallini A, Natali L, Durante M, Maserti B (1999) Mercury uptake, distribution and DNA affinity in durum wheat (*Triticum durum* Desf.) plants. Sci Total Environ 244:119–127

Choudhary D (2011) Plant growth-promotion (PGP) activities and molecular characterization of rhizobacterial strains isolated from soybean (*Glycine max* L. Merril) plants against charcoal rot pathogen, *Macrophomina phaseolina*. Biotechnol Lett 33(11):2287–2295

Clemens S, Ma JF (2016) Toxic heavy metal and metalloid accumulation in crop plants and foods. Annu Rev Plant Biol 67(1):489–512

De Flora S, Bennicelli C, Bagnasco M (1994) Genotoxicity of mercury compounds. A review. Mut Res 317(1):57–79

de Oliveira SMB, Melfi AJ, Fostier AH et al (2001) Soils as an important sink for mercury in the Amazon. Water Air Soil Pollut 126(3–4):321–337

Dessaux Y, Hinsinger P, Lemanceau P (2009) Rhizosphere: so many achievements and even more challenges. Plant Soil 321(1):1–3

Dimkpa C, Weinand T, Asch F (2009) Plant–rhizobacteria interactions alleviate abiotic stress conditions. Plant Cell Environ 32(12):1682–1694

Divya B, Kumar MD (2011) Plant–microbe interaction with enhanced bioremediation. Res J Biotechnol 6(1):72–79

Dixit R, Malaviya D, Pandiyan K et al (2015) Bioremediation of heavy metals from soil and aquatic environment: an overview of principles and criteria of fundamental processes. Sustainability 7(2):2189–2212

Duffus JH (1981) Environmental toxicology. Wiley, New York

FiskesjÖ G (1969) Some results from allium tests with organic mercury halogenides. Hereditas 62(3):314–322

Foy C, Chaney R, White M (1978) The physiology of metal toxicity in plants. Annu Rev Plant Physiol 29(1):511–566

Gao C, De Schamphelaere K, Smolders E (2016) Zinc toxicity to the alga *Pseudokirchneriella subcapitata* decreases under phosphate limiting growth conditions. Aquat Toxicol 173(1):74–82

Gupta N, Ali A (2004) Mercury volatilization by R factor systems in *Escherichia coli* isolated from aquatic environments of India. Curr Microbiol 48(2):88–96

Han FX, Su Y, Monts DL, Waggoner CA, Plodinec MJ (2006) Binding, distribution, and plant uptake of mercury in a soil from Oak Ridge, Tennessee, USA. Sci Total Environ 368(2–3):753–768

Harada M, Akagi H, Tsuda T, Kizaki T, Ohno H (1999) Methylmercury level in umbilical cords from patients with congenital Minamata disease. Sci Total Environ 234(1):59–62

Hayat R, Ali S, Amara U, Khalid R, Ahmed I (2010) Soil beneficial bacteria and their role in plant growth promotion: a review. Ann Microbiol 60(4):579–598

https://people.uwec.edu (2014) University of Wisconsin-Eau Claire. Mercury in the environment and water supply. https://people.uwec.edu/piercech/Hg/mercury_water/cycling.htm. Accessed 2 Dec 2014

Israr M, Sahi S, Datta R, Sarkar D (2006) Bioaccumulation and physiological effects of mercury in *Sesbania drummondii*. Chemosphere 65(4):591–598

Järup L (2003) Hazards of heavy metal contamination. Brit Medic Bull 68(1):167–182

Khalid B, Tinsley J (1980) Some effects of nickel toxicity on rye grass. Plant Soil 55(1):139–144

Khan A, Kuek C, Chaudhry T, Khoo C, Hayes W (2000) Role of plants, mycorrhizae and phytochelators in heavy metal contaminated land remediation. Chemosphere 41(1):197–207

Khan MS, Zaidi A, Wani P, Ahemad M, Oves M (2009) Functional diversity among plant growth-promoting rhizobacteria: current status. In: Khan MS, Zaidia A, Mussarat J (eds) Microbial strategies for crop improvement. Springer, Berlin/Heidelberg, pp 105–132

Kim P, Choi BH (1995) Selective inhibition of glutamate uptake by mercury in cultured mouse astrocytes. Yonsei Med J 36(3):299–305

Kostoff D (1939) Effect of the fungicide "Granosan" on atypical growth and chromosome doubling in plants. Nature 144(1):334

Kostoff D (1940) Atypical growth, abnormal mitosis and polyploidy induced by ethyl-mercury-chloride. J Phytopathol (Berlin) 13(1):91–96

Kumar P, Dubey R, Maheshwari D (2012) *Bacillus* strains isolated from rhizosphere showed plant growth promoting and antagonistic activity against phytopathogens. Microbiol Res 167(8):493–499

Lal S, Tabacchioni S (2009) Ecology and biotechnological potential of *Paenibacillus polymyxa*: a minireview. Ind J Microbiol 49(1):2–10

Manivasagaperumal R, Balamurugan S, Thiyagarajan G, Sekar J (2011) Effect of zinc on germination, seedling growth and biochemical content of cluster bean (*Cyamopsis tetragonoloba* (L.) Taub). Curr Bot 2(5):11–15

Marschner P, Crowley D, Rengel Z (2011) Rhizosphere interactions between microorganisms and plants govern iron and phosphorus acquisition along the root axis-model and research methods. Soil Biol Biochem 43(5):883–894

McGrath S, Zhao J, Lombi E (2002) Phytoremediation of metals, metalloids, and radionuclides. Adv Agron 75(1):1–56

Nicholson WL, Munakata N, Horneck G, Melosh HJ, Setlow P (2000) Resistance of *Bacillus* endospores to extreme terrestrial and extra-terrestrial environments. Microbiol Mol Biol Rev 64(3):548–572

Ono B, Ishii N, Fujino S, Aoyama I (1991) Role of hydrosulfide ions (HS^-) in methylmercury resistance in *Saccharomyces cerevisiae*. Appl Environ Microbiol 57(11):3183–3186

Ono B, Kijima K, Ishii N et al (1996) Regulation of sulphate assimilation in *Saccharomyces cerevisiae*. Yeast 12(11):1153–1162

Orcutt DM (2000) The physiology of plants under stress: soil and biotic factors, vol 2. Wiley, New Jersey

Ortega-Villasante C, Rellan-Alvarez R, Del Campo FF et al (2005) Cellular damage induced by cadmium and mercury in *Medicago sativa*. J Exp Bot 56(418):2239–2251

Padmavathiamma PK, Li LY (2007) Phytoremediation technology: hyper-accumulation metals in plants. Water Air Soil Pollut 184(1–4):105–126

Patra M, Sharma A (2000) Mercury toxicity in plants. Bot Rev 66(3):379–422

Patra M, Bhowmik N, Bandopadhyay B, Sharma A (2004) Comparison of mercury, lead and arsenic with respect to genotoxic effects on plant systems and the development of genetic tolerance. Environ Exp Bot 52(3):199–223

Perez-Sanz A, Millan R, Sierra MJ et al (2012) Mercury uptake by *Silene vulgaris* grown on contaminated spiked soils. J Environ Manag 95(3):S233–S237

Piehler M, Swistak J, Pinckney J, Paerl H (1999) Stimulation of diesel fuel biodegradation by indigenous nitrogen fixing bacterial consortia. Microb Ecol 38(1):69–78

Rafique A, Amin A, Latif Z (2015) Screening and characterization of mercury-resistant nitrogen fixing bacteria and their use as biofertilizers and for mercury bioremediation. Pak J Zool 47(5):1271–1277

Ray S, Gachhui R, Pahan K, Chaudhury J, Mandal A (1989) Detoxification of mercury and organomercurials by nitrogen-fixing soil bacteria. J Biosci 14(2):173–182

Rodríguez H, Fraga R, Gonzalez T, Bashan Y (2006) Genetics of phosphate solubilization and its potential applications for improving plant growth-promoting bacteria. Plant Soil 287(1–2):15–21

Rugh CL, Gragson GM, Meagher RB, Merkle SA (1998) Toxic mercury reduction and remediation using transgenic plants with a modified bacterial gene. Hortscience 33(4):618–621

Sakakibara M, Watanabe A, Inoue M, Sano S, Kaise T (2010) Phytoextraction and phytovolatilization of arsenic from As-contaminated soils by *Pteris vittata*. In: Proceedings of the annual international conference on soils, sediments, water and energy, Vol 12, Article 26

Salt DE, Blaylock M, Kumar NP et al (1995) Phytoremediation: a novel strategy for the removal of toxic metals from the environment using plants. Nat Biotechnol 13(5):468–474

Sharma D, Sharma C (1993) Chromium uptake and its effects on growth and biological yield of wheat. Cereal Res Commun 21(4):317–322

Sharma RK, Agrawal M, Marshall F (2007) Heavy metal contamination of soil and vegetables in suburban areas of Varanasi, India. Ecotoxicol Environ Saf 66(2):258–266

Sparks DL (2005) Toxic metals in the environment: the role of surfaces. Elements 1(4):193–197

Stefan M, Mihasan M, Dunca S (2008) Plant growth promoting rhizobacteria can inhibit the in vitro germination of *Glycine max* L. seeds. Analele Stiintifice ale Universitatii "Al I Cuza" Din Iasi (Serie Noua) Sectiunea 2 a Genetica si Biologie Moleculara 9 (3):35–40

Suszcynsky EM, Shann JR (1995) Phytotoxicity and accumulation of mercury in tobacco subjected to different exposure routes. Environ Toxicol Chem 14(1):61–67

Tariq S, Amin A, Latif Z (2015) PCR based DNA fingerprinting of mercury resistant and nitrogen fixing *Pseudomonas* spp. Pure Appl Biol 4(1):129–134

Tian F, Ding Y, Zhu H, Yao L, Du B (2009) Genetic diversity of siderophore-producing bacteria of tobacco rhizosphere. Braz J Microbiol 40(2):276–284

Uwah E, Ndahi N, Ogugbuaja V (2009) Study of the levels of some agricultural pollutants in soils, and water leaf (*Talinum triangulare*) obtained in Maiduguri, Nigeria. J Appl Sci Environ Sanitat 4(2):71–78

Versieren L, Evers S, De Schamphelaere K, Blust R, Smolders E (2016) Mixture toxicity and interactions of Cu, Ni, Cd and Zn to barley at low effect levels: something from nothing? Environ Toxicol Chem. doi:10.1002/etc.3380. [Epub ahead of print]

Wilson B, Pyatt F (2007) Heavy metal dispersion, persistance, and bioccumulation around an ancient copper mine situated in Anglesey, UK. Ecotoxicol Environ Saf 66(2):224–231

Yu X, Li Y, Zhang C et al (2014) Culturable heavy metal-resistant and plant growth promoting bacteria in V-Ti magnetite mine tailing soil from Panzhihua, China. PloS one 9(9):e106618

MIX
Papier aus verantwortungsvollen Quellen
Paper from responsible sources
FSC® C105338

If you have any concerns about our products,
you can contact us on
ProductSafety@springernature.com

In case Publisher is established outside the EU,
the EU authorized representative is:
Springer Nature Customer Service Center GmbH
Europaplatz 3, 69115 Heidelberg, Germany

Printed by Libri Plureos GmbH
in Hamburg, Germany